Fernando G. Lobo, Cláudio F. Lima and Zbigniew Michalewicz (Eds.)

Parameter Setting in Evolutionary Algorithms

Studies in Computational Intelligence, Volume 54

Editor-in-chief
Prof. Janusz Kacprzyk
Systems Research Institute
Polish Academy of Sciences
ul. Newelska 6
01-447 Warsaw
Poland
E-mail: kacprzyk@ibspan.waw.pl

Further volumes of this series
can be found on our homepage:
springer.com

Vol. 33. Martin Pelikan, Kumara Sastry, Erick Cantú-Paz (Eds.)
Scalable Optimization via Probabilistic Modeling, 2006
ISBN 978-3-540-34953-2

Vol. 34. Ajith Abraham, Crina Grosan, Vitorino Ramos (Eds.)
Swarm Intelligence in Data Mining, 2006
ISBN 978-3-540-34955-6

Vol. 35. Ke Chen, Lipo Wang (Eds.)
Trends in Neural Computation, 2007
ISBN 978-3-540-36121-3

Vol. 36. Ildar Batyrshin, Janusz Kacprzyk, Leonid Sheremetor, Lotfi A. Zadeh (Eds.)
Preception-based Data Mining and Decision Making in Economics and Finance, 2006
ISBN 978-3-540-36244-9

Vol. 37. Jie Lu, Da Ruan, Guangquan Zhang (Eds.)
E-Service Intelligence, 2007
ISBN 978-3-540-37015-4

Vol. 38. Art Lew, Holger Mauch
Dynamic Programming, 2007
ISBN 978-3-540-37013-0

Vol. 39. Gregory Levitin (Ed.)
Computational Intelligence in Reliability Engineering, 2007
ISBN 978-3-540-37367-4

Vol. 40. Gregory Levitin (Ed.)
Computational Intelligence in Reliability Engineering, 2007
ISBN 978-3-540-37371-1

Vol. 41. Mukesh Khare, S.M. Shiva Nagendra (Eds.)
Artificial Neural Networks in Vehicular Pollution Modelling, 2007
ISBN 978-3-540-37417-6

Vol. 42. Bernd J. Krämer, Wolfgang A. Halang (Eds.)
Contributions to Ubiquitous Computing, 2007
ISBN 978-3-540-44909-6

Vol. 43. Fabrice Guillet, Howard J. Hamilton (Eds.)
Quality Measures in Data Mining, 2007
ISBN 978-3-540-44911-9

Vol. 44. Nadia Nedjah, Luiza de Macedo Mourelle, Mario Neto Borges, Nival Nunes de Almeida (Eds.)
Intelligent Educational Machines, 2007
ISBN 978-3-540-44920-1

Vol. 45. Vladimir G. Ivancevic, Tijana T. Ivancevic
Neuro-Fuzzy Associative Machinery for Comprehensive Brain and Cognition Modeling, 2007
ISBN 978-3-540-47463-0

Vol. 46. Valentina Zharkova, Lakhmi C. Jain
Artificial Intelligence in Recognition and Classification of Astrophysical and Medical Images, 2007
ISBN 978-3-540-47511-8

Vol. 47. S. Sumathi, S. Esakkirajan
Fundamentals of Relational Database Management Systems, 2007
ISBN 978-3-540-48397-7

Vol. 48. H. Yoshida (Ed.)
Advanced Computational Intelligence Paradigms in Healthcare, 2007
ISBN 978-3-540-47523-1

Vol. 49. Keshav P. Dahal, Kay Chen Tan, Peter I. Cowling (Eds.)
Evolutionary Scheduling, 2007
ISBN 978-3-540-48582-7

Vol. 50. Nadia Nedjah, Leandro dos Santos Coelho, Luiza de Macedo Mourelle (Eds.)
Mobile Robots: The Evolutionary Approach, 2007
ISBN 978-3-540-49719-6

Vol. 51. Shengxiang Yang, Yew Soon Ong, Yaochu Jin Honda (Eds.)
Evolutionary Computation in Dynamic and Uncertain Environment, 2007
ISBN 978-3-540-49772-1

Vol. 52. Abraham Kandel, Horst Bunke, Mark Last (Eds.)
Applied Graph Theory in Computer Vision and Pattern Recognition, 2007
ISBN 978-3-540-68019-2

Vol. 53. Huajin Tang, Kay Chen Tan, Zhang Yi
Neural Networks: Computational Models and Applications, 2007
ISBN 978-3-540-69225-6

Vol. 54. Fernando G. Lobo, Cláudio F. Lima and Zbigniew Michalewicz (Eds.)
Parameter Setting in Evolutionary Algorithms, 2007
ISBN 978-3-540-69431-1

Fernando G. Lobo
Cláudio F. Lima
Zbigniew Michalewicz
(Eds.)

Parameter Setting in Evolutionary Algorithms

With 100 Figures and 24 Tables

Springer

Fernando G. Lobo
Departamento de Engenharia
Electrónica e Informática
Universidade do Algarve
Campus de Gambelas
8000-117 Faro Portugal
E-mail: flobo@ualg.pt

Cláudio F. Lima
Departamento de Engenharia
Electrónica e Informática
Universidade do Algarve
Campus de Gambelas
8000-117 Faro Portugal
E-mail: clima@ualg.pt

Zbigniew Michalewicz
School of Computer Science
University of Adelaide
SA 5005 Adelaide Australia
E-mail: zbyszek@cs.adelaide.edu.au

Library of Congress Control Number: 2006939345

ISSN print edition: 1860-949X
ISSN electronic edition: 1860-9503
ISBN-10 3-540-69431-5 Springer Berlin Heidelberg New York
ISBN-13 978-3-540-69431-1 Springer Berlin Heidelberg New York

Springer is a part of Springer Science+Business Media
springer.com

Cover design: deblik, Berlin
Typesetting by the editors using a Springer LaTeX macro package
Printed on acid-free paper SPIN: 11543152 89/SPi 5 4 3 2 1 0

Preface

One of the main difficulties that a user faces when applying an evolutionary algorithm (or, as a matter of fact, any heuristic method) to solve a given problem is to decide on an appropriate set of parameter values. Before running the algorithm, the user typically has to specify values for a number of parameters, such as population size, selection rate, and operator probabilities, not to mention the representation and the operators themselves. Over the years, there have been numerous research studies on different approaches to automate control of these parameters as well as understand their interactions.

At the 2005 Genetic and Evolutionary Computation Conference, held in Seattle, the first two editors of this book organized a workshop entitled *Parameter setting in evolutionary algorithms*. Shortly after we announced the workshop, we were approached by Janusz Kacprzyk to prepare a volume containing extended versions of some of the papers presented at the workshop, as well as contributions from other authors in the field.

We gladly accepted the invitation and Zbigniew Michalewicz joined us in the project.

The resulting work is in front of you, a book with 15 chapters covering the topic on various areas of evolutionary computation, including genetic algorithms, evolution strategies, genetic programming, estimation of distribution algorithms, and also discussing the issues of specific parameters used in parallel implementations, multi-objective EAs, and practical consideration used for real-world applications. Some of these chapters are overview oriented while others describe recent advances in the area.

In the first chapter, Ken De Jong gives us an historical overview on parameterized evolutionary algorithms, as well as his personal view on the issues related to parameter adaptation. De Jong was the first person to conduct a systematic study of the effect of parameters on the performance of genetic algorithms, and it is interesting to hear his view now that more than 30 years have passed since his 1975 PhD dissertation.

In the second chapter, Agoston Eiben et al. present a survey of the area giving special attention to on-the-fly parameter setting.

In chapter 3, Silja Meyer-Nieberg and Hans-Georg Beyer focus on self-adaptation, a technique that consists of encoding parameters into the individual's genome and evolving them together with the problem's decision variables, and that has been mainly used in the area of evolutionary programming and evolution strategies.

Chapter 4 by Dirk Thierens deals with adaptive operator allocation. These rules are often used for learning probability values of applying a given operator from a fixed set of variation operators. Thierens surveys the probability matching method, which has been incorporated in several adaptive operator allocation algorithms proposed in the literature, and proposes an alternative method called the *adaptive pursuit strategy*. The latter turns out to exhibit a better performance than the probability matching method, in a controlled, non-stationary environment.

In chapter 5, Mike Preuss and Thomas Bartz-Beielstein present the *sequential parameter optimization* (SPO), a technique based on statistical design of experiments. The authors motivate the technique and demonstrate its usefulness for experimental analysis. As a test case, the SPO procedure is applied to self-adaptive EA variants for binary coded problems.

In chapter 6, Bo Yuan and Marcus Gallagher bring a statistical technique called *Racing*, originally proposed in the *machine learning* field, to the context of choosing parameter settings in EAs. In addition, they also suggest an hybridization scheme for combining the technique with meta-EAs.

In chapter 7, Alan Piszcz and Terrence Soule discuss structure altering mutation techniques in genetic programming and observe that the parameter settings associated with the operators generally show a nonlinear response with respect to population fitness and computational effort.

In chapter 8, Michael Samples et al. present *Commander*, a software solution that assists the user in conducting parameter sweep experiments in a distributed computing environment.

In chapter 9, Fernando Lobo and Cláudio Lima provide a review of various adaptive population sizing methods that have been proposed for genetic algorithms. For each method, the major advantages and disadvantages are discussed. The chapter ends with some recommendations for those who design and compare self-adjusting population sizing mechanisms for genetic algorithms.

In chapter 10, Tian-Li Yu et al. suggest an adaptive population sizing scheme for genetic algorithms. The method has strong similarities with the work proposed by Smith and Smuda in 1993, but the components of the population sizing model are automatically estimated through the use of linkage-learning techniques.

In chapter 11, Martin Pelikan et al. present a parameter-less version of the hierarchical Bayesian optimization algorithm (hBOA). The resulting algorithm solves nearly decomposable and hierarchical problems in a quadratic or subquadratic number of function evaluations without the need of user intervention for setting parameters. The chapter also discusses how the parameter-less technique can be applied to other estimation of distribution algorithms.

In chapter 12, Kalyanmoy Deb presents a functional decomposition of an evolutionary multi-objective methodology and shows how a specific algorithm, the elitist non-dominated sorting GA (NSGA-II), was designed and implemented without the need of any additional parameter with respect to those existing in a traditional EA. Deb argues that this property of NSGA-II is one of the main reasons for its success and popularity.

In chapter 13, Erick Cantú-Paz presents theoretical models that predict the effects of the parameters in parallel genetic algorithms. The models explore the effect of communication topologies, migration rates, population sizing, and migration strategies. Although the models make assumptions about the class of problems being solved, they provide useful guidelines for practitioners who are looking for increased efficiency by means of parallelization.

In chapter 14, Zbigniew Michalewicz and Martin Schmidt summarize their experience of tuning and/or controlling various parameters of evolutionary algorithms from working on real word problems. A car distribution system is used as an example. The authors also discuss prediction and optimization issues present in dynamic environments, and explain the ideas behind *Adaptive Business Intelligence*.

The last chapter of the book, by Neal Wagner and Zbigniew Michalewicz, presents the results of recent studies investigating non-static parameter settings that are controlled by feedback from the genetic programming search process in the context of forecasting applications.

We hope you will find the volume enjoyable and inspiring; we also invite you to take part in future workshops on *Parameter setting in evolutionary algorithms*!

Faro, Portugal, *Fernando Lobo*
Faro, Portugal, *Cláudio Lima*
Adelaide, Australia, *Zbigniew Michalewicz*

November 2006

Contents

List of Contributors

Thomas Bartz-Beielstein
Dortmund University
D-44221 Dortmund, Germany
thomas.bartz-beielstein@udo.edu

Hans-Georg Beyer
Vorarlberg University of Applied Sciences
Hochschulstr. 1
A-6850 Dornbirn, Austria
hans-georg.beyer@fhv.at

Matthew J. Byom
University of Michigan
Ann Arbor, MI 48109-2143, USA
mjbyom@umich.edu

Erick Cantú-Paz
Yahoo!, Inc.
701 First Avenue
Sunnyvale, CA 94089
cantupaz@acm.org

Jason M. Daida
University of Michigan
Ann Arbor, MI 48109-2143, USA
daida@umich.edu

Kalyanmoy Deb
Indian Institute of Technology Kanpur
Kanpur, PIN 208016, India
deb@iitk.ac.in

Kenneth De Jong
George Mason University
4400 University Drive, MSN 4A5
Fairfax, VA 22030, USA
kdejong@gmu.edu

A.E. Eiben
Free University Amsterdam
The Netherlands
gusz@cs.vu.nl

Marcus Gallagher
University of Queensland
Qld. 4072, Australia
marcusg@itee.uq.edu.au

David E. Goldberg
University of Illinois at Urbana-Champaign
Urbana, IL 61801
deg@illigal.ge.uiuc.edu

Alexander Hartmann
Universität Göttingen
Friedrich-Hund-Platz 1
37077 Göttingen, Germany
hartmann@physik.uni-goettingen.de

Cláudio F. Lima
University of Algarve
Campus de Gambelas
8000-117 Faro, Portugal
clima@ualg.pt

Tz-Kai Lin
University of Missouri–St. Louis
One University Blvd.
St. Louis, MO 63121
tlkq4@studentmail.umsl.edu

Fernando G. Lobo
University of Algarve
Campus de Gambelas
8000-117 Faro, Portugal
flobo@ualg.pt

Silja Meyer-Nieberg
Universtät der Bundeswehr München
D-85577 Neubiberg, Germany
silja.meyer-nieberg@unibw.de

Zbigniew Michalewicz
University of Adelaide
Adelaide, SA 5005, Australia
zbyszek@cs.adelaide.edu.au

Martin Pelikan
University of Missouri–St. Louis
One University Blvd.
St. Louis, MO 63121
pelikan@cs.umsl.edu

Alan Piszcz
University of Idaho
Moscow, ID, 83844-1010
apiszcz@acm.org

Mike Preuss
Dortmund University
D-44221 Dortmund, Germany
mike.preuss@cs.uni-dortmund.de

Michael E. Samples
University of Michigan
Ann Arbor, MI 48109-2143, USA
msamples@umich.edu

Kumara Sastry
University of Illinois at Urbana-Champaign
Urbana, IL 61801
kumara@illigal.ge.uiuc.edu

Martin Schmidt
SolveIT Software
PO Box 3161
Adelaide, SA 5000, Australia
martin.schmidt@solveitsoftware.com

M. Schoenauer
INRIA
France
marc@lri.fr

J.E. Smith
UWE
United Kingdom
james.smith@uwe.ac.uk

Terence Soule
University of Idaho
Moscow, ID, 83844-1010
tsoule@cs.uidaho.edu

Dirk Thierens
Universiteit Utrecht
The Netherlands
dirk.thierens@cs.uu.nl

Neal Wagner
Augusta State University
Augusta, GA 30904, USA
nwagner@aug.edu

Tian-Li Yu
University of Illinois at Urbana-Champaign
Urbana, IL 61801
tianliyu@illigal.ge.uiuc.edu

Bo Yuan
University of Queensland
Qld. 4072, Australia
boyuan@itee.uq.edu.au

Parameter Setting in EAs: a 30 Year Perspective

Kenneth De Jong

Department of Computer Science
George Mason University
4400 University Drive, MSN 4A5
Fairfax, VA 22030, USA
kdejong@gmu.edu

Summary. Parameterized evolutionary algorithms (EAs) have been a standard part of the Evolutionary Computation community from its inception. The widespread use and applicability of EAs is due in part to the ability to adapt an EA to a particular problem-solving context by tuning its parameters. However, tuning EA parameters can itself be a challenging task since EA parameters interact in highly non-linear ways. In this chapter we provide a historical overview of this issue, discussing both manual (human-in-the-loop) and automated approaches, and suggesting when a particular strategy might be appropriate.

1 Introduction

More than 30 years have passed since I completed my thesis [18], and more than 40 years since the pioneering work of Rechenberg [34], Schwefel [36], Foget et al. [17] and Holland [24]. Since then the field of Evolutionary Computation (EC) has grown dramatically and matured into a well-established discipline.

One of the characteristics of the field from the very beginning was the notion of a parameterized evolutionary algorithm (EA), the behavior of which could be modified by changing the value of one or more of its parameters. This is not too surprising since EAs consist of populations of individuals that produce offspring using reproductive mechanisms that introduce variation into the population. It was clear from the beginning that changing the population size, the type and amount of reproductive variation, etc. could significantly change the behavior of an EA on a particular fitness landscape and, conversely, appropriate parameter settings for one fitness landscape might be inappropriate for others.

It is not surprising, then, that from the beginning EA practitioners have wanted to know the answers to questions like:

K. De Jong: *Parameter Setting in EAs: a 30 Year Perspective*, Studies in Computational Intelligence (SCI) **54**, 1–18 (2007)
www.springerlink.com

- Are there optimal settings for the parameters of an EA in general?
- Are there optimal settings for the parameters of an EA for a particular class of fitness landscapes?
- Are there robust settings for the parameters of and EA than produce good performance over a broad range of fitness landscapes?
- Is it desirable to dynamically change parameter values during an EA run?
- How do changes in a parameter affect the performance of an EA?
- How do landscape properties affect parameter value choices?

A look at the table of contents of this book will convince you that there are no simple answers to these questions. In this chapter I will provide my perspective on these issues and describe a general framework that I use to help understand the answers to such questions.

2 No Free Lunch Theorems

It has been shown in a variety of contexts including the areas of search, optimization, and machine learning that, unless adequate restrictions are placed on the the class of problems one is attempting to solve, there is no single algorithm that will outperform all other algorithms (see, for example, [83]). While very general and abstract, No Free Lunch (NFL) theorems provide a starting point for answering some of the questions listed in the previous section. In particular, these theorems serve as both a cautionary note to EA practitioners to be careful about generalizing their results, and as a source of encouragement to use parameter tuning as a mechanism for adapting algorithms to particular classes of problems.

It is important at this point to clarify several frequently encountered misconceptions about NFL results. The first misconception is that NFL results are often interpreted as applying only to algorithms that do not self-adapt during a problem solving run. So, for example, I might imagine a two-stage algorithm that begins by first figuring out what kind of problem it is being asked to solve and then choosing the appropriate algorithm (or parameter settings) from its repertoire, thus avoiding NFL constraints. While these approaches can extend the range of problems for which an algorithm is effective, it is clear that the ability of the first stage to identify problem types and select appropriate algorithms is itself an algorithm that is subject to NFL results.

A second misconception is the interpretation of NFL results as leaving us in a hopeless state unable to say anything general about the algorithms we develop. A more constructive interpretation of NFL results is that they present us with a challenge to define more carefully the classes of problems we are trying to solve and the algorithms we are developing to solve them. In particular, NFL results provide the context for a meaningful discussion parameter setting in EAs, including:

- What EA parameters are useful for improving performance?

- How does one choose appropriate parameter values?
- Should parameter values be fixed during a run or be modified dynamically?

These issues are explored in some detail in the following sections.

3 Parameter Setting in Simple EAs

Before we tackle more complicated EAs, it is helpful to ascertain what we know about parameter setting for simple EAs. I use the term "simple EA" to refer to algorithms in which:

- A population of individuals is evolved over time.
- The current population is used as a source of parents to produce some offspring.
- A subset of the parents and offspring are selected to "survive" into the next generation.

This is the form that the early EAs took (Evolution Strategies (ESs), Evolutionary Programming (EP), and Genetic Algorithms (GAs)), and it represents the basic formulation of many EAs used today. With any algorithm, the easiest things to consider parameterizing are the numerical elements. In the case of the early EAs, the obvious candidate was the population size. For ESs, the parent population size μ was specified independently of the offspring population size λ, while early EP and GA versions defined both populations to be of the same size controlled by a single parameter m. The values of these parameters were specified at the beginning of a run and remained unchanged throughout a run.

Somewhat more subtle were the parameters defined to control reproductive variation (the degree to which offspring resemble their parents). Early ES and EP algorithms produced offspring via *asexual* reproduction, i.e., by cloning and mutation. Hence, the amount of reproductive variation is controlled by varying the parameters of the mutation operator, one of which specifies the (expected) number of "genes" undergoing mutation. A second parameter controls how different (on average) a mutated gene is from its original form.

Since many ES and EP algorithms were evolving populations of individuals whose genes were real-valued parameters of an objective fitness function, a Gaussian mutation operator was used with a mean of zero and a variance of σ^2. It became clear early on that leaving σ fixed for the duration of an entire evolutionary run was suboptimal, and mechanisms were developed for dynamically adapting its value were developed. The earliest of these mechanisms was the "1/5th rule"' developed by Rechenberg and Schwefel [36] in which the ratio of "successful" mutations (improvements in fitness) to unsuccessful ones was monitored during the course of a run. If the ratio fell below 1/5, σ was decreased. If the ratio exceeded 1/5, σ was increased. The amount that σ was increased or decreased was a parameter as well, but set to a fixed value at the beginning of a run.

By contrast, early GA versions used an internal binary string representation, and produced reproductive variation via two-parent (sexual) recombination and a "bit-flipping" mutation operator. In this case the degree of reproductive variation was controlled by the amount of recombination and the number of bits flipped. The amount of recombination was controlled in two ways: by specifying the percentage of offspring to be produced using recombination, and the number of crossover!points to be used when performing recombination (see, for example, [18]). Interestingly, there was less concern about dynamically adapting these GA parameters, in part because the variation due to recombination decreases automatically as the population becomes more homogeneous.

So, already with the early canonical EAs we see examples of parameters whose values remained fixed through out an EA run and parameters whose values are dynamically modified during a run. Both of these approaches are examined in more detail in the following sections.

4 Static Parameter Setting Strategies

The No Free Lunch results place an obligation on the EA practitioner to understand something about the particular properties of the problems that (s)he is trying to solve that relate to particular choices of EA parameter settings. This poses a bit of a dilemma in that often an EA is being used precisely because of a lack of knowledge about the fitness landscapes being explored. As a consequence, the parameter setting strategy frequently adopted is a multiple run, human-in-the-loop approach in which various parameter settings are tried in an attempt to fine-tune an EA to a particular problem.

Since this can be a somewhat tedious process, it is not uncommon to automate this process with a simple, top-level "parameter sweep" procedure that systematically adjusts selected parameter values. However, unless care is taken, this can lead to a combinatorial explosion of parameter value combinations and require large amounts of computation time. One possible escape from this is to replace the parameter sweep procedure with a parameter optimization procedure. Which optimization procedure to use is an interesting question. The fitness landscapes defined by EA parameter tuning generally have the same unknown properties as the underlying problems that suggested the use of EAs in the first place! So, a natural consequence is the notion of a two-level EA, the top level of which is evolving the parameters of the lower level EA. The best example of this approach is the "nested" Evolution Strategy in which a top-level ES evolves the parameters for a second-level ES [35]. An interesting (and open question) is how to choose the parameters for top-level EAs!

Since exploring EA parameter space is generally very time consuming and computationally expensive, it is done in a more general "offline" setting. The idea here is to find EA parameter settings that are optimal for a class of problems using a sweep or optimization procedure on a selected "test suite" of sample problems, and then use the resulting parameter settings for all of the problems of that class when encountered in practice (see, for example, [18] or [34]). The key insight from such studies is the robustness of EAs with respect to their parameter settings. Getting "in the ball park" is generally sufficient for good EA performance. Stated another way, the EA parameter "sweet spot" is reasonably large and easy to find [18]. As a consequence most EAs today come with a default set of static parameter values that have been found to be quite robust in practice.

5 Dynamic Parameter Setting Strategies

A more difficult thing to ascertain is whether there is any advantage to be gained by dynamically changing the value of a parameter during an EA run and, if so, how to change it. The intuitive motivation is that, as evolution proceeds, it should be possible to use the accumulating information about the fitness landscape to improve future performance. The accumulating information may relate to global properties of the fitness landscape such as noise or ruggedness, or it may relate to the local properties of a particular region of the landscape. A second intuition is a sense of the need to change EA parameters as it evolves from a more diffuse global search process to a more focused converging local search process.

One way to accomplish this is to set up an *a priori* schedule of parameter changes in a fashion similar to those used in simulating annealing [16]. This is difficult to accomplish in general since it is not obvious how to predict the number of generations an EA will take to converge, and set a parameter adjustment schedule appropriately. A more successful approach is to monitor particular properties of an evolutionary run, and use changes in these properties as a signal to change parameter values. A good example of this approach is the 1/5 rule described in the previous section. More recently, the use of a covariance matrix for mutation step size adaptation has been proposed [23]. Another example is the adaptive reproductive operator selection procedure developed by Davis [17] in which reproductive operators are dynamically chosen based on their recent performance.

The purist might argue that inventing feedback control procedures for EAs is a good example of over-engineering an already sophisticated adaptive system. A better strategy is to take our inspiration from nature and design our EAs to be self-regulating. For example, individuals in the population might contain "regulatory genes" that control mutation and recombination mechanisms, and these regulatory genes would be subject to the same evolutionary processes as the rest of the genome. Historically, the ES community has used this approach as a way of independently controlling the mutation step size σ_i of each objective parameter value x_i [3]. Similar ideas have been used to control the probability of choosing a particular mutation operator for finite state machines [16]. Within the GA community Spears [76] has used a binary control gene to determine which of two crossover operators to use.

These two approaches to setting parameters dynamically, using either an *ad hoc* or a self-regulating control mechanism, have been dubbed "adaptive" and "self-adaptive" in the literature [11].

Perhaps the most interesting thing to note today, after more than 30 years of experimenting with dynamic parameter setting strategies, is that, with one exception, none of them are used routinely in every day practice. The one exception are the strategies used by the ES community for mutation step size adaptation. In my opinion this is due is the fact that it is difficult to say anything definitive and general about the performance improvements obtained through dynamic parameter setting strategies. There are two reasons for this difficulty. First, our EAs are stochastic, non-linear algorithms. This means that formal proofs are extremely difficult to obtain, and experimental studies must be carefully designed to provide statistically significant results. The second reason is that a legitimate argument can be made that comparing the performance of an EA with static parameter settings to one with dynamic settings is unfair since it is likely that the static settings were established via some preliminary parameter tuning runs that have not been included in the comparison.

My own view is that there is not much to be gained in dynamically adapting EA parameter settings when solving static optimization problems. The real payoff for dynamic parameter setting strategies is when the fitness landscapes are themselves dynamic (see, for example, [4], [5], or [32]).

6 Choosing the EA Parameters to Set

An important aspect to EA parameter setting is a deeper understanding of how changes in EA parameters affect EA performance. Without that understanding it is difficult to ascertain which parameters, if properly set, would improve performance on a particular problem. This deeper understanding can be obtained in several ways. The simplest approach is to study parameter setting in the context of a particular family of EAs (e.g., ESs, GAs, etc.). The advantage here is that they are well-studied and well-understood. For example, there is a large body of literature in the ES community regarding the role and the effects of the parent population size μ, the offspring population size λ, and mutation step size σ. Similarly, the GA community has studied population size m, various crossover operators, and mutation rates.

The disadvantage to this approach is that the EAs in use today seldom fit precisely into one of these canonical forms. Rather, they have evolved via recombinations and mutations of these original ideas. It would be helpful, then, to understand the role of various parameters at a higher level of abstraction. This is something that I have been interested in for several years. The basis for this is a unified view of simple EAs [9]. From this perspective an EA practitioner makes a variety of design choices, perhaps the most important of which is representation. Having made that choice, there are still a number of additional decisions to be made that affect EA performance, the most important of which are:

- The size of the parent population m.
- The size of the offspring population n.
- The procedure for selecting parents *p_select*.
- The procedure for producing offspring *reproduction*.
- The procedure for selecting survivors *s_select*.

Most modern EA toolkits parameterize these decisions allowing one to choose traditional parameter settings or create new and novel variations. So, for example, a canonical $(\mu + \lambda)$-ES would be specified as:

- $m = \mu$
- $n = \lambda$
- *p_select* = deterministic and uniform
- *reproduction* = clone and mutate
- *s_select* = deterministic truncation

and a canonical GA would be specified as:

- $m = n$
- *p_select* = probabilistic and fitness-proportional
- *reproduction* = clone, recombine, and mutate
- *s_select* = offspring only

It is at this level of generality that we would like to understand the effects of parameter settings. We explore this possibility in the following subsections.

6.1 Parent Population Size m

From my perspective the parent population size m can be viewed as a measure of the degree of parallel search that an EA supports, since the parent population is the basis for generating new search points. For simple landscapes like the 2-dimensional (inverted) parabola illustrated in Figure 1, little, if any, parallelism is required since any sort of simple hill climbing technique will provide reasonable performance.

By contrast, more complex, multi-peaked landscapes may require populations of 100s or even 1000s of parents in order to have some reasonable chance of finding globally optimal solutions. As a simple illustration of this, consider the 2-dimensional landscape defined by the fitness function $f(x_1, x_2) = x_1^2+x_2^2$ in which the variables x_1 and x_2 are constrained to the interval $[-10, 5]$.

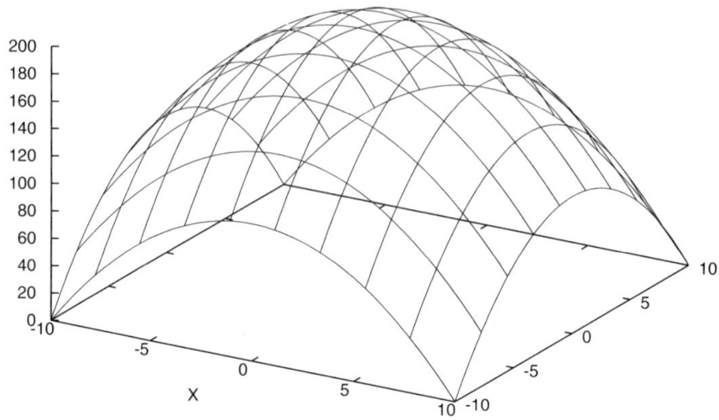

Fig. 1. A simple 2-dimensional (inverted) parabolic fitness landscape.

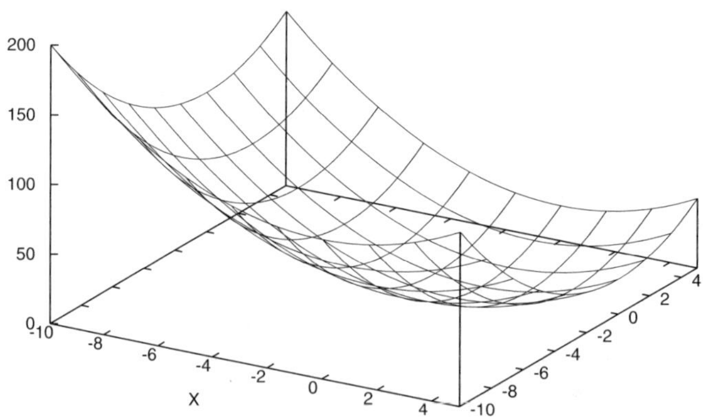

Fig. 2. A 2-dimensional parabolic objective fitness landscape with multiple peaks and a unique maximum.

This landscape has four peaks and a unique optimal fitness value of 200 at $\langle -10, -10 \rangle$ as shown in Figure 2.

Regardless of which simple EA you choose, increasing m increases the probability of finding the global optimum. Figure 3 illustrates this for a standard ES-style EA. What is plotted are best-so-far curves averaged over 100 runs for increasing values of m while keeping n fixed at 1 and using a non-adaptive Gaussian mutation operator with an average step size of 1.0.

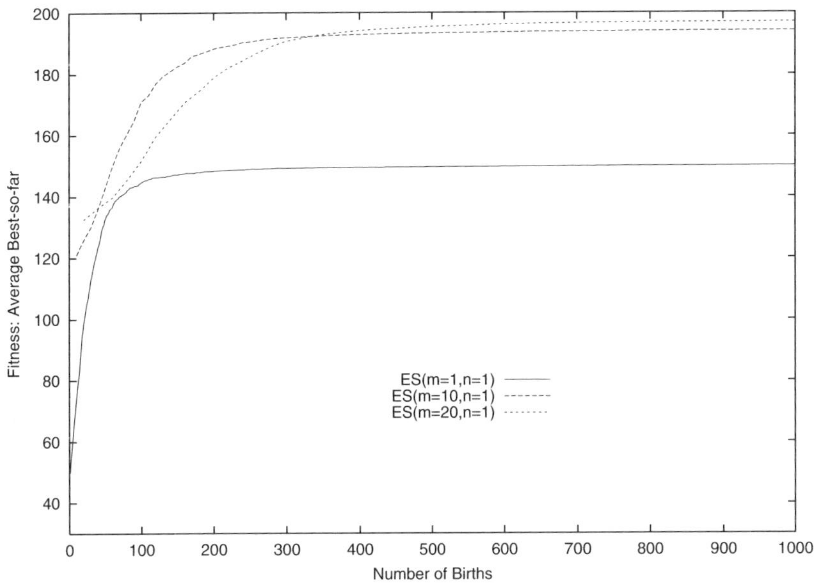

Fig. 3. The effects of parent population size on average best-so-far curves on a multi-peaked landscape.

In this particular case, we see that the *average* behavior of $ES(m, n = 1)$ improves with increasing m, but at a decreasing rate. The same thing can be shown for GA-like EAs and other non-traditional simple EAs.

For me, this is the simplest and clearest example of a relationship between fitness landscape properties and EA parameters, and provides useful insight into possible strategies for choosing parent population size. At the beginning of an EA run it is important to have sufficient parallelism to handle possible multi-modalities, while at the end of a run an EA is likely to have converged to a local area of the fitness landscape that is no more complex than Figure 1. That, in turn, suggests some sort of "annealing" strategy for parent population size that starts with a large value that decreases over time. The difficulty is in defining a computational procedure to do so effectively since it is difficult to predict *a priori* the time to convergence. One possibility is to set a fixed time limit to an EA run and define an annealing schedule for it *a priori* (e.g., [30]).

Studies of this sort confirm our intuition about the usefulness of dynamically adapting population size, but they achieve it by making (possibly) suboptimal *a priori* decisions. Ideally, one would like parent population size controlled by the current state of an EA. This has proved to be quite difficult to achieve (see, for example, [39], [2], or [12]). Part of the reason for this is that there are other factors that affect parent population size choices as well, including the interacting effects of offspring population size (discussed in the next section), the effects of noisy fitness landscapes [20], selection pressure [13], and whether generations overlap or not [37]. As a consequence, most EAs used in practice today run with a fixed population size, the value of which may be based on existing off-line studies (e.g., [34]) or more likely via manual tuning over multiple runs.

6.2 Offspring Population Size n

By contrast the offspring population size n plays a quite different role in a simple EA. The current parent population reflects where in the solution space an EA is focusing its search. The number of offspring n generated reflects the amount of exploration performed using the current parent population before integrating the newly generated offspring back into the parent population. Stated in another way, this is the classic exploration-exploitation tradeoff that all search techniques face.

This effect can be seen quite clearly if we keep the parent population size m constant and increase the offspring population size n. Figure 4 illustrates this for the same ES-like algorithm used previously with $m = 1$. Notice how increasing n on this multi-peaked landscape results in a decrease in average performance unlike what we saw in the previous section with increases in m.

This raises an interesting EA question: should the parameters m and n be coupled as they are in GAs and EP, or is it better to decouple them as they are in ESs? Having fewer parameters to worry about is a good thing, so is there anything to be gained by having two population size parameters? There is considerable evidence to support an answer of "yes", including a variety of studies involving "steady-state" GAs (e.g., [42]) that have large m and $n = 1$, and other simple EAs (e.g., [27]). However, just as we saw with parent population size, defining an effective strategy for adapting the offspring population size during an EA run is quite difficult (see, for example, [26]). As a consequence, most EAs in use today run with a fixed offspring population size, the default value of which is based on some off-line studies and possibly manually tuned over several preliminary runs.

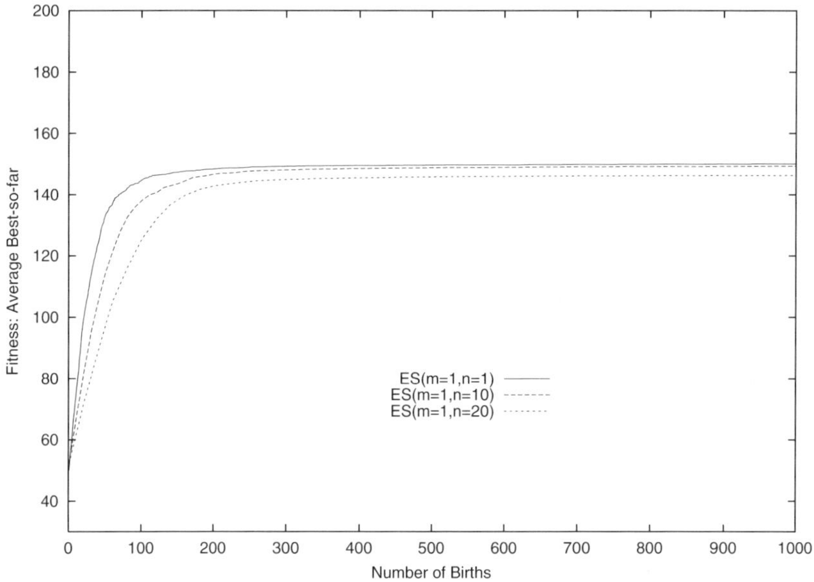

Fig. 4. The effects of offspring population size on average best-so-far curves on a multi-peaked landscape.

7 Selection

Selection procedures are used in EAs in two different contexts: as a procedure for selecting parents to produce offspring and as a procedure for deciding which individuals "survive" into the next generation. It is quite clear that the more "elitist" a selection algorithm is, the more an EA behaves like a local search procedure (i.e, a hill-climber, a "greedy" algorithm) and is less likely to converge to a global optimum. Just as we saw with parent population size m, this suggests that selection!pressure should be adapted during an EA run, weak at first to allow for more exploration and then stronger towards the end as an EA converges. However, the same difficulty arises here in developing an effective mechanism for modifying selection pressure as a function of the current state of an EA. For example, although fitness-proportional selection does vary selection pressure over time, it does so by inducing stronger pressure at the beginning of a run and very little at the end.

One source of difficulty here is that selection pressure is not as easy to "parameterize" as population size. We have a number of families of selection procedures (e.g, tournament selection, truncation selection, fitness-proportional selection, etc.) to choose from and a considerable body of literature analyzing their differences (see, for example, [19] or [15]), but deciding which family to choose or even which member of a parameterized family is still quite difficult, particularly because of the interacting effects with population size [13].

Another interesting question is whether the selection algorithm chosen for parent selection is in any way coupled with the choice made for survival selection. The answer is a qualified "yes" in the following sense: the overall selection pressure of an EA is due to the combined effects of both selection procedures. Hence, strong selection pressure for one of these (e.g, truncation selection) is generally combined with weak selection pressure (e.g., uniform selection) for the other. For example, standard ES and EP algorithms use uniform parent selection and truncation survival selection, while standard GAs use fitness-proportional parent selection and uniform survival selection.

One additional "parameterization" of selection is often made available: the choice between overlapping and non-overlapping generations. With non-overlapping models, the entire parent population dies off each generation and the offspring only compete with each other for survival. Historical examples of non-overlapping EAs include "generational GAs" and the "," version of ESs.

The alternative is an overlapping-generation model such as a steady-state GA, a $(\mu+\lambda)$-ES, or any EP algorithm. In this case, parents and offspring compete with each other for survival. The effects of this choice are quite clear. An overlapping-generation EA behaves more like a local search algorithm, while a non-overlapping-generation EA exhibits better global search properties [9]. Although an intriguing idea, I am unaware of any attempts to change this parameter dynamically.

8 Reproductive Operators

Most EAs produce offspring using two basic classes of reproductive mechanisms: an asexual (single parent) mutation operator and a sexual (more than one parent) recombination operator. Deciding if and how to parameterize this aspect of an EA is a complex issue because the effects of reproductive operators invariably interact with all of the other parameterized elements discussed so far.

At a high level of abstraction reproductive operators introduce variation into the population, counterbalancing the reduction in diversity due to selection. Intuitively, we sense that population diversity should start out high and decrease over time, reflecting a shift from a more global search to a local one. The difficulty is in finding an appropriate "annealing schedule" for population diversity. If selection is too strong, convergence to a local optimum is highly likely. If reproductive variation is too strong, the result is undirected random search. Finding a balance between exploration and exploitation has been a difficult-to-achieve goal from the beginning [18].

What is clear is that there is no one single answer. ES and EP algorithms typically match up strong selection pressure with strong reproductive variation, while GAs match up weaker selection pressure with weaker reproductive variation. Finding a balance usually involves holding one of these fixed (say, selection pressure) and tuning the other (say, reproductive variation).

This is complicated by the fact that what is really needed from reproductive operators is *useful* diversity. This was made clear early on in [33] via the notion of fitness correlation between parents and offspring, and has been shown empirically in a variety of settings (e.g., [31], [1], [28]). This in turn is a function of the properties of the fitness landscapes being explored which makes it clear that the ability to dynamically improve fitness correlation will improve EA performance. Precisely how to achieve this is less clear. One approach is to maintain a collection of plausible reproductive operators and dynamically select the ones that seem to be helping most (see, for example, [17] or [16]). Alternatively, one can focus on tuning specific reproductive operators to improve performance. This is a somewhat easier task about which considerable work has been done, and is explored in more detail in the following subsections.

8.1 Adapting Mutation

The classic one-parent reproductive mechanism is mutation that operates by cloning a parent and then providing some variation by modifying one or more genes in the offspring's genome. The amount of variation is controlled by specifying how many genes are to be modified and the manner in which genes are to be modified. These two aspects together determine both the amount and usefulness of the resulting variation.

Although easy to parameterize, the expected number of modified genes is seldom adapted dynamically. Rather, there are a number of studies that suggest a fixed value of 1 is quite robust, and is the default value for traditional GAs. By contrast, traditional ES algorithms mutate *every* gene, typically by a small amount. This seeming contradiction is clarified by noting that GAs make small changes in genotype space while the ES approach makes small changes in phenotype space.

Both of these approaches allow for dynamic adaptation. The "1/5" rule discussed earlier is a standard component of many ES algorithms. Adapting GA mutation rates is not nearly as common, and is typically done via a self-adaptive mechanism in which the mutation rate is encoded as a control gene on individual genomes (see, for example, [3]).

Mutating gene values independently can be suboptimal when their effects on fitness are coupled (i.e., epistatic non-linear interactions). Since these interactions are generally not known *a priori*, but difficult to determine dynamically during an EA run. The most successful example of this is the pair-wise covariance matrix approach used by the ES community [23]. Attempting to detect higher order interactions is computational expensive and seldom done.

8.2 Adapting Recombination

The classic two-parent reproductive mechanism is a recombination operator in which subcomponents of the parents' genomes are cloned and reassembled to create an offspring genome. For simple fixed-length linear genome representations, the recombination operators have traditionally taken the form of "crossover" operators, in which the crossover points mark the linear subsegments on the parents' genomes to be copied and reassembled. For these kinds of recombination operators, the amount of variation introduced is dependent on two factors: how many crossover points there are and how similar the parents are to each other. The interesting implication of this is that, unlike mutation, the amount of variation introduced by crossover diminishes over time as selection makes the population more homogeneous.

This dependency on the contents of the population makes it much more difficult to estimate the level of crossover-induced variation, but has the virtue of self-adapting along the lines of our intuition: more variation early in an EA run and less variation as evolution proceeds. Providing *useful* variation is more problematic in that, intuitively, one would like offspring to inherit important combinations of gene values from their parents. However, just as we saw for mutation, which combinations of genes should be inherited is seldom known *a priori*. For example, it is well-known that the 1-point crossover operator used in traditional GAs introduces a distance bias in the sense that the values of genes that are far apart on a chromosome are much less likely to be inherited together than those of genes that are close together [25]. Simply increasing the number of crossover points reduces that bias but increases the amount of reproductive variation at the same time [10]. One elegant solution to this is to switch to a parameterized version of uniform crossover. This simultaneously removes the distance bias and provides a single parameter for controlling diversity [40]. Although an intriguing possibility, I am unaware of any current EAs that dynamically adjust parameterized uniform crossover.

Alternatively, one might consider keeping a recombination operator fixed and adapting the representation in ways that improve the production of useful diversity. This possibility is discussed in the next section.

9 Adapting the Representation

Perhaps the most difficult and least understood area of EA design is that of adapting its internal representation. It is clear that choices of representation play an extremely important role in the performance of an EA, but are difficult to automate. As a field we have developed a large body of literature that helps us select representations *a priori* for particular classes of problems. However, there are only a few examples of strategies for dynamically adapting the representation during an EA run. One example is the notion of a "messy GA" [21] in which the position of genes on the chromosome are adapted over time to improve the effectiveness 1-point crossover. Other examples focus on adapting the range and/or resolution of the values of a gene (see, for example, [38] or [82]). Both of these approaches have been shown to be useful in specific contexts, but are not general enough to be included in today's EA toolkits.

10 Parameterless EAs

Perhaps the ultimate goal of these efforts is to produce an effective and general problem-solving EA with *no externally visible* parameters. The No Free Lunch discussion at the beginning of this chapter makes it clear that this will only be achieved if there are effective ways to dynamically adapt various *internal* parameters. However, as we have seen throughout this chapter, there are very few techniques that do so effectively in a general setting for even one internal parameter, much less for more than one simultaneously. The few examples of parameterless EAs that exist in the literature involve simplified EAs in particular contexts (e.g., [29]). To me this is a clear indication of the difficulty of such a task.

An approach that appears more promising is to design an EA to perform internal restarts, i.e. multiple runs, and use information from previous runs to improve performance on subsequent (internal) runs. The most notable success in this area is the CHC algorithm developed by [14]. Nested ESs are also quite effective and based on a similar ideas but still require a few external parameter settings [35].

Clearly, we still have a long way to go in achieving the goal of effective parameterless EAs.

11 Summary

The focus in this chapter has be primarily of parameter setting in simple EAs for two reasons. First, this is where most of the efforts have been over the past 30 or more years. Second, although many new and more complex EAs have been developed (e.g., spatially distributed EAs, multi-population island models, EAs with speciation and niching, etc.), these new EAs do little to resolve existing parameter tuning issues. Rather, they generally exacerbate the problem by creating new parameters that need to be set.

My sense is that, for static optimization problems, it will to continue to be the case that particular types of EAs that have been pre-tuned for particular classes of problems will continue to outperform EAs that try to adapt too many things dynamically. However, if we switch our focus to time-varying fitness landscapes, dynamic parameter adaptation will have a much stronger impact.

References

1. L. Altenberg. The schema theorem and Price's theorem. In M. Vose and D. Whitley, editors, *Foundations of Genetic Algorithms 3*, pages 23–49. Morgan Kaufmann, 1994.

2. J. Arabas, Z. Michalewitz, and J. Mulawka. Gavaps - a genetic algorithm with varying population size. In *Proceedings of the First IEEE Conference on Evolutionary Computation*, pages 73–78. IEEE Press, 1994.
3. T. Bäck. *Evolutionary Algorithms in Theory and Practice.* Oxford University Press, New York, 1996.
4. T. Bäck. On the behavior of evolutionary algorithms in dynamic fitness landscapes. In *IEEE International Conference on Evolutionary Computation*, pages 446–451. IEEE Press, 1998.
5. J. Branke. *Evolutionary Optimization in Dynamic Environments.* Kluwer, Boston, 2002.
6. L. Davis. Adapting operator probabilities in genetic algorithms. In *Third International Conference on Genetic Algorithms*, pages 61–69. Morgan Kaufmann, 1989.
7. L. Davis. *The Handbook of Genetic Algorithms.* Van Nostrand Reinhold, New York, 1991.
8. K. De Jong. *Analysis of Behavior of a Class of Genetic Adaptive Systems.* PhD thesis, University of Michigan, Ann Arbor, MI, 1975.
9. K. De Jong. *Evolutionary Computation: A Unified Approach.* MIT Press, Cambridge, MA, 2006.
10. K. De Jong and W. Spears. A formal analysis of the role of multi-point crossover in genetic algorithms. *Annals of Mathematics and Artificial Intelligence*, 5(1): 1–26, 1992.
11. A. Eiben, R. Hinterding, and Z. Michalewicz. Parameter control in evolutionary algorithms. *IEEE Transactions on Evolutionary Computation*, 3(2):124–141, 1999.
12. A. Eiben, E. Marchiori, and V. Valko. Evolutionary algorithms with on-the-fly population size adjustment. In X. Yao et al., editor, *Proceedings of PPSN VIII*, pages 41–50. Springer-Verlag, 2004.
13. A. Eiben, M. Schut, and A. de Wilde. Is self-adaptation of selection pressure and population size possible? In T. Runarsson et al., editor, *Proceedings of PPSN VIII*, pages 900–909. Springer-Verlag, 2006.
14. L. Eshelman. The CHC adaptive search algorithm. In G. Rawlins, editor, *Foundations of Genetic Algorithms 1*, pages 265–283. Morgan Kaufmann, 1990.
15. S. Ficici and J. Pollack. Game–theoretic investigation of selection methods used in evolutionary algorithms. In D. Whitley, editor, *Proceedings of CEC 2000*, pages 880–887. IEEE Press, 2000.
16. L. Fogel, P. Angeline, and D. Fogel. An evolutionary programming approach to self-adaptation on finite state machines. In J. McDonnell, R. Reynolds, and D. Fogel, editors, *Proceedings of the 4th Annual Conference on Evolutionary Programming*, pages 355–365. MIT Press, 1995.
17. L.J. Fogel, A.J. Owens, and M.J. Walsh. *Artificial Intelligence through Simulated Evolution.* John Wiley & Sons, New York, 1966.
18. D. Goldberg. *The Design of Innovation: Lessons from and for Competent Genetic Algorithms.* Kluwer, Boston, 2002.
19. D. Goldberg and K. Deb. A comparative analysis of selection schemes used in genetic algorithms. In G. Rawlins, editor, *Proceedings of the First Workshop on Foundations of Genetic Algorithms*, pages 69–92. Morgan Kaufmann, 1990.
20. D. Goldberg, K. Deb, and J. Clark. Accounting for noise in sizing of populations. In D. Whitley, editor, *Foundations of Genetic Algorithms 2*, pages 127–140. Morgan Kaufmann, 1992.

21. D. Goldberg, K. Deb, and B. Korb. Don't worry, be messy. In R. K. Belew and L. B. Booker, editors, *Proceedings of the Fourth International Conference on Genetic Algorithms*, pages 24–30. Morgan Kaufmann, 1991.
22. J. Grefenstette. Optimization of control parameters for genetic algorithms. *IEEE Transactions on Systems, Man, and Cybernetics*, 16(1):122–128, 1986.
23. N. Hansen and A. Ostermeier. Completely derandomized step-size adaptation in evolution strategies. *Evolutionary Computation*, 9(2):159–195, 2001.
24. J.H. Holland. Outline for a logical theory of adaptive systems. *JACM*, 9:297–314, 1962.
25. J.H. Holland. *Adaptation in Natural and Artificial Systems*. University of Michigan Press, Ann Arbor, MI, 1975.
26. T. Jansen, K. De Jong, and I. Wegener. On the choice of offspring population size in evolutionary algorithms. *Evolutionary Computation*, 13(4):413–440, 2005.
27. Thomas Jansen and Kenneth De Jong. An analysis of the role of offspring population size in EAs. In W. B. Langdon et al., editor, *Proceedings of the Genetic and Evolutionary Computation Conference (GECCO 2002)*, pages 238–246. Morgan Kaufman, 2002.
28. T. Jones. *Evolutionary algorithms, fitness landscapes, and search*. PhD thesis, University of New Mexico, 1995.
29. C. Lima and F. Lobo. Parameter-less optimization with the extended compact genetic algorithm and iterated local search. In K. Deb et al, editor, *Proceedings of GECCO-2004*, pages 1328–1339. Springer-Verlag, 2004.
30. S. Luke, G. Balan, and L. Panait. Population implosion in genetic programming. In *Proceedings of GECCO-2003*, volume 2724, pages 1729–1739. Springer LNCS Series, 2003.
31. B. Manderick, M. de Weger, and P. Spiessens. The genetic algorithm and the structure of the fitness landscape. In R. K. Belew and L. B. Booker, editors, *The Fourth International Conference on Genetic Algorithms and Their Applications*, pages 143–150. Morgan Kaufmann, 1991.
32. R. Morrison. *Designing Evolutionary Algorithms for Dynamic Environments*. Springer-Verlag, Berlin, 2004.
33. George Price. Selection and covariance. *Nature*, 227:520–521, 1970.
34. I. Rechenberg. Cybernatic solution path of an experimental problem. In *Library Translation 1122*. Royal Aircraft Establishment, Farnborough, 1965.
35. I. Rechenberg. *Evolutionsstrategie '94*. Frommann-Holzboog, Stuttgart, 1994.
36. H.-P. Schwefel. *Evolutionsstrategie und numerishe Optimierung*. PhD thesis, Technical University of Berlin, Berlin, Germany, 1975.
37. H.-P. Schwefel. *Evolution and Optimum Seeking*. John Wiley & Sons, New York, 1995.
38. C. Shaefer. The ARGOT strategy: adaptive representation genetic optimizer technique. In J. Grefenstette, editor, *Proceedings of the Second International Conference on Genetic Algorithms*, pages 50–58. Lawrence Erlbaum, 1987.
39. R. Smith. Adaptively resizing populations: an algorithm and analysis. In S. Forrest, editor, *Proceedings of the Fifth International Conference on Genetic Algorithms and their Applications*, page 653. Morgan Kaufmann, 1993.
40. W. Spears and K. De Jong. On the virtues of parameterized uniform crossover. In R. K. Belew and L. B. Booker, editors, *International Conference on Genetic Algorithms*, volume 4, pages 230–236. Morgan Kaufmann, 1991.

41. W.M. Spears. Adapting crossover in evolutionary algorithms. In J. McDonnell, R. Reynolds, and D.B. Fogel, editors, *Proceedings of the Fourth Conference on Evolutionary Programming*, pages 367–384. MIT Press, 1995.
42. D. Whitley. The Genitor algorithm and selection pressure: Why rank-based allocation of reproductive trials is best. In J.D. Schaffer, editor, *Proceedings of the Third International Conference on Genetic Algorithms*, pages 116–121. Morgan Kaufmann, 1989.
43. D. Whitley, K. Mathias, and P. Fitzhorn. Delta coding: an iterative search strategy for genetic algorithms. In R. K. Belew and L. B. Booker, editors, *Proceeding of the Fourth International Conference on Genetic Algorithms*, pages 77–84. Morgan Kaufmann, 1991.
44. D. Wolpert and W. Macready. No free lunch theorems for optimization. *IEEE Transactions on Evolutionary Computation*, 1:67–82, 1997.

Parameter Control in Evolutionary Algorithms

A.E. Eiben[1], Z. Michalewicz[2], M. Schoenauer[3], and J.E. Smith[4]

[1] Free University Amsterdam, The Netherlands gusz@cs.vu.nl
[2] University of Adelaide, Australia zbyszek@cs.adelaide.edu.au
[3] INRIA, France marc@lri.fr
[4] UWE, United Kingdom james.smith@uwe.ac.uk

Summary. The issue of setting the values of various parameters of an evolutionary algorithm is crucial for good performance. In this paper we discuss how to do this, beginning with the issue of whether these values are best set in advance or are best changed during evolution. We provide a classification of different approaches based on a number of complementary features, and pay special attention to setting parameters on-the-fly. This has the potential of adjusting the algorithm to the problem while solving the problem.

This paper is intended to present a survey rather than a set of prescriptive details for implementing an EA for a particular type of problem. For this reason we have chosen to interleave a number of examples throughout the text. Thus we hope to both clarify the points we wish to raise as we present them, and also to give the reader a feel for some of the many possibilities available for controlling different parameters.

1 Introduction

Finding the appropriate setup for an evolutionary algorithm is a long standing grand challenge of the field [22, 25]. The main problem is that the description of a specific EA contains its components, such as the choice of representation, selection, recombination, and mutation operators, thereby setting a framework while still leaving quite a few items undefined. For instance, a simple GA might be given by stating it will use binary representation, uniform crossover, bit-flip mutation, tournament selection, and generational replacement. For a full specification, however, further details have to be given, for instance, the population size, the probability of mutation p_m and crossover p_c, and the tournament size. These data – called the **algorithm parameters** or **strategy parameters** – complete the definition of the EA and are necessary to produce an executable version. The values of these parameters greatly determine whether the algorithm will find an optimal or near-optimal solution, and whether it will find such a solution efficiently. Choosing the right parameter values is, however, a hard task.

A.E. Eiben et al.: *Parameter Control in Evolutionary Algorithms*, Studies in Computational Intelligence (SCI) **54**, 19–46 (2007)
www.springerlink.com

Globally, we distinguish two major forms of setting parameter values: **parameter tuning** and **parameter control**. By parameter tuning we mean the commonly practised approach that amounts to finding good values for the parameters *before* the run of the algorithm and then running the algorithm using these values, which remain fixed during the run. Later on in this section we give arguments that any static set of parameters having the values fixed during an EA run seems to be inappropriate. Parameter control forms an alternative, as it amounts to starting a run with initial parameter values that are changed *during* the run.

Parameter tuning is a typical approach to algorithm design. Such tuning is done by experimenting with different values and selecting the ones that give the best results on the test problems at hand. However, the number of possible parameters and their different values means that this is a very time-consuming activity. Considering four parameters and five values for each of them, one has to test $5^4 = 625$ different setups. Performing 100 independent runs with each setup, this implies 62,500 runs just to establish a good algorithm design.

The technical drawbacks to parameter tuning based on experimentation can be summarised as follows:

- Parameters are not independent, but trying all different combinations systematically is practically impossible.
- The process of parameter tuning is time consuming, even if parameters are optimised one by one, regardless of their interactions.
- For a given problem the selected parameter values are not necessarily optimal, even if the effort made for setting them was significant.

This picture becomes even more discouraging if one is after a "generally good" setup that would perform well on a range of problems or problem instances. During the history of EAs considerable effort has been spent on finding parameter values (for a given type of EA, such as GAs), that were good for a number of test problems. A well-known early example is that of [18], determining recommended values for the probabilities of single-point crossover and bit mutation on what is now called the DeJong test suite of five functions. About this and similar attempts [34, 62], it should be noted that genetic algorithms used to be seen as robust problem solvers that exhibit approximately the same performance over a wide range of problems [33, page 6]. The contemporary view on EAs, however, acknowledges that specific problems (problem types) require specific EA setups for satisfactory performance [12]. Thus, the scope of "optimal" parameter settings is necessarily narrow. There are also theoretical arguments that any quest for generally good EA, thus generally good parameter settings, is lost a priori, such as the No Free Lunch theorem [83].

To elucidate another drawback of the parameter tuning approach recall how we defined it: finding good values for the parameters before the run of the algorithm and then running the algorithm using these values, *which remain fixed during the run.* However, a run of an EA is an intrinsically dynamic, adaptive process. The use of rigid parameters that do not change their values is thus in contrast to this spirit. Additionally, it is intuitively obvious, and has been empirically and theoretically demonstrated, that different values of parameters might be optimal at different stages of the evolutionary process [6, 7, 8, 16, 39, 45, 63, 66, 68, 72, 73, 78, 79].

To give an example, large mutation steps can be good in the early generations, helping the exploration of the search space, and small mutation steps might be needed in the late generations to help fine-tune the suboptimal chromosomes. This implies that the use of static parameters itself can lead to inferior algorithm performance.

A straightforward way to overcome the limitations of static parameters is by replacing a parameter p by a function $p(t)$, where t is the generation counter (or any other measure of elapsed time). However, as indicated earlier, the problem of finding optimal static parameters for a particular problem is already hard. Designing optimal dynamic parameters (that is, functions for $p(t)$) may be even more difficult. Another possible drawback to this approach is that the parameter value $p(t)$ changes are caused by a "blind" deterministic rule triggered by the progress of time t, without taking any notion of the actual progress in solving the problem, i.e., without taking into account the current state of the search. A well-known instance of this problem occurs in simulated annealing [49] where a so-called cooling schedule has to be set before the execution of the algorithm.

Mechanisms for modifying parameters during a run in an "informed" way were realised quite early in EC history. For instance, evolution strategies changed mutation parameters on-the-fly by Rechenberg's 1/5 success rule using information on the ratio of successful mutations. Davis experimented within GAs with changing the crossover rate based on the progress realised by particular crossover operators [16]. The common feature of these and similar approaches is the presence of a human-designed feedback mechanism that utilises actual information about the search process for determining new parameter values.

Yet another approach is based on the observation that finding good parameter values for an evolutionary algorithm is a poorly structured, ill-defined, complex problem. This is exactly the kind of problem on which EAs are often considered to perform better than other methods. It is thus a natural idea to use an EA for tuning an EA to a particular problem. This could be done using two EAs: one for problem solving and another one – the so-called meta-EA – to tune the first one [32, 34, 48]. It could also be done by using only one EA that tunes itself to a given problem, while solving that problem. Self-adaptation, as introduced in Evolution Strategies for varying the mutation parameters, falls within this category. In the next section we discuss various options for changing parameters, illustrated by an example.

2 A case study: Evolution Strategies

The history of Evolution Strategies (ES) is a typical case study for parameter tuning, as it went through several successive steps pertaining to many of the different approaches listed so far.

Typical ES work in a real-valued search space (typically $\mathbb{R}^n$, or a subset of $\mathbb{R}^n$, for some integer n).

2.1 Gaussian Mutations

The main operator (and almost the trademark) of ES is the Gaussian mutation, that adds centered normally distributed noise to the variables of the individuals. The most general Gaussian distribution in $\mathbb{R}^n$ is the multivariate normal distribution $\mathcal{N}(m, C)$, with mean m and **covariance matrix** C, a $n \times n$ positive definite matrix, that has the following Probability Distribution Function

$$\Phi(X) = \frac{\exp(-\frac{1}{2}(X-m)^t C^{-1}(X-m))}{\sqrt{(2\pi)^n |C|}}$$

where $|C|$ is the determinant of C.

It is then convenient to write the mutation of a vector $X \in \mathbb{R}^n$ as

$$X \rightarrow X + \sigma N(0, C)$$

i.e. to distinguish a scaling factor σ, also called the **step-size**, from the directions of the Gaussian distribution given by the covariance matrix C.

For example, the simplest case of Gaussian mutation assumes that C is the identity matrix in $\mathbb{R}^n$ (the diagonal matrix that has only 1s on the diagonal). In this case, all coordinates will be mutated independently, and will have added to them some Gaussian noise with variance σ^2.

Tuning an ES algorithm therefore amounts to tuning the step-size and the covariance matrix – or simply tuning the step-size in the simple case mentioned above.

2.2 Adapting the Step-Size

The step-size of the Gaussian mutation gives the scale of the search. To make things clear, suppose that you are minimizing x^2 in one dimension, running a (1+1)-ES (one parent gives birth to one offspring, and the best of both is the next parent) with a fixed step-size σ. Then the average distance between parent and successful offspring is proportional to σ. This has two consequences: first, starting from distance d_0 from the solution, it will take an average of d_0/σ steps to reach a region close to the optimum. On the other hand, when hovering around the optimum, the precision you can hope is again proportional to σ. Those arguments naturally lead to the optimal **adaptive** setting of the step-size for the sphere function: σ should be proportional to the distance to the optimum. Details can be found in the studies of the so-called *progress rate*: early work was done by Schwefel [66], completed and extended by Beyer and recent work by Auger [5] gave a formal global convergence proof of this... impractical algorithm: indeed, the distance to the optimum is not known real situations!

But another piece of information is always available to the algorithm: the *success rate* (the proportion of successful mutations, where the offspring is better than the parent). This can indirectly give information about the step-size: this was Rechenberg's main idea to propose the first practical method for an **adaptive** step-size, the so-called *one-fifth rule*: if the success rate over some time window is larger than the success rate when the step-size is optimal (0.2, or one-fifth!), the the step-size should be increased (the algorithm is making too many small steps); on the other hand, if the success rate is smaller than 0.2, then the step-size should be decreased (the algorithm is constantly missing the target, because it shoots too far). Though formally derived from studies on the sphere function and the corridor function (a bounded linear function), the one-fifth rule was generalized to any function.

However, there are many situations where the one-fifth rule can be mislead. Moreover, it does not in any way handle the case of non-isotropic functions, where a non-diagonal covariance matrix is mandatory. Hence, it is no longer used today.

2.3 Self-Adaptive ES

The next big step in ESs was the invention of the **self-adaptive** mutation: the parameters of the mutation (both the step-size and the covariance matrix) are attached to each individual, and are subject to mutation, too. Those personal mutation parameters range from a single step-size, leading to *isotropic* mutation, where all coordinates are mutated independently with the same variance, to the *non-isotropic* mutation, that use a vector of n "standard deviations" σ_i, equivalent to a diagonal matrix C with σ_i on the diagonal, and to the *correlated* mutations, where a full covariance matrix is attached to each individual. Mutating an individual then amounts to first mutating the mutation parameters themselves, and then mutating the variables using the new mutation parameters. Details can be found in [66, 13].

The rationale for SA-ES are that algorithm relies on the selection step to keep in the population not only the best fit individuals, but also the individuals with the best mutation parameters – according to the region of the search space they are in. Indeed, although the selection acts based on the fitness, the underlying idea beneath Self-Adaptive ES (SA-ES) is that if two individuals starts with the same fitness, the offspring of one that has "better" mutation parameters will reach regions of higher fitness faster than the offspring of the other: selection will hence keep the ones with the good mutation parameters. This is what has often been stated as *"mutation parameters are optimized for free"* by the evolution itself. And indeed, SA-ES have long been the state-of-the-art in parametric optimization [9].

But what are "good" mutation parameters? The issue has already been discussed for the step-size in previous section, and similar arguments can be given for the covariance matrix itself. Replace the sphere model (min $\sum x_i^2 \equiv X^t X$) with an *elliptic* function (min $\frac{1}{2} X^t H X$ for some positive definite matrix H). Then it is clear that the mutation should progress slower along the directions of steepest descent of H: the covariance matrix should be proportional to H_{-1}. And whereas the step-size actually self-adapts to quasi-optimal values [9, 19], the covariance matrix that is learned by the *correlated* SA-ES is not the actual inverse of the Hessian [4].

2.4 CMA-ES: a Clever Adaptation

Another defect of SA-ES is the relative slowness of adaptation of the mutation parameters: even for the simple case of the step-size, if the initial value is not the optimal one (proportional to the distance to the optimum in the case of the sphere function), it takes some time to reach that optimal value and to start being efficient.

This observation led Hansen and Ostermeier to propose deterministic schedules to adapt the mutation parameters in ES, hence heading back to an **adaptive** method for parameter tuning. Their method was first limited to the step-size [38], and later addressed the adaptation of the full covariance matrix [36]. The complete *Covariance Matrix Adaptation* (CMA-ES) algorithm was finally detailed (and its parameters carefully tuned) in [37] and an improvement for the update of the covariance matrix was proposed in [35].

The basic idea in CMA-ES is to use the path followed by the algorithm to deterministically update the different mutation parameters, and a simplified view is given by the following: suppose that the algorithm has made a series of steps in colinear directions; then the step-size should be increased, to allow larger steps and increase speed. Similar ideas undermine the covariance matrix update(s).

Indeed, using such clever learning method, CMA-ES proved to outperform most other stochastic algorithms for parametric optimization, as witnessed by its success in the 2005 contest that took place at CEC'2005.

2.5 Lessons Learned

This brief summary of ES history witnesses that

- Static parameters are not only hard but can be impossible to tune: there doesn't exist any good static value for the step-size in Gaussian mutation
- Adaptive methods use some information about the current state of the search, and are as good as the information they get: the *success rate* is a very raw information, and lead to the "easy-to-defeat" one-fifth rule, while

CMA-ES uses high-level information to cleverly update all the parameters of the most general Gaussian mutation
- Self-adaptive methods are efficient methods when applicable, i.e. when the only available selection (based on the fitness) can prevent bad parameters from proceeding to future generations. They outperform basic static and adaptive methods, but are outperformed by clever adaptive methods.

3 Case Study: Changing the Penalty Coefficients

Let us assume we deal with a numerical optimisation problem to minimise

$$f(\overline{x}) = f(x_1, \ldots, x_n),$$

subject to some inequality and equality constraints

$$g_i(\overline{x}) \leq 0, i = 1, \ldots, q,$$

and

$$h_j(\overline{x}) = 0, j = q+1, \ldots, m,$$

where the domains of the variables are given by lower and upper bounds $l_i \leq x_i \leq u_i$ for $1 \leq i \leq n$.

For such a numerical optimisation problem we may consider an evolutionary algorithm based on a floating-point representation, where each individual $\overline{x}$ in the population is represented as a vector of floating-point numbers $\overline{x} = \langle x_1, \ldots, x_n \rangle$.

In the previous section we described different ways to modify a parameter controlling mutation. Several other components of an EA have natural parameters, and these parameters are traditionally tuned in one or another way. Here we show that other components, such as the evaluation function (and consequently the fitness function) can also be parameterised and thus varied. While this is a less common option than tuning mutation (although it is practised in the evolution of variable-length structures for parsimony pressure [84]), it may provide a useful mechanism for increasing the performance of an evolutionary algorithm.

When dealing with constrained optimisation problems, penalty functions are often used. A common technique is the method of static penalties [60], which requires fixed user-supplied penalty parameters. The main reason for its widespread use is that it is the simplest technique to implement: it requires only the straightforward modification of the evaluation function as follows:

$$eval(\overline{x}) = f(\overline{x}) + W \cdot penalty(\overline{x}),$$

where f is the objective function, and $penalty(\overline{x})$ is zero if no violation occurs, and is positive,[5] otherwise. Usually, the *penalty* function is based on the

[5] For minimisation problems.

distance of a solution from the feasible region, or on the effort to "repair" the solution, i.e., to force it into the feasible region. In many methods a set of functions f_j $(1 \leq j \leq m)$ is used to construct the penalty, where the function f_j measures the violation of the jth constraint in the following way:

$$f_j(\overline{x}) = \begin{cases} \max\{0, g_j(\overline{x})\} & \text{if} \quad 1 \leq j \leq q, \\ |h_j(\overline{x})| & \text{if} \quad q+1 \leq j \leq m. \end{cases} \tag{1}$$

W is a user-defined weight, prescribing how severely constraint violations are weighted. In the most traditional penalty approach the weight W does not change during the evolution process. We sketch three possible methods of changing the value of W.

First, we can replace the static parameter W by a dynamic parameter, e.g., a function $W(t)$. Just as for the mutation parameter σ, we can develop a heuristic that modifies the weight W over time. For example, in the method proposed by Joines and Houck [46], the individuals are evaluated (at the iteration t) by a formula, where

$$eval(\overline{x}) = f(\overline{x}) + (C \cdot t)^{\alpha} \cdot penalty(\overline{x}),$$

where C and α are constants. Since

$$W(t) = (C \cdot t)^{\alpha},$$

the penalty pressure grows with the evolution time provided $1 \leq C, \alpha$.

Second, let us consider another option, which utilises feedback from the search process. One example of such an approach was developed by Bean and Hadj-Alouane [14], where each individual is evaluated by the same formula as before, but $W(t)$ is updated in every generation t in the following way:

$$W(t+1) = \begin{cases} (1/\beta_1) \cdot W(t) & \text{if } \overline{b}^i \in \mathcal{F} \quad\quad \text{for all} \quad t-k+1 \leq i \leq t, \\ \beta_2 \cdot W(t) & \text{if } \overline{b}^i \in \mathcal{S}-\mathcal{F} \text{ for all} \quad t-k+1 \leq i \leq t, \\ W(t) & \text{otherwise.} \end{cases}$$

In this formula, $\mathcal{S}$ is the set of all search points (solutions), $\mathcal{F} \subseteq \mathcal{S}$ is a set of all *feasible* solutions, $\overline{b}^i$ denotes the best individual in terms of the function *eval* in generation i, $\beta_1, \beta_2 > 1$, and $\beta_1 \neq \beta_2$ (to avoid cycling). In other words, the method decreases the penalty component $W(t+1)$ for the generation $t+1$ if all best individuals in the last k generations were feasible (i.e., in $\mathcal{F}$), and increases penalties if all best individuals in the last k generations were infeasible. If there are some feasible and infeasible individuals as best individuals in the last k generations, $W(t+1)$ remains without change.

Third, we could allow self-adaptation of the weight parameter, similarly to the mutation step sizes in the previous section. For example, it is possible to extend the representation of individuals into

$$\langle x_1, \ldots, x_n, W \rangle,$$

where W is the weight. The weight component W undergoes the same changes as any other variable x_i (e.g., Gaussian mutation and arithmetic recombination).

To illustrate this method,which is analogous to using a separate σ_i for each x_i, we need to redefine the evaluation function. Let us first introduce penalty functions for each constraint as per Eq. (1). Clearly, these penalties are all non-negative and are at zero if no constraints are violated. Then consider a vector of weights $\overline{w} = (w_1, \ldots, w_m)$, and define

$$eval(\overline{x}) = f(\overline{x}) + \sum_{j=1}^{m} w_j f_j(\overline{x}),$$

as the function to be minimised and also extend the representation of individuals into

$$\langle x_1, \ldots, x_n, w_1, \ldots, w_m \rangle.$$

Variation operators can then be applied to both the $\overline{x}$ and the $\overline{w}$ part of these chromosomes, realising a self-adaptation of the constraint weights, and thereby the fitness function.

It is important to note the crucial difference between self-adapting mutation step sizes and constraint weights. Even if the mutation step sizes are encoded in the chromosomes, the evaluation of a chromosome is *independent* from the actual σ values. That is,

$$eval(\langle \overline{x}, \overline{\sigma} \rangle) = f(\overline{x}),$$

for any chromosome $\langle \overline{x}, \overline{\sigma} \rangle$. In contrast, if constraint weights are encoded in the chromosomes, then we have

$$eval(\langle \overline{x}, \overline{w} \rangle) = f_{\overline{w}}(\overline{x}),$$

for any chromosome $\langle \overline{x}, W \rangle$. This could enable the evolution to "cheat" in the sense of making improvements by minimising the weights instead of optimising f and satisfying the constraints. Eiben et al. investigated this issue in [22] and found that using a specific tournament selection mechanism neatly solves this problem and enables the EA to solve constraints.

3.1 Summary

In the previous sections we illustrated how the mutation operator and the evaluation function can be controlled (adapted) during the evolutionary process. The latter case demonstrates that not only can the traditionally adjusted components, such as mutation, recombination, selection, etc., be controlled by parameters, but so can other components of an evolutionary algorithm. Obviously, there are many components and parameters that can be changed and tuned for optimal algorithm performance. In general, the three options we sketched for the mutation operator and the evaluation function are valid for any parameter of an evolutionary algorithm, whether it is population size, mutation step, the penalty coefficient, selection pressure, and so forth.

The mutation example above also illustrates the phenomenon of the **scope** of a parameter. Namely, the mutation step size parameter can have different domains of influence, which we call scope. Using the $\langle x_1, \ldots, x_n, \sigma_1, \ldots, \sigma_n \rangle$ model, a particular mutation step size applies only to one variable of a single individual. Thus, the parameter σ_i acts on a subindividual, or component, level. In the $\langle x_1, \ldots, x_n, \sigma \rangle$ representation, the scope of σ is one individual, whereas the dynamic parameter $\sigma(t)$ was defined to affect all individuals and thus has the whole population as its scope.

These remarks conclude the introductory examples of this section. We are now ready to attempt a classification of parameter control techniques for parameters of an evolutionary algorithm.

4 Classification of Control Techniques

In classifying parameter control techniques of an evolutionary algorithm, many aspects can be taken into account. For example:

1. *What* is changed? (e.g., representation, evaluation function, operators, selection process, mutation rate, population size, and so on)
2. *How* the change is made (i.e., deterministic heuristic, feedback-based heuristic, or self-adaptive)
3. *The evidence* upon which the change is carried out (e.g., monitoring performance of operators, diversity of the population, and so on)
4. *The scope/level* of change (e.g., population-level, individual-level, and so forth)

In the following we discuss these items in more detail.

4.1 *What* is Changed?

To classify parameter control techniques from the perspective of what component or parameter is changed, it is necessary to agree on a list of all major components of an evolutionary algorithm, which is a difficult task in itself. For that purpose, let us assume the following components of an EA:

- Representation of individuals
- Evaluation function
- Variation operators and their probabilities
- Selection operator (parent selection or mating selection)
- Replacement operator (survival selection or environmental selection)
- Population (size, topology, etc.)

Note that each component can be parameterised, and that the number of parameters is not clearly defined. For example, an offspring $\overline{v}$ produced by an arithmetical crossover of k parents $\overline{x}_1, \ldots, \overline{x}_k$ can be defined by the following formula:

$$\overline{v} = a_1\overline{x}_1 + \ldots + a_k\overline{x}_k,$$

where $a_1, \ldots, a_k$, and k can be considered as parameters of this crossover. Parameters for a population can include the number and sizes of subpopulations, migration rates, and so on for a general case, when more then one population is involved. Despite the somewhat arbitrary character of this list of components and of the list of parameters of each component, we will maintain the "what-aspect" as one of the main classification features, since this allows us to locate where a specific mechanism has its effect.

4.2 *How* are Changes Made?

As discussed and illustrated in the two earlier case studies, methods for changing the value of a parameter (i.e., the "how-aspect") can be classified into one of three categories.

- **Deterministic parameter control**
  This takes place when the value of a strategy parameter is altered by some deterministic rule. This rule modifies the strategy parameter in a fixed, predetermined (i.e., user-specified) way without using any feedback from the search. Usually, a time-varying schedule is used, i.e., the rule is used when a set number of generations have elapsed since the last time the rule was activated.
- **Adaptive parameter control**
  This takes place when there is some form of feedback from the search that serves as inputs to a mechanism used to determine the direction or magnitude of the change to the strategy parameter. The assignment of the value of the strategy parameter may involve credit assignment, based on the quality of solutions discovered by different operators/parameters, so that the updating mechanism can distinguish between the merits of competing strategies. Although the subsequent action of the EA may determine whether or not the new value persists or propagates throughout the population, the important point to note is that the updating mechanism used to control parameter values is externally supplied, rather than being part of the "standard" evolutionary cycle.
- **Self-adaptive parameter control**
  The idea of the evolution of evolution can be used to implement the self-adaptation of parameters (see [10] for a good review). Here the parameters to be adapted are encoded into the chromosomes and undergo mutation and recombination. The better values of these encoded parameters lead to better individuals, which in turn are more likely to survive and produce offspring and hence propagate these better parameter values. This is an

important distinction between adaptive and self-adaptive schemes: in the latter the mechanisms for the credit assignment and updating of different strategy parameters are entirely implicit, i.e., they are the selection and variation operators of the evolutionary cycle itself.

This terminology leads to the taxonomy illustrated in Fig. 1.

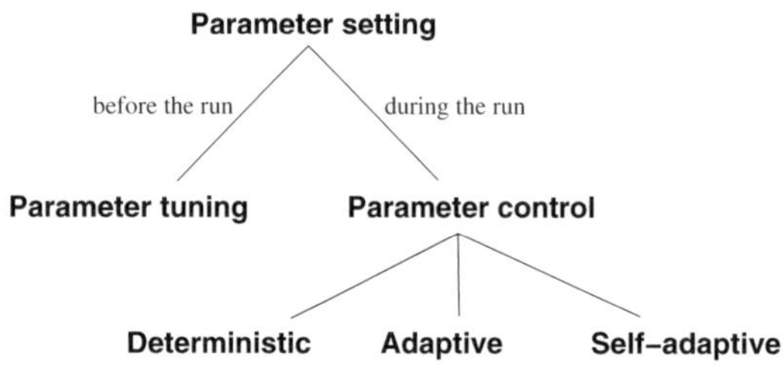

Fig. 1. Global taxonomy of parameter setting in EAs

Some authors have introduced a different terminology. Angeline [2] distinguished "absolute" and "empirical" rules, which correspond to the "uncoupled" and "tightly-coupled" mechanisms of Spears [76]. Let us note that the uncoupled/absolute category encompasses deterministic and adaptive control, whereas the tightly-coupled/empirical category corresponds to self-adaptation. We feel that the distinction between deterministic and adaptive parameter control is essential, as the first one does not use any feedback from the search process. However, we acknowledge that the terminology proposed here is not perfect either. The term "deterministic" control might not be the most appropriate, as it is not determinism that matters, but the fact that the parameter-altering transformations take no input variables related to the progress of the search process. For example, one might *randomly* change the mutation probability after every 100 generations, which is not a deterministic process. The name "fixed" parameter control might provide an alternative that also covers this latter example. Also, the terms "adaptive" and "self-adaptive" could be replaced by the equally meaningful "explicitly adaptive" and "implicitly adaptive" controls, respectively. We have chosen to use "adaptive" and "self-adaptive" for the widely accepted usage of the latter term.

4.3 What *Evidence* Informs the Change?

The third criterion for classification concerns the evidence used for determining the change of parameter value [67, 74]. Most commonly, the progress of the search is monitored, e.g., by looking at the performance of operators, the diversity of the population, and so on. The information gathered by such a monitoring process is used as feedback for adjusting the parameters. From this perspective, we can make further distinction between the following two cases:

- **Absolute evidence**
  We speak of absolute evidence when the value of a strategy parameter is altered by some rule that is applied when a predefined event occurs. The difference from deterministic parameter control lies in the fact that in deterministic parameter control a rule fires by a deterministic trigger (e.g., time elapsed), whereas here feedback from the search is used. For instance, the rule can be applied when the measure being monitored hits a previously set threshold – this is the event that forms the evidence. Examples of this type of parameter adjustment include increasing the mutation rate when the population diversity drops under a given value [53], changing the probability of applying mutation or crossover according to a fuzzy rule set using a variety of population statistics [52], and methods for resizing populations based on estimates of schemata fitness and variance [75]. Such mechanisms require that the user has a clear intuition about how to steer the given parameter into a certain direction in cases that can be specified in advance (e.g., they determine the threshold values for triggering rule activation). This intuition may be based on the encapsulation of practical experience, data-mining and empirical analysis of previous runs, or theoretical considerations (in the order of the three examples above), but all rely on the implicit assumption that changes that were appropriate to make on *another* search of *another* problem are applicable to *this* run of the EA on *this* problem.
- **Relative evidence**
  In the case of using relative evidence, parameter values are compared according to the fitness of the offspring that they produce, and the better values get rewarded. The direction and/or magnitude of the change of the strategy parameter is not specified deterministically, but relative to the performance of other values, i.e., it is necessary to have more than one value present at any given time. Here, the assignment of the value of the strategy parameter involves credit assignment, and the action of the EA may determine whether or not the new value persists or propagates throughout the population. As an example, consider an EA using more crossovers with crossover rates adding up to 1.0 and being reset based on the crossovers performance measured by the quality of offspring they create. Such methods may be controlled adaptively, typically using "bookkeeping" to monitor performance and a user-supplied update procedure [16, 47, 64], or self-adaptively [7, 29, 51, 65, 71, 76] with the selection operator acting indirectly on operator or parameter frequencies via their association with "fit" solutions.

4.4 What is the *Scope* of the Change?

As discussed earlier, any change within any component of an EA may affect a gene (parameter), whole chromosomes (individuals), the entire population, another component (e.g., selection), or even the evaluation function. This is the aspect of the scope or level of adaptation [2, 40, 74]. Note, however, that the scope or level is not an independent dimension, as it usually depends on the component of the EA where the change takes place. For example, a change of the mutation step size may affect a gene, a chromosome, or the entire population, depending on the particular implementation (i.e., scheme used), but a change in the penalty coefficients typically affects the whole population. In this respect the scope feature is a secondary one, usually depending on the given component and its actual implementation.

It should be noted that the issue of the scope of the parameter might be more complicated than indicated in Sect. 3.1. First of all, the scope depends on the interpretation mechanism of the given parameters. For example, an individual might be represented as

$$\langle x_1, \ldots, x_n, \sigma_1, \ldots, \sigma_n, \alpha_1, \ldots, \alpha_{n(n-1)/2} \rangle,$$

where the vector $\overline{\alpha}$ denotes the covariances between the variables $\sigma_1, \ldots, \sigma_n$. In this case the scope of the strategy parameters in $\overline{\alpha}$ is the whole individual, although the notation might suggest that they act on a subindividual level.

The next example illustrates that the same parameter (encoded in the chromosomes) can be interpreted in different ways, leading to different algorithm variants with different scopes of this parameter. Spears [76], following [30], experimented with individuals containing an extra bit to determine whether one-point crossover or uniform crossover is to be used (bit 1/0 standing for one-point/uniform crossover, respectively). Two interpretations were considered. The first interpretation was based on a pairwise operator choice: If both parental bits are the same, the corresponding operator is used; otherwise, a random choice is made. Thus, this parameter in this interpretation acts at an individual level. The second interpretation was based on the bit distribution over the whole population: If, for example, 73% of the population had bit 1, then the probability of one-point crossover was 0.73. Thus this parameter under this interpretation acts on the population level. Spears noted that there was a definite impact on performance, with better results arising from the individual level scheme, and more recently Smith [69] compared three versions of a self-adaptive recombination operator, concluding that the component-level version significantly outperformed the individual or population-level versions.

However, the two interpretations of Spears' scheme can be easily combined. For instance, similar to the first interpretation, if both parental bits are the same, the corresponding operator is used, but if they differ, the operator is selected according to the bit distribution, just as in the second interpretation. The scope/level of this parameter in this interpretation is neither individual nor population, but rather both. This example shows that the notion of scope can be ill-defined and very complex. This, combined with the arguments that the scope or level entity is primarily a feature of the given parameter and only secondarily a feature of adaptation itself, motivates our decision to exclude it as a major classification criterion.

4.5 Summary

In conclusion, the main criteria for classifying methods that change the values of the strategy parameters of an algorithm during its execution are:

1. What component/parameter is changed?
2. How is the change made?
3. Which evidence is used to make the change?

Our classification is thus three-dimensional. The *component* dimension consists of six categories: representation, evaluation function, variation operators (mutation and recombination), selection, replacement, and population. The other dimensions have respectively three (deterministic, adaptive, self-adaptive) and two categories (absolute, relative). Their possible combinations are given in Table 1. As the table indicates, deterministic parameter control with relative evidence is impossible by definition, and so is self-adaptive parameter control with absolute evidence. Within the adaptive scheme both options are possible and are indeed used in practice.

	Deterministic	Adaptive	Self-adaptive
Absolute	+	+	–
Relative	–	+	+

Table 1. Refined taxonomy of parameter setting in EAs: types of parameter control along the type and evidence dimensions. The – entries represent meaningless (nonexistent) combinations

5 Examples of Varying EA Parameters

Here we review some illustrative examples from the literature concerning all major components. For a more comprehensive overview the reader is referred to [22].

5.1 Representation

The choice of representation forms an important distinguishing feature between different streams of evolutionary computing. From this perspective GAs and ES can be distinguished from (historical) EP and GP according to the data structure used to represent individuals. In the first group this data structure is linear, and its length is fixed, that is, it does not change during a run of the algorithm. For (historical) EP and GP this does not hold: finite state machines and parse trees are nonlinear structures, and their size (the number of states, respectively nodes) and shape can change during a run. It could be argued that this implies an intrinsically adaptive representation in traditional EP and GP. On the other hand, the main structure of the finite state machines does not change during the search in traditional EP, nor do the function and terminal sets in GP (without automatically defined functions, ADFs). If one identifies "representation" with the basic syntax (plus the encoding mechanism), then the differently sized and shaped finite state machines, respectively trees, are only different expressions in this unchanging syntax. Based on this view we do not consider the representations in traditional EP and GP intrinsically adaptive.

We illustrate variable representations with the delta coding algorithm of Mathias and Whitley [82], which effectively modifies the encoding of the function parameters. The motivation behind this algorithm is to maintain a good balance between fast search and sustaining diversity. In our taxonomy it can be categorised as an adaptive adjustment of the representation based on absolute evidence.

The GA is used with multiple restarts; the first run is used to find an *interim solution*, and subsequent runs decode the genes as distances (*delta values*) from the last interim solution. This way each restart forms a new hypercube with the interim solution at its origin. The resolution of the delta values can also be altered at the restarts to expand or contract the search space. The restarts are triggered when population diversity (measured by the Hamming distance between the best and worst strings of the current population) is not greater than one. The sketch of the algorithm showing the main idea is given in Fig. 2.

Note that the number of bits for δ can be increased if the same solution `INTERIM` is found. This technique was further refined in [57, 58] to cope with deceptive problems.

5.2 Evaluation Function

Evaluation functions are typically not varied in an EA because they are often considered as part of the problem to be solved and not as part of the problem-solving algorithm. In fact, an evaluation function forms the bridge between the two, so both views are at least partially true. In many EAs the evaluation function is derived from the (optimisation) problem at hand with a simple transformation of the objective function. In the class of constraint satisfaction problems, however, there is no objective function in the problem definition [20]. Rather, these are normally posed as decision problems with an Boolean outcome ϕ denoting whether a given assignment of variables represents a valid

```
BEGIN
  /* given a starting population and genotype-phenotype encoding */
  WHILE (  HD > 1 ) DO
    RUN_GA with k bits per object variable;
  OD
  REPEAT UNTIL (  global termination is satisfied ) DO
    save best solution as INTERIM;
    reinitialise population with new coding;
    /*  k-1 bits as the distance δ to the object value in  */
    /*  INTERIM and one sign bit */
    WHILE (  HD > 1 ) DO
      RUN_GA with this encoding;
    OD
  OD
END
```

Fig. 2. Outline of the delta coding algorithm

solution. One possible approach using EAs is to treat these as minimisation problems where the evaluation function is defined as the amount of constraint violation by a given candidate solution. This approach, commonly known as the penalty approach, can be formalised as follows. Let us assume that we have constraints c_i ($i = \{1, \ldots, m\}$) and variables v_j ($j = \{1, \ldots, n\}$) with the same domain S. The task is to find one variable assignment $\bar{s} \in S$ satisfying all constraints. Then the penalties can be defined as follows:

$$f(\bar{s}) = \sum_{i=1}^{m} w_i \times \chi(\bar{s}, c_i),$$

where

$$\chi(\bar{s}, c_i) = \begin{cases} 1 \text{ if } \bar{s} \text{ violates } c_i, \\ 0 \text{ otherwise.} \end{cases}$$

Obviously, for each $\bar{s} \in S$ we have that $\phi(\bar{s}) = \mathit{true}$ if and only if $f(\bar{s}) = 0$, and the weights specify how severely the violation of a certain constraint is penalised. The setting of these weights has a large impact on the EA performance, and ideally w_i should reflect how hard c_i is to satisfy. The problem is that finding the appropriate weights requires much insight into the given problem instance, and therefore it might not be practicable.

The stepwise adaptation of weights (SAW) mechanism, introduced by Eiben and van der Hauw [26] as an improved version of the weight adaptation mechanism of Eiben, Raué, and Ruttkay [23, 24], provides a simple and effective way to set these weights. The basic idea behind the SAW mechanism is that constraints that are not satisfied after a certain number of steps (fitness evaluations) must be difficult, and thus must be given a high weight (penalty). SAW-ing changes the evaluation function adaptively in an EA by periodically checking the best individual in the population and raising the weights of those constraints this individual violates. Then the run continues with the new evaluation function. A nice feature of SAW-ing is that it liberates the user from seeking good weight settings, thereby eliminating a possible source of error. Furthermore, the used weights reflect the difficulty of constraints for *the given algorithm* on the *given problem instance* in *the given stage of the search* [27]. This property is also valuable since, in principle, different weights could be appropriate for different algorithms.

5.3 Mutation

A large majority of work on adapting or self-adapting EA parameters concerns variation operators: mutation and recombination (crossover). As we discussed above, the 1/5 rule of Rechenberg constitutes a classical example for adaptive mutation step size control in ES. In the same paper we also showed that self-adaptive control of mutation step sizes is traditional in ES.

Hesser and Männer [39] derived theoretically optimal schedules within GAs for deterministically changing p_m for the counting-ones function. They suggest:

$$p_m(t) = \sqrt{\frac{\alpha}{\beta}} \times \frac{\exp\left(\frac{-\gamma t}{2}\right)}{\lambda\sqrt{L}},$$

where α, β, γ are constants, L is the chromosome length, λ is the population size, and t is the time (generation counter). This is a purely deterministic parameter control mechanism.

A self-adaptive mechanism for controlling mutation in a bit-string GA is given by Bäck [6]. This technique works by extending the chromosomes by an additional 20 bits that together encode the individuals' own p_m. Mutation then works by:

1. Decoding these bits first to p_m
2. Mutating the bits that encode p_m with mutation probability p_m
3. Decoding these (changed) bits to p'_m
4. Mutating the bits that encode the solution with mutation probability p'_m

This approach is highly self-adaptive since even the rate of variation of the search parameters is given by the encoded value, as opposed to the use of an external parameters such as learning rates for step-sizes. More recently Smith [70] showed theoretical predictions, verified experimentally, that this scheme gets "stuck" in suboptimal regions of the search space with a low, or zero, mutation rate attached to each member of the population. He showed that a more robust problem-solving mechanism can simply be achieved by ignoring the first step of the algorithm above, and instead using a fixed learning rate as the probability of applying bitwise mutation to the encoding of the strategy parameters in the second step.

5.4 Crossover

The classical example for adapting crossover rates in GAs is Davis's adaptive operator fitness. The method adapts the rates of crossover operators by rewarding those that are successful in creating better offspring. This reward is diminishingly propagated back to operators of a few generations back, who helped setting it all up; the reward is a shift up in probability at the cost of other operators [17]. This, actually, is very close in spirit to the "implicit bucket brigade" credit assignment principle used in classifier systems [33].

The GA using this method applies several crossover operators simultaneously within the same generation, each having its own crossover rate $p_c(op_i)$. Additionally, each operator has its "local delta" value d_i that represents the strength of the operator measured by the advantage of a child created by using that operator with respect to the best individual in the population. The local deltas are updated after every use of operator i. The adaptation mechanism recalculates the crossover rates after K generations. The main idea is to redistribute 15% of the probabilities biased by the accumulated operator strengths, that is, the local deltas. To this end, these d_i values are normalised so that their sum equals 15, yielding d_i^{norm} for each i. Then the new value for each $p_c(op_i)$ is 85% of its old value and its normalised strength:

$$p_c(op_i) = 0.85 \cdot p_c(op_i) + d_i^{norm}.$$

Clearly, this method is adaptive based on relative evidence.

5.5 Selection

It is interesting to note that neither the parent selection nor the survivor selection (replacement) component of an EA has been commonly used in an adaptive manner, even though there are selection methods whose parameters can be easily adapted. For example, in linear ranking there is a parameter s representing the expected number of offspring to be allocated to the best individual. By changing this parameter within the range of $[1 \ldots 2]$ the selective pressure of the algorithm can be varied easily. Similar possibilities exist for tournament selection, where the tournament size provides a natural parameter.

Most existing mechanisms for varying the selection pressure are based on the so-called **Boltzmann** selection mechanism, which changes the selection pressure during evolution according to a predefined "cooling schedule" [55]. The name originates from the Boltzmann trial from condensed matter physics, where a minimal energy level is sought by state transitions. Being in a state i the chance of accepting state j is

$$P[\text{accept } j] = \begin{cases} 1 & \text{if } E_i \geq E_j, \\ \exp\left(\frac{E_i - E_j}{K_b \cdot T}\right) & \text{if } E_i < E_j, \end{cases}$$

where E_i, E_j are the energy levels, K_b is a parameter called the Boltzmann constant, and T is the temperature. This acceptance rule is called the Metropolis criterion.

We illustrate variable selection pressure in the survivor selection (replacement) step by **simulated annealing** (SA). SA is a generate-and-test search technique based on a physical, rather than a biological analogy [1]. Formally, however, SA can be envisioned as an evolutionary process with population size of 1, undefined (problem-dependent) representation and mutation, and a specific survivor selection mechanism. The selective pressure changes during the course of the algorithm in the Boltzmann style. The main cycle in SA is given in Fig. 3.

```
BEGIN
  /* given a current solution i ∈ S */
  /* given a function to generate the set of neighbours N_i of i */
  generate j ∈ N_i;
  IF (f(i) < f(j)) THEN
    set i = j;
    ELSE
      IF ( exp((f(i)-f(j))/c_k) > random[0,1)) THEN
        set i = j;
      FI
    ESLE
  FI
END
```

Fig. 3. Outline of the simulated annealing algorithm

In this mechanism the parameter c_k, the temperature, decreases according to a predefined scheme as a function of time, making the probability of accepting inferior solutions smaller and smaller (for minimisation problems). From an evolutionary point of view, we have here a (1+1) EA with increasing selection pressure.

A successful example of applying Boltzmann acceptance is that of Smith and Krasnogor [50], who used it in the local search part of a memetic algorithm (MA), with the temperature inversely related to the fitness diversity of the population. If the population contains a wide spread of fitness values, the "temperature" is low, so only fitter solutions found by local search are likely to be accepted, concentrating the search on good solutions. However, when the spread of fitness values is low, indicating a converged population which is a common problem in MAs, the "temperature" is higher, making it more likely that an inferior solution will be accepted, thus reintroducing diversity and offering a potential means of escaping from local optima.

5.6 Population

An innovative way to control the population size is offered by Arabas et al. [3, 59] in their GA with variable population size (GAVaPS). In fact, the population size parameter is removed entirely from GAVaPS, rather than adjusted on-the-fly. Certainly, in an evolutionary algorithm the population always has a size, but in GAVaPS this size is a derived measure, not a controllable parameter. The main idea is to assign a lifetime to each individual when it is created, and then to reduce its remaining lifetime by one in each consecutive generation. When the remaining lifetime becomes zero, the individual is removed from the population. Two things must be noted here. First, the lifetime allocated to a newborn individual is biased by its fitness: fitter individuals are allowed to live longer. Second, the expected number of offspring of an individual is proportional to the number of generations it survives. Consequently, the resulting system favours the propagation of good genes.

Fitting this algorithm into our general classification scheme is not straightforward because it has no explicit mechanism that sets the value of the population size parameter. However, the procedure that implicitly determines how many individuals are alive works in an adaptive fashion using information about the status of the search. In particular, the fitness of a newborn individual is related to the fitness of the present generation, and its lifetime is allocated accordingly. This amounts to using relative evidence.

5.7 Varying Several Parameters Simultaneously

One of the studies explicitly devoted to adjusting more parameters (and also on more than one level) is that of Hinterding et al. on a "self-adaptive GA" [41]. This GA uses self-adaptation for mutation rate control, plus relative-based adaptive control for the population size.[6] The mechanism for controlling mutation is similar to that from Bäck [6], (Sect. 5.3), except that mutating the bits encoding the mutation strength is not based on the bits in question, but is done by a universal mechanism fixed for all individuals and all generations. In other words, the self-adaptive mutation parameter is only used for the genes encoding a solution. As for the population size, the GA works with three subpopulations: a small, a medium, and a large one, P1, P2, and P3, respectively

[6] Strictly speaking, the authors' term "self-adaptive GA" is only partially correct. However, this paper is from 1996, and the contemporary terminology distinguishing dynamic, adaptive, and self-adaptive schemes as we do it here was only published in 1999 [22].

(the initial sizes respectively being 50, 100, and 200). These populations are evolved in parallel for a given number of fitness evaluations (an epoch) independently by the same GA setup. After each epoch, the subpopulations are resized based on some heuristic rules, maintaining a lower and an upper bound (10 and 1000) and keeping P2 always the medium-sized subpopulation. There are two categories of rules. Rules in the first category are activated when the fitnesses in the subpopulations converge and try to move the populations apart. For instance, if P2 and P3 have the same fitness, the size of P3 is doubled. Rules from another set are activated when the fitness values are distinct at the end of an epoch. These rules aim at maximising the performance of P2. An example of one such rule is: if the performance of the subpopulations ranks them as P2 $<$ P3 $<$ P1 then size(P3) = (size(P2) + size(P3))/2. In our taxonomy, this population size control mechanism is adaptive, based on relative evidence.

Lis and Lis [54] also offer a parallel GA setup to control the mutation rate, the crossover rate, and the population size during a run. The idea here is that for each parameter a few possible values are defined in advance, say *lo, med, hi*, and only these values are allowed in any of the GAs, that is, in the subpopulations evolved in parallel. After each epoch the performances of the applied parameter values are compared by averaging the fitnesses of the best individuals of those GAs that use a given value. If the winning parameter value is:

1. *hi*, then all GAs shift one level up concerning this parameter in the next epoch;
2. *med*, then all GAs use the same value concerning this parameter in the next epoch;
3. *lo*, then all GAs shift one level down concerning this parameter in the next epoch.

Clearly, the adjustment mechanism for all parameters here is adaptive, based on relative evidence.

Mutation, crossover, and population size are all controlled on-the-fly in the GA "without parameters" of Bäck et al. in [11]. Here, the self-adaptive mutation from [6] (Sect. 5.3) is adopted without changes, a new self-adaptive technique is invented for regulating the crossover rates of the individuals, and the GAVaPS lifetime idea (Sect. 5.6) is adjusted for a steady-state GA model. The crossover rates are included in the chromosomes, much like the mutation rates. If a pair of individuals is selected for reproduction, then their individual crossover rates are compared with a random number $r \in [0, 1]$ and an individual is seen as ready to mate if its $p_c > r$. Then there are three possibilities:

1. If both individuals are ready to mate then uniform crossover is applied, and the resulting offspring is mutated.
2. If neither is ready to mate then both create a child by mutation only.

3. If exactly one of them is ready to mate, then the one not ready creates a child by mutation only (which is inserted into the population immediately through the steady-state replacement), the other is put on the hold, and the next parent selection round picks only one other parent.

This study differs from those discussed before in that it explicitly compares GA variants using only one of the (self-)adaptive mechanisms and the GA applying them all. The experiments show remarkable outcomes: the completely (self-)adaptive GA wins, closely followed by the one using only the adaptive population size control, and the GAs with self-adaptive mutation and crossover are significantly worse. These results suggest that putting effort into adapting the population size could be more effective than trying to adjust the variation operators. This is truly surprising considering that traditionally the on-line adjustment of the variation operators has been pursued and the adjustment of the population size received relatively little attention. The subject certainly requires more research.

6 Discussion

Summarising this paper a number of things can be noted. First, parameter control in an EA can have two purposes. It can be done to avoid suboptimal algorithm performance resulting from suboptimal parameter values set by the user. The basic assumption here is that the applied control mechanisms are intelligent enough to do this job better than the user could, or that they can do it approximately as good, but they liberate the user from doing it. Either way, they are beneficial. The other motivation for controlling parameters on-the-fly is the assumption that the given parameter can have a different "optimal" value in different phases of the search. If this holds, then there is simply no optimal static parameter value; for good EA performance one must vary this parameter.

The second thing we want to note is that making a parameter (self adaptive) does not necessarily mean that we have an EA with fewer parameters. For instance, in GAVaPS the population size parameter is eliminated at the cost of introducing two new ones: the minimum and maximum lifetime of newborn individuals. If the EA performance is sensitive to these new parameters then such a parameter replacement can make things worse. This problem also occurs on another level. One could say that the procedure that allocates lifetimes in GAVaPS, the probability redistribution mechanism for adaptive crossover rates (Sect. 5.4), or the function specifying how the σ values are mutated in ES are also (meta) parameters. It is in fact an assumption that these are intelligently designed and their effect is positive. In many cases there are more possibilities, that is, possibly well-working procedures one can design. Comparing these possibilities implies experimental (or theoretical) studies very much like comparing different parameter values in a classical setting.

Here again, it can be the case that algorithm performance is not so sensitive to details of this (meta) parameter, which fully justifies this approach.

Finally, let us place the issue of parameter control in a larger perspective. Over the last 20 years the EC community shifted from believing that EA performance is to a large extent independent from the given problem instance to realising that it is. In other words, it is now acknowledged that EAs need more or less fine-tuning to specific problems and problem instances. Ideally, it should be the algorithm that performs the necessary problem-specific adjustments. Parameter control as discussed here is a step towards this.

References

1. E.H.L. Aarts and J. Korst. *Simulated Annealing and Boltzmann Machines.* Wiley, Chichester, UK, 1989.
2. P.J. Angeline. Adaptive and self-adaptive evolutionary computations. In *Computational Intelligence*, pages 152–161. IEEE Press, 1995.
3. J. Arabas, Z. Michalewicz, and J. Mulawka. GAVaPS – a genetic algorithm with varying population size. In ICEC-94 [42], pages 73–78.
4. A. Auger. *Contributions théoriques et numériques à l'optimisation continue par algorithmes évolutionnaires.* PhD thesis, Université Paris 6, December 2004. in French.
5. A. Auger, C. Le Bris, and M. Schoenauer. Dimension-independent convergence rate for non-isotropic $(1, \lambda) - es$. In *proceedings of GECCO2003*, pages 512–524, 2003.
6. T. Bäck. The interaction of mutation rate, selection and self-adaptation within a genetic algorithm. In Männer and Manderick [56], pages 85–94.
7. T. Bäck. Self adaptation in genetic algorithms. In F.J. Varela and P. Bourgine, editors, *Toward a Practice of Autonomous Systems: Proceedings of the 1st European Conference on Artificial Life*, pages 263–271. MIT Press, Cambridge, MA, 1992.
8. T. Bäck. Optimal mutation rates in genetic search. In Forrest [31], pages 2–8.
9. T. Bäck. *Evolutionary Algorithms in Theory and Practice.* New-York:Oxford University Press, 1995.
10. T. Bäck. Self-adaptation. In T. Bäck, D.B. Fogel, and Z. Michalewicz, editors, *Evolutionary Computation 2: Advanced Algorithms and Operators*, chapter 21, pages 188–211. Institute of Physics Publishing, Bristol, 2000.
11. T. Bäck, A.E. Eiben, and N.A.L. van der Vaart. An empirical study on GAs "without parameters". In M. Schoenauer, K. Deb, G. Rudolph, X. Yao, E. Lutton, J.J. Merelo, and H.-P. Schwefel, editors, *Proceedings of the 6th Conference on Parallel Problem Solving from Nature*, number 1917 in Lecture Notes in Computer Science, pages 315–324. Springer, Berlin, Heidelberg, New York, 2000.
12. T. Bäck, D.B. Fogel, and Z. Michalewicz, editors. *Handbook of Evolutionary Computation.* Institute of Physics Publishing, Bristol, and Oxford University Press, New York, 1997.
13. T. Bäck, M. Schütz, and S. Khuri. A comparative study of a penalty function, a repair heuristic and stochastic operators with set covering problem. In *Proceedings of Evolution Artificial 1995*, number 1063 in LNCS. Springer-Verlag, 1995.

14. J.C. Bean and A.B. Hadj-Alouane. A dual genetic algorithm for bounded integer problems. Technical Report 92-53, University of Michigan, 1992.
15. R.K. Belew and L.B. Booker, editors. *Proceedings of the 4th International Conference on Genetic Algorithms.* Morgan Kaufmann, San Francisco, 1991.
16. L. Davis. 'Adapting operator probabilities in genetic algorithms. In Schaffer [61], pages 61–69.
17. L. Davis, editor. *Handbook of Genetic Algorithms.* Van Nostrand Reinhold, 1991.
18. K.A. De Jong. *An Analysis of the Behaviour of a Class of Genetic Adaptive Systems.* PhD thesis, University of Michigan, 1975.
19. K. Deb and H.-G. Beyer. Self-adaptive genetic algorithms with simulated binary crossover. *Evolutionary Computation.*
20. A.E. Eiben. Evolutionary algorithms and constraint satisfaction: Definitions, survey, methodology, and research directions. In L. Kallel, B. Naudts, and A. Rogers, editors, *Theoretical Aspects of Evolutionary Computing*, pages 13–58. Springer, Berlin, Heidelberg, New York, 2001.
21. A.E. Eiben, R. Hinterding, and Z. Michalewicz. Parameter control in evolutionary algorithms. *IEEE Transactions on Evolutionary Computation*, 3(2):124–141, 1999.
22. A.E. Eiben, B. Jansen, Z. Michalewicz, and B. Paechter. Solving CSPs using self-adaptive constraint weights: how to prevent EAs from cheating. In Whitley et al. [81], pages 128–134.
23. A.E. Eiben, P.-E. Raué, and Z. Ruttkay. GA-easy and GA-hard constraint satisfaction problems. In M. Meyer, editor, *Proceedings of the ECAI-94 Workshop on Constraint Processing*, number 923 in LNCS, pages 267–284. Springer, Berlin, Heidelberg, New York, 1995.
24. A.E. Eiben and Z. Ruttkay. Self-adaptivity for constraint satisfaction: Learning penalty functions. In ICEC-96 [43], pages 258–261.
25. A.E. Eiben and J.E. Smith. *Introduction to Evolutionary Computation.* Springer, 2003.
26. A.E. Eiben and J.K. van der Hauw. Solving 3-SAT with adaptive genetic algorithms. In ICEC-97 [44], pages 81–86.
27. A.E. Eiben and J.I. van Hemert. SAW-ing EAs: adapting the fitness function for solving constrained problems. In D. Corne, M. Dorigo, and F. Glover, editors, *New Ideas in Optimization*, Chapter 26, pages 389–402. McGraw Hill, London, 1999.
28. L.J. Eshelman, editor. *Proceedings of the 6th International Conference on Genetic Algorithms.* Morgan Kaufmann, San Francisco, 1995.
29. D.B. Fogel. *Evolutionary Computation.* IEEE Press, 1995.
30. D.B. Fogel and J.W. Atmar. Comparing genetic operators with Gaussian mutations in simulated evolutionary processes using linear systems. *Biological Cybernetics*, 63(2):111–114, 1990.
31. S. Forrest, editor. *Proceedings of the 5th International Conference on Genetic Algorithms.* Morgan Kaufmann, San Francisco, 1993.
32. B. Friesleben and M. Hartfelder. Optimisation of genetic algorithms by genetic algorithms. In R.F. Albrecht, C.R. Reeves, and N.C. Steele, editors, *Artifical Neural Networks and Genetic Algorithms*, pages 392–399. Springer, Berlin, Heidelberg, New York, 1993.
33. D.E. Goldberg. *Genetic Algorithms in Search, Optimization and Machine Learning.* Addison-Wesley, 1989.

34. J.J. Grefenstette. Optimisation of control parameters for genetic algorithms. *IEEE Transaction on Systems, Man and Cybernetics*, 16(1):122–128, 1986.
35. N. Hansen, S. Müller, and P. Koumoutsakos. Reducing the Time Complexity of the Derandomized Evolution Strategy with Covariance Matrix Adaptation (CMA-ES). *Evolution Computation*, 11(1), 2003.
36. N. Hansen and A. Ostermeier. Adapting arbitrary normal mutation distributions in evolution strategies: The covariance matrix adaption. In *ICEC96*, pages 312–317. IEEE Press, 1996.
37. N. Hansen and A. Ostermeier. Completely derandomized self-adaptation in evolution strategies. *Evolutionary Computation*, 9(2):159–195, 2001.
38. N. Hansen, A. Ostermeier, and A. Gawelczyk. On the adaptation of arbitrary normal mutation distributions in evolution strategies: The generating set adaptation. In Eshelman [28], pages 57–64.
39. J. Hesser and R. Manner. Towards an optimal mutation probablity in genetic algorithms. In H.-P. Schwefel and R. Männer, editors, *Proceedings of the 1st Conference on Parallel Problem Solving from Nature*, number 496 in Lecture Notes in Computer Science, pages 23–32. Springer, Berlin, Heidelberg, New York, 1991.
40. R. Hinterding, Z. Michalewicz, and A.E. Eiben. Adaptation in evolutionary computation: A survey. In ICEC-97 [44].
41. R. Hinterding, Z. Michalewicz, and T.C. Peachey. Self-adaptive genetic algorithm for numeric functions. In Voigt et al. [80], pages 420–429.
42. *Proceedings of the First IEEE Conference on Evolutionary Computation*. IEEE Press, Piscataway, NJ, 1994.
43. *Proceedings of the 1996 IEEE Conference on Evolutionary Computation*. IEEE Press, Piscataway, NJ, 1996.
44. *Proceedings of the 1997 IEEE Conference on Evolutionary Computation*. IEEE Press, Piscataway, NJ, 1997.
45. A. Jain and D.B. Fogel. Case studies in applying fitness distributions in evolutionary algorithms. II. comparing the improvements from crossover and gaussian mutation on simple neural networks. In X. Yao and D.B. Fogel, editors, *Proc. of the 2000 IEEE Symposium on Combinations of Evolutionary Computation and Neural Networks*, pages 91–97. IEEE Press, 2000.
46. J.A. Joines and C.R. Houck. On the use of non-stationary penalty functions to solve nonlinear constrained optimisation problems with ga's. In ICEC-94 [42], pages 579–584.
47. B.A. Julstrom. What have you done for me lately?: Adapting operator probabilities in a steady-state genetic algorithm. In Eshelman [28], pages 81–87.
48. Y. Kakuza, H. Sakanashi, and K. Suzuki. Adaptive search strategy for genetic algorithms with additional genetic algorithms. In Männer and Manderick [56], pages 311–320.
49. S. Kirkpatrick, C. Gelatt, and M. Vecchi. Optimization by simulated anealing. *Science*, 220:671–680, 1983.
50. N. Krasnogor and J.E. Smith. A memetic algorithm with self-adaptive local search: TSP as a case study. In Whitley et al. [81], pages 987–994.
51. N. Krasnogor and J.E. Smith. Emergence of profitable search strategies based on a simple inheritance mechanism. In Spector et al. [77], pages 432–439.
52. M. Lee and H. Takagi. Dynamic control of genetic algorithms using fuzzy logic techniques. In Forrest [31], pages 76–83.

53. J. Lis. Parallel genetic algorithm with dynamic control parameter. In ICEC-96 [43], pages 324–329.
54. J. Lis and M. Lis. Self-adapting parallel genetic algorithm with the dynamic mutation probability, crossover rate, and population size. In J. Arabas, editor, *Proceedings of the First Polish Evolutionary Algorithms Conference*, pages 79–86. Politechnika Warszawska, Warsaw, 1996.
55. S.W. Mahfoud. Boltzmann selection. In Bäck et al. [12], pages C2.5:1–4.
56. R. Männer and B. Manderick, editors. *Proceedings of the 2nd Conference on Parallel Problem Solving from Nature*. North-Holland, Amsterdam, 1992.
57. K.E. Mathias and L.D. Whitley. Remapping hyperspace during genetic search: Canonical delta folding. In L.D. Whitley, editor, *Foundations of Genetic Algorithms 2*, pages 167–186. Morgan Kaufmann, San Francisco, 1993.
58. K.E. Mathias and L.D. Whitley. Changing representations during search: A comparative study of delta coding. *Evolutionary Computation*, 2(3):249–278, 1995.
59. Z. Michalewicz. *Genetic Algorithms + Data Structures = Evolution Programs*. Springer, Berlin, Heidelberg, New York, 3rd edition, 1996.
60. Z. Michalewicz and M. Schoenauer. Evolutionary algorithms for constrained parameter optimisation problems. *Evolutionary Computation*, 4(1):1–32, 1996.
61. J.D. Schaffer, editor. *Proceedings of the 3rd International Conference on Genetic Algorithms*. Morgan Kaufmann, San Francisco, 1989.
62. J.D. Schaffer, R.A. Caruana, L.J. Eshelman, and R. Das. A study of control parameters affecting online performance of genetic algorithms for function optimisation. In Schaffer [61], pages 51–60.
63. J.D. Schaffer and L.J. Eshelman. On crossover as an evolutionarily viable strategy. In Belew and Booker [15], pages 61–68.
64. D. Schlierkamp-Voosen and H. Mühlenbein. Strategy adaptation by competing subpopulations. In Y. Davidor, H.-P. Schwefel, and R. Männer, editors, *Proceedings of the 3rd Conference on Parallel Problem Solving from Nature*, number 866 in Lecture Notes in Computer Science, pages 199–209. Springer, Berlin, Heidelberg, New York, 1994.
65. H.-P. Schwefel. *Numerische Optimierung von Computer-Modellen mittels der Evolutionsstrategie*, volume 26 of *ISR*. Birkhaeuser, Basel/Stuttgart, 1977.
66. H.-P. Schwefel. *Numerical Optimisation of Computer Models*. Wiley, New York, 1981.
67. J.E. Smith. *Self Adaptation in Evolutionary Algorithms*. PhD thesis, University of the West of England, Bristol, UK, 1998.
68. J.E. Smith. Modelling GAs with self-adaptive mutation rates. In Spector et al. [77], pages 599–606.
69. J.E. Smith. On appropriate adaptation levels for the learning of gene linkage. *J. Genetic Programming and Evolvable Machines*, 3(2):129–155, 2002.
70. J.E. Smith. Parameter perturbation mechanisms in binary coded gas with self-adaptive mutation. In Rowe, Poli, DeJong, and Cotta, editors, *Foundations of Genetic Algorithms 7*, pages 329–346. Morgan Kaufmann, San Francisco, 2003.
71. J.E. Smith and T.C. Fogarty. Adaptively parameterised evolutionary systems: Self adaptive recombination and mutation in a genetic algorithm. In Voigt et al. [80], pages 441–450.
72. J.E. Smith and T.C. Fogarty. Recombination strategy adaptation via evolution of gene linkage. In ICEC-96 [43], pages 826–831.

73. J.E. Smith and T.C. Fogarty. Self adaptation of mutation rates in a steady state genetic algorithm. In ICEC-96 [43], pages 318–323.
74. J.E. Smith and T.C. Fogarty. Operator and parameter adaptation in genetic algorithms. *Soft Computing*, 1(2):81–87, 1997.
75. R.E. Smith and E. Smuda. Adaptively resizing populations: Algorithm, analysis and first results. *Complex Systems*, 9(1):47–72, 1995.
76. W.M. Spears. Adapting crossover in evolutionary algorithms. In J.R. McDonnell, R.G. Reynolds, and D.B. Fogel, editors, *Proceedings of the 4th Annual Conference on Evolutionary Programming*, pages 367–384. MIT Press, Cambridge, MA, 1995.
77. L. Spector, E. Goodman, A. Wu, W.B. Langdon, H.-M. Voigt, M. Gen, S. Sen, M. Dorigo, S. Pezeshk, M. Garzon, and E. Burke, editors. *Proceedings of the Genetic and Evolutionary Computation Conference (GECCO-2001)*. Morgan Kaufmann, San Francisco, 2001.
78. C.R. Stephens, I. Garcia Olmedo, J. Moro Vargas, and H. Waelbroeck. Self-adaptation in evolving systems. *Artificial life*, 4:183–201, 1998.
79. G. Syswerda. A study of reproduction in generational and steady state genetic algorithms. In G. Rawlins, editor, *Foundations of Genetic Algorithms*, pages 94–101. Morgan Kaufmann, San Francisco, 1991.
80. H.-M. Voigt, W. Ebeling, I. Rechenberg, and H.-P. Schwefel, editors. *Proceedings of the 4th Conference on Parallel Problem Solving from Nature*, number 1141 in Lecture Notes in Computer Science. Springer, Berlin, Heidelberg, New York, 1996.
81. D. Whitley, D. Goldberg, E. Cantu-Paz, L. Spector, I. Parmee, and H.-G. Beyer, editors. *Proceedings of the Genetic and Evolutionary Computation Conference (GECCO-2000)*. Morgan Kaufmann, San Francisco, 2000.
82. L.D. Whitley, K.E. Mathias, and P. Fitzhorn. Delta coding: An iterative search strategy for genetic algorithms,. In Belew and Booker [15], pages 77–84.
83. D.H. Wolpert and W.G. Macready. No Free Lunch theorems for optimisation. *IEEE Transactions on Evolutionary Computation*, 1(1):67–82, 1997.
84. B. Zhang and H. Mühlenbeim. Balancing accuracy and parsimony in genetic programming. *Evolutionary Computing*, 3(3):17–38, 1995.

Self-Adaptation in Evolutionary Algorithms

Silja Meyer-Nieberg[1] and Hans-Georg Beyer[2]

[1] Department for Computer Science,
Universerstät der Bundeswehr München,
D-85577 Neubiberg, Germany
silja.meyer-nieberg@unibw.de
[2] Research Center Process and Product Engineering,
Department of Computer Science,
Vorarlberg University of Applied Sciences,
Hochschulstr. 1, A-6850 Dornbirn, Austria
hans-georg.beyer@fhv.at

Summary. In this chapter, we will give an overview over self-adaptive methods in evolutionary algorithms. Self-adaptation in its purest meaning is a state-of-the-art method to adjust the setting of control parameters. It is called self-adaptive because the algorithm controls the setting of these parameters itself – embedding them into an individual's genome and evolving them. We will start with a short history of adaptation methods. The section is followed by a presentation of classification schemes for adaptation rules. Afterwards, we will review empirical and theoretical research of self-adaptation methods applied in genetic algorithms, evolutionary programming, and evolution strategies.

1 Introduction

Evolutionary algorithms (EA) operate on basis of populations of individuals. Their performance depends on the characteristics of the population's distribution. Self-Adaptation aims at biasing the distribution towards appropriate regions of the search space – maintaining sufficient diversity among individuals in order to enable further evolvability.

Generally, this is achieved by adjusting the setting of control parameters. Control parameters can be of various forms – ranging from mutation rates, recombination probabilities, and population size to selection operators (see e.g. [6]).

The goal is not only to find suitable adjustments but to do this efficiently. The task is further complicated by taking into account that the optimizer is faced by a dynamic problem since a parameter setting that was optimal at the beginning of an EA-run might become unsuitable during the evolutionary process. Thus, there is generally the need for a steady modification or adaptation of the control parameters during the run of an EA.

S. Meyer-Nieberg and H.-G. Beyer: *Self-Adaptation in Evolutionary Algorithms*, Studies in Computational Intelligence (SCI) **54**, 47–75 (2007)
www.springerlink.com

We will consider the principle of self-adaptation which is explicitly used in evolutionary programming (EP) [27, 26] and evolution strategies (ES) [49, 55] while it is rarely used in genetic algorithms (GA) [35, 36].

Individuals of a population have a set of object parameters that serves as a representative of possible solutions. The basic idea of explicit self-adaptation consists in incorporating the strategy parameters into the individual's genome and evolving them alongside with the object parameters.

In this paper, we will give an overview over the self-adaptive behavior of evolutionary algorithms. We will start with a short overview over the historical development of adaptation mechanisms in evolutionary computation. In the following part, i.e., Section 2.2, we will introduce classification schemes that are used to group the various approaches. Afterwards, self-adaptive mechanisms will be considered. The overview is started by some examples – introducing self-adaptation of the strategy parameter and of the crossover operator. Several authors have pointed out that the concept of self-adaptation may be extended. Section 3.2 is devoted to such ideas. The mechanism of self-adaptation has been examined in various areas in order to find answers to the question under which conditions self-adaptation works and when it could fail. In the remaining sections, therefore, we present a short overview over some of the research done in this field.

2 Adaptation and Self-Adaptation

2.1 A Short History of Adaptation in Evolutionary Algorithms

In this section, we will shortly review the historical development of adaptation mechanisms. The first proposals to adjust the control parameters of a computation automatically date back to the early days of evolutionary computation.

In 1967, Reed, Toombs, and Barricelli [51] experimented with the evolution of probabilistic strategies playing a simplified poker game. Half of a player's genome consisted of strategy parameters determining, e.g., the probabilities for mutation or the probabilities for crossover with other strategies. Interestingly, it was shown for a play for which a known optimal strategy existed that the evolutionary simulation realized nearly optimal plans.

Also in 1967, Rosenberg [52] proposed to adapt crossover probabilities. Concerning genetic algorithms, Bagley [9] considered incorporating the control parameters into the representation of an individual. Although Bagley's suggestion is one of the earliest proposals of applying classical self-adaptive methods, self-adaptation as usually used in ES appeared relatively late in genetic algorithms. In 1987, Schaffer and Morishima [54] introduced the self-adaptive *punctuated crossover* adapting the number and location of crossover points. Some years later, a first method to self-adapt the mutation operator was suggested by Bäck [4, 4]. He proposed a self-adaptive mutation rate in genetic algorithms similar to evolution strategies.

The idea of using a meta-GA can be found quite early. Here, an upper-level GA tries to tune the control parameters of a lower-level algorithm which tries in turn to solve the original problem. The first suggestion stems from Weinberg [71] giving rise to the work by Mercer and Sampson [45].

Concerning evolution strategies, the need to adapt the mutation strength (or strengths) appropriately during the evolutionary process was recognized in Rechenberg's seminal book *Evolutionsstrategie* [63].

He proposed the well-known, which was originally developed for (1+1)-ES. It relies on counting the successful and unsuccessful mutations for a certain number of generations. If more than 1/5th of mutations leads to an improvement the mutation strength is increased and decreased otherwise. The aim was to stay in the so-called *evolution window* guaranteeing nearly optimal progress.

In addition to the 1/5th rule, Rechenberg [63] also proposed to couple the evolution of the strategy parameters with that of the object parameters. The strategy parameters were randomly changed. The idea of (explicit) was born. To compare the performance of this *learning population* with that of an ES using the 1/5th rule, Rechenberg conducted some experiments on the sphere and corridor model. The learning population exhibited a higher convergence speed and even more important it proved to be applicable in cases where it is improper to use the 1/5th rule. Self-adaptation thus appeared as a more universally usable method.

Since then various methods for adapting control parameters in evolutionary algorithms have been developed – ranging from adapting crossover probabilities in genetic algorithms to a direct adaptation of the distribution [16].

Schwefel [71, 72] introduced a self-adaptive method for changing the strategy parameters in evolution strategies which is today commonly associated with the term self-adaptation. In its most general form, the full covariance matrix of a general multidimensional normal distribution is adapted. A similar method of adapting the strategy parameters was offered by Fogel et al. [25] in the area of evolutionary programming – the so-called meta-EP operator for changing the mutation strength.

A more recent technique, the cumulative path-length control, stems from Ostermeier, Hansen, and Gawelczyk [48]. One of the aims is to derandomize the adaptation of the strategy parameters. The methods developed, the cumulative step-size adaptation (CSA) as well as the covariance matrix adaptation (CMA) [30], make use of an *evolution path*, $\mathbf{p}^{(g+1)} = (1-c)\mathbf{p}^{(g)} + \sqrt{c(2-c)}\mathbf{z}_{\text{sel}}^{(g+1)}$, which cumulates the selected mutation steps. To illustrate this concept, consider an evolution path where purely random selection is applied. Since the mutations are normally distributed, the cumulated evolution path is given by $\mathbf{u}^{(g)} = \sum_{k=1}^{g} \sigma \mathcal{N}^{(k)}(\mathbf{0}, \mathbf{1})$, where $\mathcal{N}(\mathbf{0}, \mathbf{1})$ is a random vector with identically independently distributed $\mathcal{N}(0, 1)$ normally distributed components with zero mean and variance 1. The length of $u^{(g)}$ is χ-distributed with expectation $\overline{u} = \sigma\overline{\chi}$. Fitness based selection changes the situation. Too large mutation steps result in a selection of smaller mutations. Thus, the path-length is smaller than $\overline{u}$ and the step size should be decreased. If on the other hand the path-length is larger than the expected $\overline{u}$, the step-size should be increased. CSA is also used in the CMA-algorithm. However, additionally the CMA adapts the whole covariance matrix [30] and as such it represents the state-of-the-art in real-coded evolutionary optimization algorithms.

2.2 A Taxonomy of Adaptation

As we have seen, various methods for changing and adapting control parameters of evolutionary algorithms exist and adaptation can take place on different levels.

Mainly, there are two taxonomy schemes [2, 22] which group adaptive computations into distinct classes – distinguishing by the type of adaptation, i.e., how the parameter is changed, and by the level of adaptation, i.e. where the changes occur. The classification scheme of Eiben, Hinterding, and Michaelewicz [22] extends and broadens the concepts introduced by Angeline in [2].

Let us start with Angeline's classification [2]. Considering the type of adaptation, adaptive evolutionary computations are divided into algorithms with *absolute update rules* and *empirical update rules*.

If an *absolute update rule* is applied, a statistic is computed. This may be done by sampling over several generations or by sampling the population. Based on the result, it is decided by means of a deterministic and fixed rule if and how the operator is to be changed. Rechenberg's 1/5th-rule [63] is one well-known example of this group.

In contrast to this, evolutionary algorithms with *empirical update rules* control the values of the strategy parameters themselves. The strategy parameter may be interpreted as an incorporated part of the individual's genome, thus being subject to "genetic variations". In the case the strategy parameter variation leads to an individual with a sufficiently good fitness, it is selected and "survives". Individuals with appropriate strategy parameters should – on average – have good fitness values and thus a higher chance of survival than those with badly tuned parameters. Thus, the EA should be able to self-control the parameter change.

As Smith [63] points out, the difference of the algorithms lies in the nature of the transition function. The transition function maps the set of parameters at generation t on that at $t+1$. In the case of absolute update rules, it is defined externally. In the case of self-adaptive algorithms, the transition function is a result of the operators and is defined by the algorithm itself.

Both classes of adaptive evolutionary algorithms can be further subdivided based on the level the adaptive parameters operate on. Angeline distinguished between population-, individual-, and component-level adaptive parameters.

Population-level adaptive parameters are changed globally for the whole population. Examples are for instance the mutation strength and the covariance matrix adaptation in CSA and CMA evolution strategies.

Adaptation on the individual-level changes the control parameters of an individual and these changes only affect that individual. The probability for crossover in GA is for instance adapted in [54] on the level of individuals.

Finally, component-level adaptive methods affect each component of an individual separately. Self-Adaptation in ES with correlated mutations (see Section 3.1) belongs to this type.

Angeline's classification was broadened by Eiben, Hinterding, and Michaelewicz [22]. Adaptation schemes are again classified firstly by the type of adaptation and secondly – as in [2] – by the level of adaptation. Considering the different levels of adaptation a fourth level, *environment level adaptation*, was introduced to take into account cases where the responses of the environment are not static.

Concerning the adaptation type, in [22] the algorithms are first divided into *static*, i.e., no changes of the parameters occur, and *dynamic* algorithms. The term "dynamic adaptation" is used to classify any algorithm where the strategy parameters according to some rule, i.e., without external control. Based on the *mechanism of adaptation* three subclasses are distinguished: *deterministic*, *adaptive*, and finally *self-adaptive* algorithms. The latter comprise the same class of algorithms as in [2].

A deterministic adaptation is used if the control parameter is changed according to a deterministic rule *without* taking into account any present information by the evolutionary algorithm itself. Examples of this adaptation class are the time-dependent change of the mutation rates proposed by Holland [36] and the cooling schedule in simulated annealing like selection schemes.

Algorithms with an adaptive dynamic adaptation rule take feedback from the EA itself into account and change the control parameters accordingly. Again, a well known member of this class is Rechenberg's 1/5th-rule. Further examples (see [22]) include Davis' adaptive operator fitness [15] and Julstrom's adaptive mechanism [38]. In both cases, the usage probability of an operator depends on its success or performance.

3 Self-Adaptation: The Principles

3.1 Self-Adapted Parameters: Some Examples

Self-Adaptation of Strategy Parameters

The technique most commonly associated with the term was introduced by Rechenberg [63] and Schwefel [71, 57] in the area of evolution strategies and independently by Fogel [25] for evolutionary programming. The control parameters considered here apply to the mutation process and parameterize the mutation distribution. The mutation is usually given by a normally distribution random vector, i.e. $\mathbf{Z} \sim \mathbf{N}(\mathbf{0}, \mathbf{C})$. The entries c_{ij} of the covariance matrix $\mathbf{C}$ are given by $c_{ii} = var(Z_i)$ or by $c_{ij} = cov(Z_i, Z_j)$ if $j \neq i$. The density function reads

$$p_Z(Z_1, \ldots, Z_N) = \frac{\mathrm{e}^{-\frac{1}{2}\mathbf{Z}^{\mathrm{T}}\mathbf{C}^{-1}\mathbf{Z}}}{\sqrt{(2\pi)^N \det(\mathbf{C})}}, \tag{1}$$

where N is the dimensionality of the search space. The basic step in the self-adaptation mechanism consists of a mutation of the mutation parameters themselves. In contrast to the additive change of the object variables, the mutation of the mutation strength strengths (i.e., the standard deviations $\sqrt{c_{ii}}$ in (1)) is realized by a multiplication with a random variable. The resulting mutation parameters are then applied in the variation of the object parameters. It should be mentioned here that concerning evolution strategies, the concept of self-adaptation was originally developed for non-recombinative $(1, \lambda)$-ES. Later on it was transferred to multi-parent strategies.

Figure 1 illustrates the basic mechanism of a multi-parent $(\mu/\rho, \lambda)$-ES with σ-self-adaptation. At generation g the ES maintains a population of μ candidate solutions – with the strategy parameters used in their creation. Based on that parent population, λ offspring are created via variation. The variation process usually comprises recombination and mutation. For each offspring, ρ parents are chosen for the recombination. First, the strategy parameters are changed. The strategy parameters of the chosen ρ parents are recombined and the result is mutated afterwards. The change of the object parameters occurs in the next step. Again, the parameters are first recombined and then mutated. In the mutation process, the newly created strategy parameter is used. After that, the fitness of the offspring is calculated. Finally, the μ-best individuals are chosen according to their fitness values as the next parental population. Two selection schemes are generally distinguished: "comma" and "plus"-selection. In the former case, only the offspring population is considered. In the latter, the new parent population is chosen from the old parent population and the offspring population.

Depending on the form of $\mathbf{C}$, different mutation distributions have to be taken into account. Considering the simplest case $\mathbf{Z} = \sigma\mathcal{N}(\mathbf{0}, \mathbf{I})$, the mutation of σ is given by

$$\sigma' = \sigma\mathrm{e}^{\tau\epsilon} \tag{2}$$

and using the new σ', the mutation of the object parameters reads

$$x_i' = x_i + \sigma'\mathcal{N}(0, 1). \tag{3}$$

The ϵ in Eq. (2) is a random number, often chosen as

$$\epsilon \sim \mathcal{N}(0, 1), \tag{4}$$

thus, producing log-normally distributed σ' variants. This way of choosing ϵ is also referred to as the "log-normal mutation rule". Equation (2) contains a new strategy specific parameter – the *learning rate* τ to be fixed. The general recommendation is to choose $\tau \propto 1/\sqrt{N}$, which has been shown to be optimal with respect to the convergence speed on the sphere [11].

```
BEGIN
  g:=0
  INITIALIZATION(P_μ^(0) := {(y_m^(0), σ_m^(0), F(y_m^(0)))})
  REPEAT
    FOR EACH OF THE λ OFFSPRING DO
      P_ρ:=REPRODUCTION(P_μ^(g))
      σ'_l :=RECOMB_σ(P_ρ);
      σ_l :=MUTATE_σ(σ'_r);
      y'_l :=RECOMB_y(P_ρ);
      y_l :=MUTATE_y(x'_l, σ_l);
      F_l := F(y_l);
    END
    P_λ^(g):={(y_l, σ_l, F_l)}
    CASE ","-SELECTION: P_μ^(g+1):=SELECT(P_λ^(g))
    CASE "+"-SELECTION: P_μ^(g+1):=SELECT(P_μ^(g+1),P_λ^(g))
    g:=g+1
  UNTIL stop;
END
```

Fig. 1. The $(\mu/\rho, \lambda)$-σSA-ES.

If different mutation strengths are used for each dimension, i.e., $Z_i = \sigma_i \mathcal{N}(0,1)$, the update rule

$$\sigma_i' = \sigma_i \exp\big(\tau' \mathcal{N}(0,1) + \tau \mathcal{N}_i(0,1)\big) \tag{5}$$
$$x_i' = x_i + \sigma_i' \mathcal{N}(0,1) \tag{6}$$

has been proposed. It is recommended [57] to choose the learning rates $\tau' \propto 1/\sqrt{2N}$ and $\tau \propto 1/\sqrt{2\sqrt{N}}$. The approach can also be extended to allow for correlated mutations [6]. Here, rotation angles α_i need to be taken into account leading to the update rule

$$\sigma_i' = \sigma_i \exp\big(\tau' \mathcal{N}(0,1) + \tau \mathcal{N}_i(0,1)\big) \tag{7}$$
$$\alpha_i' = \alpha_i + \beta \mathcal{N}_i(0,1) \tag{8}$$
$$\mathbf{x}' = \mathbf{x} + \mathcal{N}\big(0, \mathbf{C}(\sigma', \alpha)\big) \tag{9}$$

where $\mathbf{C}$ is the covariance matrix [6]. The parameter β is usually [6] chosen as 0.0873.

In EP [25], a different mutation operator, called *meta-EP*, is used

$$\sigma_i' = \sigma_i \Big(1 + \alpha \mathcal{N}(0,1)\Big) \tag{10}$$
$$x_i' = x_i + \sigma_i' \mathcal{N}(0,1). \tag{11}$$

Both operators lead to similar results – provided that the parameters τ and α are sufficiently small.

The log-normal operator, Eqs. (2), (3), and the meta-EP operator introduced above are not the only possibilities. Self-Adaptation seems to be relatively robust as to the choice of the distribution. Another possible operator is given by $\epsilon = \pm\delta$, where $+|\delta|$ and $-|\delta|$ are generated with the same probability of 1/2. That is, the resulting probability distribution function (pdf) of δ is a two-point distribution giving rise to the so-called two-point rule. It is usually implemented using $\delta = 1/\tau \ln(1+\beta)$, thus, leading with (2) to

$$\sigma_i' = \begin{cases} \sigma_i(1+\beta), & \text{if } u \leq 0.5 \\ \sigma_i/(1+\beta), & \text{if } u > 0.5 \end{cases} \tag{12}$$

where u is a random variable uniformly distributed on $(0,1]$.

Another choice was proposed by Yao and Liu [73]. They substituted the normal distribution of the meta-EP operator with a Cauchy-distribution. Their new algorithm, called fast evolutionary programming, performed well on a set of test functions and appeared to be preferable in the case of multi-modal functions. The Cauchy-distribution is similar to the normal distribution but has a far heavier tail. Its moments are undefined. In [42], Lee and Yao proposed to use a Lévy-distribution. Based on results on a suite of test functions, they argued that using Lévy-distributions instead of normal distribution may lead to higher variations and a greater diversity. Compared to the Cauchy-distribution, the Lévy-distribution allows for a greater flexibility since the Cauchy-distribution appears as a special case of the Lévy-distribution.

Self-Adaptation of Recombination Operators

Crossover is traditionally regarded as the main search mechanism in genetic algorithm and most efforts to self-adapt the characteristics of this operator stem from this area.

Schaffer and Morishima [54] introduced *punctuated crossover* which adapts the positions where crossover occurs. An individual's genome is complemented with a bitstring encoding crossover points. A position in this crossover map is changed in the same manner as its counterpart in the original genome. Schaffer and Morishima reported that punctuated crossover performed better than one-point crossover. Spears [67] points out, however, that the improvement of the performance might not necessarily be due to self-adaptation but due to the difference between crossover with more than one crossover point and one-point crossover.

Spears [67] self-adapted the form of the crossover operator using an additional bit to decide whether two-point or uniform crossover should be used for creating the offspring. Again, it should be noted that Spears attributes the improved performance not to the self-adaptation process itself but rather to the increased diversity that is offered to the algorithm.

Smith and Fogarty [62] introduced the so-called LEGO-algorithm, a linkage evolving genetic algorithm. The objects which are adapted are *blocks*, i.e.

linked neighboring genes. Each gene has two additional bits which indicate whether it is linked to its left and right neighbors. Two neighboring genes are then called *linked* if the respective bits are set. Offspring are created via successive tournaments. The positions of an offspring are filled from left to right by a competition between parental blocks. The blocks have to be eligible, i.e., they have to start at the position currently considered. The fittest block is copied as a whole and then the process starts anew. More than two parents may contribute in the creation of an offspring. Mutation also extends to the bits indicating linked genes.

3.2 A Generalized Concept of Self-Adaptation

In [6], two of self-adaptation have been identified: Self-adaptation aims at biasing the population distribution to more appropriate regions of the search space by making use of an indirect link between good strategy values and good object variables. Furthermore, self-adaptation relies on a population's diversity. While the adaptation of the operator ensures a good convergence speed, the degree of diversity determines the convergence reliability. More generally speaking, self-adaptation controls the relationship between parent and offspring population, i.e., the transmission function (see e.g. Altenberg [1]). The control can be direct by manipulating control parameters in the genome or more implicit. In the following, we will see that self-adaptation of strategy parameters can be put into a broader context.

Igel and Toussaint [37] considered the effects of neutral genotype-phenotype mapping. They pointed out that neutral genome parts give an algorithm the ability to "vary the search space distribution independent of phenotypic variation". This may be regarded as one of the main benefits of neutrality. Neutrality induces a redundancy in the relationship between genotype-phenotype. But neutral parts can influence the *exploration distribution* and thus the population's search behavior. This use of neutrality is termed generalized self-adaptation. It also comprises the classical form of self-adaptation since the strategy parameters adapted in classical self-adaptation belong to the neutral part of the genome.

More formally, generalized self-adaptation is defined as "adaptation of the exploration distribution $P_{\mathcal{P}}^{(t)}$ by exploiting neutrality – i.e. independent of changing phenotypes in the population, of external control, and of changing the genotype-phenotype mapping" [37]. Igel and Toussaint showed additionally that neutrality cannot be seen generally as a disadvantage. Although the state space is increased, it might not necessarily lead to a significant degradation of the performance.

In [28], Glickman and Sycara referred to an *implicit self-adaptation* caused by a non-injective genotype-phenotype mapping. Again there are variations of the genome that do not alter the fitness value but influence the transmission function which induces a similar effect.

Beyer and Deb [18] pointed out that in well-designed real-coded GAs, the parent offspring transmission function is controlled by the characteristics of the parent population. Thus, the GA performs an implicit form of self-adaptation. In contrast to the explicit self-adaptation in ES, an individual's genome does not contain any control parameters. Deb and Beyer [20] examined the dynamic behavior of real-coded genetic algorithm (RCGA) that apply simulated binary crossover (SBX) [17, 21]. In SBX, two parents x^1 and x^2 create two offspring y^1 and y^2 according to

$$\begin{aligned} y_i^1 &= 1/2\Big((1-\beta_i)x_i^1 + (1+\beta_i)x_i^2\Big) \\ y_i^2 &= 1/2\Big((1+\beta_i)x_i^1 + (1-\beta_i)x_i^2\Big). \end{aligned} \tag{13}$$

The random variable β has the density

$$p(\beta) = \begin{cases} 1/2(\eta+1)\beta^{\eta}, & \text{if } 0 \leq \beta \leq 1 \\ 1/2(\eta+1)\beta^{-\eta-2}, & \text{if } \beta > 1 \end{cases}. \tag{14}$$

The authors pointed out that these algorithms show self-adaptive behavior although an individual's genome does not contain any control parameters. Well-designed crossover operators create offspring depending on the difference in parent solutions. The spread of children solutions is in proportion to the spread of the parent solutions, and near parent solutions are more likely to be chosen as children solutions than solutions distant from parents [20]. Thus, the diversity in the parental population controls that of the offspring population. Self-adaptation in evolution strategies has similar properties. In both cases, offspring closer to the parents have a higher probability to be created than individuals further away. While the implicit self-adaptability of real-coded crossover operators is well understood today, it is interesting to point out that even the standard one- or k-point crossover operators operating on binary strings do have this property: Due to the mechanics of these operators, bit positions which are common in both parents are transferred to the offspring. However, the other positions are randomly filled. From this point of view, crossover can be seen as a *self-adaptive mutation operator*, which is in contrast to the building block hypothesis [29] usually offered to explain the working of crossover in binary GAs.

3.3 Demands on the Operators: Real-coded Algorithms

Let us start with rules [12, 11, 13] for the design of mutation operators that stem from analyzes of implementations and theoretical considerations in evolution strategies: *reachability*, *scalability*, and *unbiasedness*. They state that every finite state must be reachable, the mutation operator must be tunable in order to adapt to the fitness landscape (scalability), and it must not introduce a bias on the population. The latter is required to hold also for the recombination operator [12, 40]. The demand of unbiasedness becomes clear when considering that the behavior of an EA can be divided into two phases: Exploitation of the search space by selecting good solutions (reproduction) and exploration of the search space by means of variation. Only the former generally makes use of fitness information, whereas ideally the latter should only rely on search space information of the population. Thus, under a variation operator, the expected population mean should remain unchanged, i.e., the variation operators should not bias the population. This requirement, firstly made explicit in [12] may be regarded as a basic design principle for variation operators in EAs. The basic work [12] additionally proposed design principles with respect to the changing behavior of the population variance. Generally, selection changes the population variance. In order to avoid premature convergence, the variation operator must counteract that effect of the reproduction phase to some extend. General rules how to do this are, of course, nearly impossible to give but some *minimal* requirements can be proposed concerning the behavior on certain fitness landscapes [12].

For instance, Deb and Beyer [12] postulated that the population variance should increase exponentially with the generation number on flat or linear fitness functions. As pointed out by Hansen [31] this demand might not be sufficient. He proposed a linear increase of the expectation of the logarithm of the variance. Being based on the desired behavior in flat fitness landscapes, Beyer and Deb [12] advocate applying variation operators that increase the population variance also in the general case of unimodal fitness functions. While the variance should be decreased if the population brackets the optimum, this should not be done by the variation operator. Instead, this task should be left to the selection operator. In the case of crossover operators in real-coded genetic algorithms (RCGA), similar guidelines have been proposed by Kita and Yamamura [40]. They suppose that the distribution of the parent population indicates an appropriate region for further search. As before, the first guideline states that the statistics of the population should not be changed. This applies here to the mean as well as to the variance-covariance matrix. Additionally, the crossover operator should lead to as much diversity in the offspring population as possible. It is noted that the first guideline may be violated since the selection operator typically reduces the variance. Therefore, it might be necessary to increase the present search region.

4 Self-Adaptation in EAs: Theoretical and Empirical Results

4.1 Genetic Algorithms

In this section, we will review the empirical and theoretical research that has been done in order to understand the working of self-adaptive EAs and to evaluate their performance.

Self-Adaptation of the Crossover Operator

Real-Coded GAs in Flat Fitness Landscapes

Beyer and Deb [19] analyzed three crossover operators commonly used in real-coded GAs, i.e., the simulated binary crossover (SBX) by Deb and Agrawala [17], the blend crossover operator (BLX) of Eshelman and Schaffer [23], and the *fuzzy recombination* of Voigt et al. [70].

In [19], expressions for the mean and the variance of the offspring population in relation to the parent population are derived. The aim is to examine if and under which conditions the postulates proposed in Section 3.3 are fulfilled. The fitness environments considered are flat fitness landscapes and the sphere. As mentioned before in Section 3.3, the self-adaptation should not change the population mean in the search space, i.e., it should not introduce a bias, but it should – since a flat fitness function is considered – increase the population variance and this exponentially fast.

It was shown in [19] that the crossover operator leaves the population mean unchanged regardless of the distribution of the random variable. Concerning the population variance, an exponential change can be asserted. Whether the variance expands or contracts depends on the population size and on the second moment of the random variable. Thus, a relationship between the population size and the distribution parameters of the random variables can be derived which ensures an expanding population.

Kita [39] investigated real-coded genetic algorithms using UNDX-crossover (unimodal normal distribution) and performed a comparison with evolution strategies. Based on empirical results, he pointed out that both appear to work reasonably well although naturally some differences in their behavior can be observed. The ES for example widens the search space faster when the system is far away from an optimum. On the other hand, the RCGA appears to have a computational advantage in high-dimensional search spaces compared to an ES which adapts the rotation angles of the covariance matrix according to Eqs. (7)–(9). Kita used a $(15, 100)$-ES with the usual recommendations for setting the learning rates.

Self-Adaptation of the Mutation Rate in Genetic Algorithms

Traditionally, the crossover (recombination) operator is regarded as the main variation operator in genetic algorithms, whereas the mutation operator was originally proposed as a kind of "background operator" [36] endowing the algorithm with the potential ability to explore the whole search space. Actually, there are good reasons to consider this as a reasonable recommendation in GAs with genotype-phenotype mapping from $\mathbb{B}^\ell \rightarrow \mathbb{R}^n$. As has been shown in [12], standard crossover of the genotypes does not introduce a bias on the population mean in the phenotype space. Interestingly, this does *not* hold for bit-flip mutations. That is, mutations in the genotype space result in a biased phenotypic population mean – thus violating the postulates formulated in [12]. On the other hand, over the course of the years it was observed that for GAs on (pseudo) boolean functions (i.e., the problem specific search space is the $\mathbb{B}^{\ell}$) the mutation operator might also be an important variation operator to explore the search space (see e.g. [68]). Additionally, it was found that the optimal mutation rate or mutation probability does not only depend on the function to be optimized but also on the search space dimensionality and the current state of the search (see e.g. [5]).

A mechanism to self-adapt the mutation rate was proposed by Bäck [4, 4] for GA using the standard ES approach. The mutation rate is encoded as a bit-string and becomes part of the individual's genome. As it is common practice, the mutation rate is mutated first which requires its decoding to $[0, 1]$. The decoded mutation rate is used to mutate the positions in the bit-string of the mutation rate itself. The mutated version of the mutation probability is then decoded again in order to be used in the mutation of the object variables.

Several investigations have been devoted to the mechanism of self-adaptation in genetic algorithms. Most of the work is concentrated on empirical studies which are directed to possible designs of mutation operators trying to identify potential benefits and drawbacks.

Bäck [4] investigated the asymptotic behavior of the encoded mutation rate by neglecting the effects of recombination and selection. The evolution of the mutation rate results in a Markov chain[3]. Zero is an absorbing state of this chain which shows the convergence of the simplified algorithm.

The author showed empirically that in the case of GA with an extinctive selection scheme[4] self-adaptation is not only possible but can be beneficial [4]. For the comparison, three high-dimensional test functions (two unimodal, one multimodal) were used.

In [4], a self-adaptive GA optimizing the bit-counting function was examined. Comparing its performance with a GA that applies an optimal deterministic schedule to tune the mutation strength, it was shown that the self-adaptive algorithm realizes nearly optimal mutation rates.

The encoding of the mutation rate as a bit-string was identified in [8] as an obstacle for the self-adaptation mechanism. To overcome this problem, Bäck and Schütz [8] extended the genome with a real-coded mutation rate $p \in]0, 1[$. Several requirements have to be fulfilled. The expected change of p should be zero and small changes should occur with a higher probability than large ones. Also, there is the symmetry requirement that a change by a factor c should have the same probability as by $1/c$. In [8], a logistic normal distribution with parameter γ was used. The algorithm was compared with a GA without any adaptation and with a GA that uses a deterministic time-dependent schedule. Two selection schemes were considered. The GA with

[3] A Markov chain is a stochastic process which possesses the Markov property, i.e., the future behavior depends on the present state but not on the past.

[4] A selection scheme is extinctive iff at least one individual is not selected (see [4]).

the deterministic schedule performed best on the test-problems chosen with the self-adaptive GA in second place. Unfortunately, the learning rate γ was found to have a high impact.

Considering the originally proposed algorithm [4], Smith [65] showed that prematurely reduced mutation rates can occur. He showed that this can be avoided by using a fixed learning rate for the bitwise mutation of the mutation rate.

In 1996, Smith and Fogarty [74] examined empirically a self-adaptive steady state $(\mu + 1)$-GA finding that self-adaptation may improve the performance of a GA. The mutation rate was encoded again as a bit-string and several encoding methods were applied. Additionally, the impact of crossover in combination with a self-adaptive mutation rate was investigated. The self-adaptive GA appeared to be relative robust with respect to changes of the encoding or crossover.

In [66], the authors examined the effect of self-adaptation when the crossover operator and the mutation rate are both simultaneously adapted. It appeared that at least on the fitness functions considered synergistic effects between the two variation operators come into play.

Trying to model the behavior of self-adaptive GAs, Smith [75] developed a model to predict the mean fitness of the population. In the model several simplifications are made. The mutation rate is only allowed to assume q different values. Because of this, Smith also introduced a new scheme for mutating the mutation rate. The probability of changing the mutation rate is given by $P_a = z(q-1)/q$, where z is the so-called *innovation rate.*

In [77], Stone and Smith compared a self-adaptive GA using the log-normal operator with a GA with discrete self-adaptation, i.e., a GA implementing the model proposed in [75]. To this end, they evaluated the performance of a self-adaptive GA with continuous self-adaptation and that of their model on a set of five test functions. Stone and Smith found that the GA with discrete self-adaptation behaves more reliably whereas the GA with continuous self-adaptation may show stagnation. They attributed this behavior to the effect that the mutation rate gives the probability of bitwise mutation. As a result, smaller differences between mutation strengths are lost and more or less the same amount of genes are changed. The variety the log-normal operator provides cannot effectively be carried over to the genome and the likelihood of large changes is small. In addition, they argued that concerning the discrete self-adaptation a innovation rate of one is connected with an explorative behavior of the algorithm. This appears more suitable for multimodal problems whereas smaller innovation rates are preferable for unimodal functions.

4.2 Evolution Strategies and Evolutionary Programming

Research on self-adaptation in evolution strategies has a long tradition. The first theoretical in-depth analysis has been presented by Beyer [10]. It focused on the conditions under which a convergence of the self-adaptive algorithm can be ensured. Furthermore, it also provided an estimate of the convergence order.

The evolutionary algorithm leads to a stochastic process which can be described by a Markov chain. The random variables chosen to describe the system's behavior are the object vector (or its distance to the optimizer, respectively) and the mutation strength.

There are several approaches to analyze the Markov chain. The first [14, 3] considers the chain directly whereas the second [59, 60, 34] analyzes induced supermartingales. The third [11, 18] uses a model of the Markov chain in order to determine the dynamic behavior.

Convergence Results using Markov Chains

Bienvenüe and François [14] examined the global convergence of adaptive and self-adaptive $(1, \lambda)$-evolution strategies on spherical functions. To this end, they investigated the induced stochastic process $z_t = \|x_t\|/\sigma_t$. The parameter σ_t denotes the mutation strength, whereas x_t stands for the object parameter vector.

They showed that (z_t) is a homogeneous Markov chain, i.e., z_t only depends on z_{t-1}. This also confirms an early result obtained in [10] that the evolution of the mutation strength can be decoupled from the evolution of $\|x_t\|$. Furthermore, they showed that (x_t) converges or diverges log-linearly – provided that the chain (z_t) is Harris-recurrent[5].

Auger [3] followed that line of research focusing on $(1, \lambda)$-ES optimizing the sphere model. She analyzed a general model of an $(1, \lambda)$-ES with

$$\begin{aligned} x_{t+1} &= \arg\min \left\{ f(x_t + \sigma_t \eta_t^1 \xi_t^1), \ldots, f(x_t + \sigma_t \eta_t^\lambda \xi_t^\lambda) \right\} \\ \sigma_{t+1} &= \sigma_t \eta^*(x_t),\ \eta^* \text{ given by } x_{t+1} = x_t + \sigma_t \eta^*(x_t) \xi^*(x_t), \end{aligned} \qquad (15)$$

i.e., σ_{t+1} is the mutation strength which accompanies the best offspring. The function f is the sphere and η and ξ are random variables. Auger proved that the Markov chain given by $z_t = x_t/\sigma_t$ is Harris-recurrent and positive if some additional assumptions on the distributions are met and the offspring number λ is chosen appropriately. As a result, a law of large numbers can be applied and $1/t\ \ln(\|x_t\|)$ and $1/t\ \ln(\sigma_t)$ converge almost surely[6] to the same

[5] Let N_A be the number of passages in the set A. The set A is called Harris-recurrent if $P_z(N_A = \infty) = 1$ for $z \in A$. (Or in other words if the process starting from z visits A infinitely often with probability one.) A process (z_t) is Harris-recurrent if a measure ψ exists such that (z_t) is ψ-irreducible and for all A with $\psi(A) > 0$, A is Harris-recurrent (see e.g. [47]). A Markov process is called φ-irreducible if a measure φ exists so that for every set A with $\varphi(A) > 0$ and every state x holds: The return time probability to set A starting from x is greater than zero. A process is then called ψ-irreducible if it is φ-irreducible and ψ is a maximal irreduciblity measure fulfilling some additional propositions (see e.g. [47])

[6] A sequence of random variables x_t defined on the probability space $(\Omega, \mathcal{A}, P)$ converges almost surely to a random variable x if $P(\{\omega \in \Omega |\ \lim_{t\to\infty} x_t(\omega) =$

quantity – the convergence rate. This ensures either log-linearly convergence or divergence of the ES – depending on the sign of the limit. Auger further showed that the Markov chain (z_t) is also geometrically ergodic (see e.g. [47]) so that the Central Limit Theorem can be applied. As a result, it is possible to derive a confidence interval for the convergence rate. This is a necessary ingredient, because the analysis still relies on Monte-Carlo simulations in order to obtain the convergence rate (along with its confidence interval) numerically for the real $(1, \lambda)$-ES. In order to perform the analysis, it is required that the random variable ξ is symmetric and that both random variables ξ and η must be absolutely continuous with respect to the Lebesgue-measure. Furthermore, the density p_ξ is assumed to be continuous almost everywhere, $p_\xi \in L^\infty(\mathbb{R})$, and zero has to be in the interior of the support of the density, i.e., $0 \in \mathring{\operatorname{supp}}\, p_\xi$. Additionally, it is assumed that $1 \in \mathring{\operatorname{supp}}\, p_\eta$ and that $\mathrm{E}[|\ln(\eta)|] < \infty$ holds. The requirements above are met by the distribution functions normally used in practice, i.e., the log-normal distribution (mutation strength) and normal distribution (object variable). In order to show the Harris-recurrence, the positivity, and the geometric ergodicity, Forster-Lyapunov drift conditions (see e.g. [47]) need to be obtained. To this end, new random variables are to be introduced

$$\hat{\eta}(\lambda)\hat{\xi}(\lambda) = \min\left\{\eta^1\xi^1, \ldots, \eta^\lambda\xi^\lambda\right\}. \tag{16}$$

They denote the minimal change of the object variable. For the drift conditions a number α is required. Firstly, α has to ensure that the expectations $\mathrm{E}[|\xi|^\alpha]$ and $\mathrm{E}[(1/\eta)^\alpha]$ are finite. Provided that also $\mathrm{E}[|1/\hat{\eta}(\lambda)|^\alpha] < 1$, α can be used to give a drift condition V. More generally stated, α has to decrease the reduction velocity of the mutation strength associated with the best offspring of λ trials sufficiently. Thus, additional conditions concerning α and the offspring number λ are introduced leading to the definition of the sets

$$\Gamma_0 = \{\gamma > 0 : \mathrm{E}\left[|1/\eta|^\gamma\right] < \infty \text{ and } \mathrm{E}\left[|\xi|^\gamma\right] < \infty\} \tag{17}$$

and

$$\Lambda = \bigcup_{\alpha\in\Gamma_0} \Lambda_\alpha = \bigcup_{\alpha\in\Gamma_0} \{\lambda \in N : \mathrm{E}\left[1/\hat{\eta}(\lambda)^\alpha\right] < 1\}. \tag{18}$$

Finally, the almost sure convergence of $1/t\ln(\|x_t\|)$ and $1/t\ln(\sigma_t)$ can be shown for all $\lambda \in \Lambda$. It is not straightforward to give expression for Λ or Λ_α in the general case although Λ_α can be numerically obtained for a given α. Only if the densities of η and ξ have bounded support, it can be shown that Λ_α is of the form $\Lambda_\alpha = \{\lambda : \lambda \geq \lambda_0\}$.

$x(\omega)\}) = 1$. Therefore, events for which the sequence does not converge have probability zero.

Convergence Theory with Supermartingales

Several authors [59, 60, 34] use the concept of martingales or supermartingales[7] to show the convergence of an ES or to give an estimate of the convergence velocity. As before, the random variables most authors are interested in are the object variable and the mutation strength.

Semenov [59] and Semenov and Terkel [60] examined the convergence and the convergence velocity of evolution strategies. To this end, they consider the stochastic Lyapunov function V_t of a stochastic process X_t. By showing the convergence of the Lyapunov function, the convergence of the original stochastic process follows under certain conditions.

From the viewpoint of probability theory, the function V_t may be regarded as a supermartingale. Therefore, a more general framework in terms of convergence of supermartingales can be developed. The analysis performed in [60] consists of two independent parts. The first concerns the conditions that imply almost surely convergence of supermartingales to a limit set. The second part (see also [59]) proposes demands on supermartingales which allow for an estimate of the convergence velocity. Indirectly, this also gives an independent convergence proof.

The adaptation of the general framework developed for supermartingales to the situation of evolution strategies requires the construction of an appropriate stochastic Lyapunov function. Because of the complicated nature of the underlying stochastic process, the authors did not succeed in the rigorous mathematical treatment of the stochastic process. Similar to the Harris-recurrent Markov chain approach, the authors had to resort to Monte-Carlo simulations in order to show that the necessary conditions are fulfilled.

In [59] and [60], $(1, \lambda)$-ES are considered where the offspring are generated according to

$$\begin{aligned} \sigma_{t,l} &= \sigma_t e^{\vartheta_{t,l}} \\ x_{t,l} &= x_t + \sigma_{t,l}\zeta_{t,l} \end{aligned} \tag{19}$$

and the task is to optimize $f(x) = -|x|$. The random variables $\vartheta_{t,l}$ and $\zeta_{t,l}$ are uniformly distributed with $\vartheta_{t,l}$ assuming values in $[-2, 2]$ whereas $\zeta_{t,l}$ is defined on $[-1, 1]$. For this problem, it can be shown that the object variable and the mutation strength converge almost surely to zero – provided that there are at least three offspring. Additionally, the convergence velocity of the mutation strength and the distance to the optimizer is bounded from above by $\exp(-at)$ which holds asymptotically almost surely.

Hart, DeLaurentis, and Ferguson [34] also used supermartingales in their approach. They considered a simplified $(1, \lambda)$-ES where the mutations are modeled by discrete random variables. This applies to the mutations of the

---

[7] A random process X_t is called a supermartingale if $\mathrm{E}[|X_t|] < \infty$ and $\mathrm{E}[X_{t+1}|\mathcal{F}_t] \leq X_t$ where $\mathcal{F}_t$ is e.g. the σ-algebra that is induced by X_t.

object variables as well as to those of the mutation strengths. Offspring are generated according to

$$\begin{aligned} \sigma_{t,l} &= \sigma_t D \\ x_{t,l} &= x_t + \sigma_{t,l} B. \end{aligned} \tag{20}$$

The random variable D assumes three values $\{\gamma, 1, \eta\}$ with $\gamma < 1 < \eta$. The random variable B takes a value of either $+1$ or -1 with probability $1/2$ each. Under certain assumptions, the strategy converges almost surely to the minimum x^* of a function $f : \mathbb{R} \rightarrow \mathbb{R}$ which is assumed to be strictly monotonically increasing for $x > x^*$ and strictly monotonically decreasing for $x < x^*$.

As a second result, the authors proved that their algorithm fails to locate the global optimum of an example multimodal function with probability one. We will return to this aspect of their analysis in Section 5.

Instead of using a Lyapunov function as Semenov and Terkel did, they introduced a random variable that is derived from the (random) object variable and the mutation strength. It can be shown that this random variable is a nonnegative supermartingale if certain requirements are met. In that case, it can be shown that the ES converges almost surely to the optimal solution if the offspring number is sufficiently high.

The techniques introduced in [34] can be applied to the multi-dimensional case [33] provided that the fitness function is separable, i.e., $g(x) = \sum_{k=0}^{N} g_k(x_k)$ and the g_k fulfill the conditions for f. The authors considered an ES-variant where only one dimension is changed in each iteration. The dimension k is chosen uniformly at random. Let $X_{\lambda,k}^t$ and $\Sigma_{\lambda,k}^t$ denote the stochastic process that results from the algorithm. It can be shown that $X_{\lambda,1}^t, \ldots, X_{\lambda,N}^t$ are independent of each other. This also holds for $\Sigma_{\lambda,1}^t, \ldots, \Sigma_{\lambda,N}^t$. Therefore, the results of the one-dimensional analysis can be directly transferred.

Although the analysis in [34, 33] provides an interesting alternative, it is restricted to very special cases: Due to the kind of mutations used, the convergence results in [34, 33] are, however, not practically relevant if the number of offspring exceeds six.

Step-by-step Approach: The Evolution Equations

In 1996, Beyer [10] was the first to provide a theoretical framework for the analysis of self-adaptive EAs. He used approximate equations to describe the dynamics of self-adaptive evolution strategies. Let the random variable $r^{(g)} = \|X^{(g)} - \hat{X}\|$ denote the distance to the optimizer and $\varsigma^{(g)}$ the mutation strength. The dynamics of an ES can be interpreted as a Markov process as we have already seen. But generally, the transition kernels for

$$\begin{pmatrix} r^{(g)} \\ \varsigma^{(g)} \end{pmatrix} \rightarrow \begin{pmatrix} r^{(g+1)} \\ \varsigma^{(g+1)} \end{pmatrix} \tag{21}$$

cannot be analytically determined. One way to analyze the system is therefore to apply a step by step approach extracting the important features of the dynamic process and thus deriving approximate equations.

The change of the random variables can be divided into two parts. While the first denotes the expected change, the second covers the stochastic fluctuations

$$\begin{aligned} r^{(g+1)} &= r^{(g)} - \varphi(r^{(g)}, \varsigma^{(g)}) + \epsilon_R(r^{(g)}, \varsigma^{(g)}) \\ \varsigma^{(g+1)} &= \varsigma^{(g)}\Big(1 + \psi(r^{(g)}\varsigma^{(g)})\Big) + \epsilon_\sigma(r^{(g)}, \varsigma^{(g)}). \end{aligned} \tag{22}$$

The expected changes φ and ψ of the variables are termed *progress rate* if the distance is considered and *self-adaptation response* in the case of the mutation strength.

The distributions of the fluctuation terms ϵ_R and ϵ_σ are approximated using Gram-Charlier series', usually cut off after the first term. Thus, the stochastic term is approximated using a normal distribution. The variance can be derived considering the evolution equations. Therefore, the second moments have to be taken into account leading to the second-order progress rate and to the second-order self-adaptation response.

To analyze the self-adaptation behavior of the system, expressions for the respective progress rate and self-adaptation response have to be found. Generally, no closed analytical solution can be derived. Up to now, only results for $(1,2)$-ES [11] using two-point mutations could be obtained. Therefore, several simplifications have to be introduced. For instance, if the log-normal operator is examined, the most important simplification is to consider $\tau \to 0$. The so derived expressions are then verified by experiments.

Self-Adaptation on the Sphere Model

It has been shown in [11] that an $(1,\lambda)$-ES with self-adaptation converges to the optimum log-linearly. Also the usual recommendation of choosing the learning rate proportionally to $1/\sqrt{N}$, where N is the search space dimensionality, is indeed approximately optimal. In the case of $(1,\lambda)$-ES, the dependency of the progress on the learning rate is weak provided that $\tau \geq c/\sqrt{N}$ where c is a constant. As a result, it is not necessary to have N-dependent learning parameters.

As has been shown in [11], the time to adapt an ill-fitted mutation strength to the fitness landscape is proportionally to $1/\tau^2$. Adhering to the scaling rule $\tau \propto 1/\sqrt{N}$ results in an adaptation time that linearly increases with the search space dimensionality. Therefore, it is recommended to work with a *generation-dependent* or constant learning rate τ, respectively, if N is large.

The maximal progress rate that can be obtained in experiments is always smaller than the theoretical maximum predicted by the progress rate theory (without considering the stochastic process dynamics). The reason for this is that the fluctuations of the mutation strength degrade the performance.

The average progress rate is deteriorated by a loss part stemming from the variance of the strategy parameter. The theory developed in [11] is able to predict this effect qualitatively.

If recombination is introduced in the algorithm the behavior of the ES changes qualitatively. Beyer and Grünz [30] showed that multi-recombinative ES that use intermediate/dominant recombination do not exhibit the same robustness with respect to the choice of the learning rate as $(1, \lambda)$-ES. Instead their progress in the stationary state has a clearly defined optimum. Nearly optimal progress is only attainable for a relatively narrow range of the learning rate τ. If the learning rate is chosen sub-optimally, the performance of the ES degrades but the ES still converges to the optimum with a log-linear rate. The reason for this behavior [46] is due to the different effects recombination has on the distance to the optimizer (i.e. on the progress rate) and on the mutation strength. An intermediate recombination of the object variables reduces the harmful parts of the mutation vector also referred to as "genetic repair effect". Thus, it reduces the loss part of the progress rate. This enables the algorithm to work with higher mutation strengths. However, since the strategy parameters are necessarily selected before recombination takes place, the self-adaptation response cannot reflect the after selection genetic repair effect and remains relatively inert to the effect of recombination.

Flat and Linear Fitness Landscapes

In [12], the behavior of multi-recombinative ES on flat and linear fitness landscapes was analyzed. Accepting the variance postulates proposed in [12] (see Section 3.3) the question arises whether the standard ES variation operators comply to these postulates, i.e., whether the strategies are able to increase the population variance in flat and linear fitness landscapes. Several common recombination operators and mutation operators were examined such as intermediate/dominant recombination of the object variables and intermediate/geometric recombination of the strategy parameters. The mutation rules applied for changing the mutation strength are the log-normal and the two-point distribution.

The analysis started with flat fitness landscapes which are selection neutral. Thus, the evolution of the mutation strength and the evolution of the object variables can be fully decoupled and the population variance can be easily computed. Beyer and Deb showed that if intermediate recombination is used for the object variables, the ES is generally able to increase the population variance exponentially fast. The same holds for dominant recombination. However, there is a memory of the old population variances that gradually vanishes. Whether this is a beneficial effect has not been investigated up to now.

In the case of linear fitness functions, only the behavior of $(1, \lambda)$-ES has been examined so far. It has been shown that the results obtained in [11] for the sphere model can be transferred to the linear case if $\sigma^* := \sigma N/R \to 0$ is considered because the sphere degrades to a hyperplane. As a result, it can be shown that the expectation of the mutation strength increases exponentially if log-normal or two-point operators are used.

5 Problems and Limitations of Self-Adaptation

Most of the research done so far seems to be centered on the effects of self-adapting the mutation strengths. Some of the problems that were reported are related to divergence of the algorithm (see e.g. Kursawe [41]) and premature convergence. Premature convergence may occur if the mutation strength and the population variance are decreased too fast. This generally results in a convergence towards a suboptimal solution. While the problem is well-known, it appears that only a few theoretical investigations have been done. However, premature convergence is not a specific problem of self-adaptation. Rudolph [53] analyzed an $(1+1)$-ES applying Rechenberg's 1/5th-rule. He showed for a test problem that the ES's transition to the global optimum cannot be ensured when the ES starts at a local optimum and if the step-sizes are decreased to fast.

Stone and Smith [77] investigated the behavior of GA on multimodal functions applying Smith's discrete self-adaptation algorithm. Premature convergence was observed for low innovation rates and high selection pressure which causes a low diversity of the population. Diversity can be increased by using high innovation rates. Stone and Smith additionally argued that one copy of the present strategy parameter should be kept while different choices are still introduced. This provides a suitable relation between exploration and exploitation.

Liang et al. [43, 44] considered the problem of a prematurely reduced mutation strength. They started with an empirical investigation on the loss of step size control for EP on five test functions [43]. The EP used a population size of $\mu = 100$ and a tournament size of $q = 10$. Stagnation of the search occurred even for the sphere model. As they argued [43, 44], this might be due to the selection of an individual with a mutation strength far too small in one dimension but with a high fitness value. This individual propagates its ill-adapted mutation strength to all descendants and, therefore, the search stagnates.

In [44], Liang et al. examined the probability of loosing the step size control. To simplify the calculations, a $(1+1)$-EP was considered. Therefore, the mutation strength changes whenever a successful mutation happens. A loss of step size control occurs if after κ successful mutations the mutation strength is smaller than an arbitrarily small positive number ϵ. The probability of such an event can be computed. It depends on the initialization of the mutation strength, the learning parameter, on the number of successful mutations, and on ϵ. As the authors showed, the probability of loosing control of the step size increases with the number of successful mutations.

A reduction of the mutation strength should occur if the EP is already close to the optimum. However, if the reduction of the distance to the optimizer cannot keep pace with that of the mutation strength, the search stagnates. This raises the question whether the operators used in this EP implementation comply to the design principles postulated in [19] (compare Section 3.3 in this paper). An analysis of the EP behavior in flat or linear fitness landscapes might reveal the very reason for this failure. It should be also noted that similar premature convergence behaviors of self-adaptive ES are rarely observed. A way to circumvent such behavior is to introduce a lower bound for the step size. Fixed lower bounds are considered in [43]. While this surely prevents premature convergence of the EP, it does not take into account the fact that the ideal lower bound of the mutation strength depends on the actual state of the search.

In [44], two schemes are considered proposing a dynamic lower bound (DLB) of the mutation strength. The first is based on the success rate reminiscent of Rechenberg's 1/5th-rule. The lower bound is adapted on the population level. A high success rate leads to an increase of the lower bound, a small success decreases it. The second DLB-scheme is called "mutation step size based". For all dimensions, the average of the mutation strengths of the successful offspring is computed. The lower bound is then obtained as the median of the averages. These two schemes appear to work well on most fitness functions of the benchmark suite. On functions with many local optima, however, both methods exhibit difficulties.

As mentioned before, Hart, Delaurentis, and Ferguson [34] analyzed an evolutionary algorithm with discrete random variables on a multi-modal function. They showed the existence of a bimodal function for which the algorithm does not converge to the global optimizer with probability one if it starts close to the local optimal solution.

Won and Lee [72] addressed a similar problem although in contrast to Hart, DeLaurentis, and Ferguson they proved sufficient conditions for premature convergence avoidance of a $(1+1)$-ES on a one-dimensional bimodal function. The mutations are modeled using Cauchy-distributed random variables and the two-point operator is used for changing the mutation strengths themselves.

Glickman and Sycara [28] identified possible causes for premature reduction of the mutation strength. They investigated the evolutionary search behavior of an EA without any crossover on a complex problem arising from the training of neural networks with recurrent connections.

What they have called *bowl effect* may occur if the EA is close to a local minimum. Provided that the mutation strength is below a threshold then the EA is confined in a local attractor and it cannot find any better solution. As a result, small mutation strengths will be preferred.

A second cause is attributed to the selection strength. Glickman and Sycara suspect that if the selection strength is high, high mutation rates have a better chance of survival compared to using low selection strength: A large mutation rate increases the variance. This is usually connected with a higher chance of degradation as compared to smaller mutation rates. But if an improvement occurs it is likely to be considerably larger than those achievable with small mutation rates. If only a small percentage of the offspring is accepted, there is a chance that larger mutation strengths "survive". Thus, using a high selection strength might be useful in safeguarding against premature stagnation. In their experiments, though, Glickman and Sycara could not observe a significant effect. They attributed this in part to the fact that the search is only effective for a narrow region of the selection strength. Recently, Hansen [31] resumed the investigation of the self-adaptive behavior of multiparent evolution strategies on linear fitness functions started in [19]. Hansen's analysis is aimed at revealing the causes why self-adaptation usually works adequately on linear fitness functions. He offered conditions under which the control mechanism of self-adaptation fails, i.e., that the EA does not increase the step size as postulated in [19]. The expectation of the mutation strength is not measured directly. Instead, a function h is introduced the expectation of which is unbiased under the variation operators. The question that now remains to be answered is whether the selection will increase the expectation of $h(\sigma)$. In other words, is the effect of an increase of the expectation a consequence of selection (and therefore due to the link between good object vectors and good strategy values) or is it due to a bias introduced by the recombination/mutation-operators chosen?

Hansen proposed two properties an EA should fulfill: First, the descendants' object vectors should be point-symmetrically distributed after mutation and recombination. Additionally, the distribution of the strategy parameters given the object vectors after recombination and mutation has to be identical for all symmetry pairs around the point-symmetric center. Evolution strategies with intermediate multi-recombination fulfill this symmetry assumption. Their descendants' distribution is point-symmetric around the recombination centroid.

Secondly, Hansen offered a so-called *σ-stationarity assumption*. It postulates the existence of a monotonically increasing function h whose expectation is left unbiased by recombination and mutation. Therefore, $\mathrm{E}[h(\mathcal{S}_k^{\sigma_{i;\lambda}|i=1,\ldots,\mu})] = 1/\mu \sum_{i=1}^{\mu} h(\sigma_{i;\lambda})$ must hold for all offspring. The term $\mathcal{S}_k^{\sigma_{i;\lambda}|i=1,\ldots,\mu}$ denotes the mutation strength of an offspring k created by recombination and mutation.

Hansen showed that if an EA fulfills the assumptions made above, self-adaptation does not change the expectation of $h(\sigma)$ if the offspring number is twice the number of parents.

The theoretical analysis was supplemented by an empirical investigation of the self-adaptation behavior of some evolution strategies examining the effect of several recombination schemes on the object variables and on the strategy parameter. It was shown that an ES which applies intermediate recombination to the object variables and to the mutation strength increases the expectation of $\log(\sigma)$ for all choices of the parent population size. On the other hand, evolution strategies that fulfill the symmetry and the stationarity assumption, increase the expectation of $\log(\sigma)$ if $\lambda < \mu/2$, keep it constant for $\lambda = \mu/2$ and decrease it for $\lambda > \mu/2$.

Intermediate recombination of the mutation strengths results in an increase of the mutation strength. This is beneficial in the case of linear problems and usually works as desired in practice, but it might also have unexpected effects in other cases.

6 Outlook

Self-adaptation usually refers to an adaptation of control parameters which are incorporated into the individual's genome. These are subject to variation and selection - thus evolving together with the object parameters. Stating it more generally, a self-adaptive algorithm controls the transmission function between parent and offspring population by itself without any external control. Thus, the concept can be extended to include algorithms where the representation of an individual is augmented with genetic information that does not code information regarding the fitness but influences the transmission function instead.

Interestingly, real-coded genetic algorithms where the diversity of the parent population controls that of the offspring may be regarded as self-adaptive. Surprisingly, even binary GAs with crossover operators as 1-point or k-point crossover share self-adaptive properties to a certain extend.

Self-Adaptation is commonly used in the area of evolutionary programming and evolution strategies. Here, generally the mutation strength or the full covariance matrix is adapted. Analyses done so far focus mainly on the convergence to the optimal solution. Nearly all analyses use either a simplified model of the algorithm or have to resort to numerical calculations in their study. The results obtained are similar: On simple fitness functions considered, conditions can be derived that ensure the convergence of the EA to local optimal solutions. The convergence is usually log-linear.

The explicit use of self-adaptation techniques is rarely found in genetic algorithm and, if at all, mainly used to adopt the mutation rate. Most of the studies found are directed at finding suitable ways to introduce self-adaptive behavior in GA. As we have pointed out, however, crossover in binary standard GAs does provide a rudimentary form of self-adaptive behavior. Therefore, the mutation rate can be often kept at a low level provided that the population size is reasonably large. However, unlike the clear goals in real-coded search spaces, it is by no means obvious to formulate desired behaviors the self-adaptation should realize in binary search spaces. This is in contrast to some real-coded genetic algorithms where it can be shown mathematically that they can exhibit self-adaptive behavior in simple fitness landscapes.

It should be noted that self-adaptation techniques are not the means to solve all adaptation problems in EAs. Concerning evolution strategies, multi-recombinative self-adaptation strategies show sensitive behavior with respect to the choices of the external learning rate τ. As a result, an optimal or a nearly optimal mutation strength is not always realized.

More problematic appear divergence or premature convergence to a suboptimal solution. The latter is attributed to a too fast reduction of the mutation strength. Several reasons for that behavior have been proposed although not rigorously investigated up to now. However, from our own research we have found that the main reason for a possible failure is due to the opportunistic way how self-adaptation uses the selection information obtained from just one generation. Self-adaptation rewards short-term gains. In its current form, it cannot look ahead. As a result, it may exhibit the convergence problems mentioned above.

Regardless of the problems mentioned, self-adaptation is a state-of-the-art adaptation technique with a high degree of robustness, especially in real-coded search spaces and in environments with uncertain or noisy fitness information. It also bears a large potential for further developments both in practical applications and in theoretical as well as empirical evolutionary computation research.

Acknowledgements

This work was supported by the Deutsche Forschungsgemeinschaft (DFG) through the Collaborative Research Center SFB 531 at the University of Dortmund and by the Research Center for Process- and Product-Engineering at the Vorarlberg University of Applied Sciences.

References

1. L. Altenberg. The evolution of evolvability in genetic programming. In K. Kinnear, editor, *Advances in Genetic Programming*, pages 47–74. MIT Press, Cambridge, MA, 1994.
2. P.J. Angeline. Adaptive and self-adaptive evolutionary computations. In M. Palaniswami and Y. Attikiouzel, editors, *Computational Intelligence: A Dynamic Systems Perspective*, pages 152–163. IEEE Press, 1995.
3. A. Auger. Convergence results for the $(1, \lambda)$-SA-ES using the theory of ϕ-irreducible Markov chains. *Theoretical Computer Science*, 334:35–69, 2005.
4. T. Bäck. The interaction of mutation rate, selection, and self-adaptation within a genetic algorithm. In R. Männer and B. Manderick, editors, *Parallel Problem Solving from Nature, 2*, pages 85–94. North Holland, Amsterdam, 1992.
5. T. Bäck. Optimal mutation rates in genetic search. In S. Forrest, editor, *Proceedings of the Fifth International Conference on Genetic Algorithms*, pages 2–8, San Mateo (CA), 1993. Morgan Kaufmann.
6. T. Bäck. Self-adaptation. In T. Bäck, D. Fogel, and Z. Michalewicz, editors, *Handbook of Evolutionary Computation*, pages C7.1:1–C7.1:15. Oxford University Press, New York, 1997.

7. Th. Bäck. Self-adaptation in genetic algorithms. In F. J. Varela and P. Bourgine, editors, *Toward a Practice of Autonomous Systems: proceedings of the first European conference on Artificial Life*, pages 263–271. MIT Press, 1992.
8. Th. Bäck and M. Schütz. Intelligent mutation rate control in canonical genetic algorithms. In *ISMIS*, pages 158–167, 1996.
9. J. D. Bagley. *The Behavior of Adaptive Systems Which Employ Genetic and Correlation Algorithms.* PhD thesis, University of Michigan, 1967.
10. H.-G. Beyer. Toward a theory of evolution strategies: Self-adaptation. *Evolutionary Computation*, 3(3):311–347, 1996.
11. H.-G. Beyer. *The Theory of Evolution Strategies.* Natural Computing Series. Springer, Heidelberg, 2001.
12. H.-G. Beyer and K. Deb. On self-adaptive features in real-parameter evolutionary algorithms. *IEEE Transactions on Evolutionary Computation*, 5(3):250–270, 2001.
13. H.-G. Beyer and H.-P. Schwefel. Evolution strategies: A comprehensive introduction. *Natural Computing*, 1(1):3–52, 2002.
14. A. Bienvenüe and O. François. Global convergence for evolution strategies in spherical problems: Some simple proofs and d difficulties. *Theoretical Computer Science*, 308:269–289, 2003.
15. L. Davis. Adapting operator probabilities in genetic algorithms. In J. D. Schaffer, editor, *Proc. 3rd Int'l Conf. on Genetic Algorithms*, pages 61–69, San Mateo, CA, 1989. Morgan Kaufmann.
16. M. W. Davis. The natural formation of gaussian mutation strategies in evolutionary programming. In *roceedings of the Third Annual Conference on Evolutionary Programming*, San Diego, CA, 1994. Evolutionary Programming Society.
17. K. Deb and R. B. Agrawal. Simulated binary crossover for continuous search space. *Complex Systems*, 9:115–148, 1995.
18. K. Deb and H.-G. Beyer. Self-adaptation in real-parameter genetic algorithms with simulated binary crossover. In W. Banzhaf, J. Daida, A.E. Eiben, M.H. Garzon, V. Honavar, M. Jakiela, and R.E. Smith, editors, *GECCO-99: Proceedings of the Genetic and Evolutionary Computation Conference*, pages 172–179, San Francisco, CA, 1999. Morgan Kaufmann.
19. K. Deb and H.-G. Beyer. Self-adaptive genetic algorithms with simulated binary crossover. Series CI 61/99, SFB 531, University of Dortmund, March 1999.
20. K. Deb and H.-G. Beyer. Self-adaptive genetic algorithms with simulated binary crossover. *Evolutionary Computation*, 9(2):197–221, 2001.
21. K. Deb and M. Goyal. A robust optimization procedure for mechanical component design based on genetic adaptive search. *Transactions of the ASME: Journal of Mechanical Desiogn*, 120(2):162–164, 1998.
22. A. E. Eiben, R. Hinterding, and Z. Michalewicz. Parameter control in evolutionary algorithms. *IEEE Transactions on Evolutionary Computation*, 3(2):124–141, 1999.
23. L. J. Eshelman and J. D. Schaffer. Real-coded genetic algorithms and interval schemata. In L. D. Whitley, editor, *Foundations of Genetic Algorithms, 2*, pages 187–202. Morgan Kaufmann, San Mateo, CA, 1993.
24. D. B. Fogel. *Evolving Artificial Intelligence.* PhD thesis, University of California, San Diego, 1992.
25. D. B. Fogel, L. J. Fogel, and J. W. Atma. Meta-evolutionary programming. In R.R. Chen, editor, *Proc. of 25th Asilomar Conference on Signals, Systems & Computers*, pages 540–545, Pacific Grove, CA, 1991.

26. L. J. Fogel, A. J. Owens, and M. J. Walsh. *Artificial Intelligence through Simulated Evolution*. Wiley, New York, 1966.
27. L.J. Fogel. Autonomous automata. *Industrial Research*, 4:14–19, 1962.
28. M. Glickman and K. Sycara. Reasons for premature convergence of self-adapting mutation rates. In *Proc. of the 2000 Congress on Evolutionary Computation*, pages 62–69, Piscataway, NJ, 2000. IEEE Service Center.
29. D.E. Goldberg. *Genetic Algorithms in Search, Optimization, and Machine Learning*. Addison Wesley, Reading, MA, 1989.
30. L. Grünz and H.-G. Beyer. Some observations on the interaction of recombination and self-adaptation in evolution strategies. In P.J. Angeline, editor, *Proceedings of the CEC'99 Conference*, pages 639–645, Piscataway, NJ, 1999. IEEE.
31. N. Hansen. Limitations of mutative σ-self-adaptation on linear fitness functions. *Evolutionary Computation*, 2005. accepted for publication.
32. N. Hansen and A. Ostermeier. Completely derandomized self-adaptation in evolution strategies. *Evolutionary Computation*, 9(2):159–195, 2001.
33. W. E. Hart and J. M. DeLaurentis. Convergence of a discretized self-adapative evolutionary algorithm on multi-dimensional problems. submitted.
34. W.E. Hart, J.M. DeLaurentis, and L.A. Ferguson. On the convergence of an implicitly self-adaptive evolutionary algorithm on one-dimensional unimodal problems. *IEEE Transactions on Evolutionary Computation*, 2003. To appear.
35. J. H. Holland. Outline for a logical theory of adaptive systems. *JACM*, 9:297–314, 1962.
36. J. H. Holland. *Adaptation in Natural and Artificial Systems*. The University of Michigan Press, Ann Arbor, 1975.
37. Ch. Igel and M. Toussaint. Neutrality and self-adaptation. *Natural Computing: an international journal*, 2(2):117–132, 2003.
38. B. A. Julstrom. Adaptive operator probabilities in a genetic algorithm that applies three operators. In *SAC*, pages 233–238, 1997.
39. H. Kita. A comparison study of self-adaptation in evolution strategies and real-coded genetic algorithms. *Evolutionary Computation*, 9(2):223–241, 2001.
40. H. Kita and M. Yamamura. A functional specialization hypothesis for designing genetic algorithms. In *Proc. IEEE International Conference on Systems, Man, and Cybernetics '99*, pages 579–584, Piscataway, New Jersey, 1999. IEEE Press.
41. F. Kursawe. *Grundlegende empirische Untersuchungen der Parameter von Evolutionsstrategien — Metastrategien*. Dissertation, Fachbereich Informatik, Universität Dortmund, 1999.
42. Ch.-Y. Lee and X. Yao. Evolutionary programming using mutations based on the levy probability distribution. *Evolutionary Computation, IEEE Transactions on*, 8(1):1–13, Feb. 2004.
43. K.-H. Liang, X. Yao, Ch. N. Newton, and D. Hoffman. An experimental investigation of self-adaptation in evolutionary programming. In V. W. Porto, N. Saravanan, Waagen D. E., and A. E. Eiben, editors, *Evolutionary Programming*, volume 1447 of *Lecture Notes in Computer Science*, pages 291–300. Springer, 1998.
44. K.-H. Liang, X. Yao, and Ch.S. Newton. Adapting self-adaptive parameters in evolutionary algorithms. *Artificial Intelligence*, 15(3):171 – 180, November 2001.
45. R. E. Mercer and J. R. Sampson. Adaptive search using a reproductive metaplan. *Kybernetes*, 7:215–228, 1978.

46. S. Meyer-Nieberg and H.-G. Beyer. On the analysis of self-adaptive recombination strategies: First results. In B. McKay et al., editors, *Proc. 2005 Congress on Evolutionary Computation (CEC'05), Edinburgh, UK*, pages 2341–2348, Piscataway NJ, 2005. IEEE Press.
47. S. P. Meyn and R.L. Tweedie. *Markov Chains and Stochastic Stability*. Springer, 1993.
48. A. Ostermeier, A. Gawelczyk, and N. Hansen. A derandomized approach to self-adaptation of evolution strategies. *Evolutionary Computation*, 2(4):369–380, 1995.
49. I. Rechenberg. Cybernetic solution path of an experimental problem. *Royal Aircraft Establishment, Farnborough*, page Library Translation 1122, 1965.
50. I. Rechenberg. *Evolutionsstrategie: Optimierung technischer Systeme nach Prinzipien der biologischen Evolution*. Frommann-Holzboog Verlag, Stuttgart, 1973.
51. J. Reed, R. Toombs, and N.A. Barricelli. Simulation of biological evolution and machine learning. i. selection of self-reproducing numeric patterns by data processing machines, effects of hereditary control, mutation type and crossing. *Journal of Theoretical Biology*, 17:319–342, 1967.
52. R.S. Rosenberg. *Simulation of genetic populations with biochemical properties*. Ph.d. dissertation, Univ. Michigan, Ann Arbor, MI, 1967.
53. G. Rudolph. Self-adaptive mutations may lead to premature convergence. *IEEE Transactions on Evolutionary Computation*, 5(4):410–414, 2001.
54. J.D. Schaffer and A. Morishima. An adaptive crossover distribution mechanism for genetic algorithms. In J.J. Grefenstette, editor, *Genetic Algorithms and their Applications: Proc. of the Second Int'l Conference on Genetic Algorithms*, pages 36–40, 1987.
55. H.-P. Schwefel. Kybernetische Evolution als Strategie der exprimentellen Forschung in der Strömungstechnik. Master's thesis, Technical University of Berlin, 1965.
56. H.-P. Schwefel. Adaptive Mechanismen in der biologischen Evolution und ihr Einfluß auf die Evolutionsgeschwindigkeit. Technical report, Technical University of Berlin, 1974. Abschlußbericht zum DFG-Vorhaben Re 215/2.
57. H.-P. Schwefel. *Numerische Optimierung von Computer-Modellen mittels der Evolutionsstrategie*. Interdisciplinary systems research; 26. Birkhäuser, Basel, 1977.
58. H.-P. Schwefel. *Numerical Optimization of Computer Models*. Wiley, Chichester, 1981.
59. M.A. Semenov. Convergence velocity of evolutionary algorithm with self-adaptation. In *GECCO 2002*, pages 210–213, 2002.
60. M.A. Semenov and D.A. Terkel. Analysis of convergence of an evolutionary algorithm with self-adaptation using a stochastic lyapunov function. *Evolutionary Computation*, 11(4):363–379, 2003.
61. J. Smith and T. C. Fogarty. Self-adaptation of mutation rates in a steady state genetic algorithm. In *Proceedings of 1996 IEEE Int'l Conf. on Evolutionary Computation (ICEC '96)*, pages 318–323. IEEE Press, NY, 1996.
62. J. Smith and T.C. Fogarty. Recombination strategy adaptation via evolution of gene linkage. In *Proc. of the 1996 IEEE International Conference on Evolutionary Computation*, pages 826–831. IEEE Publishers, 1996.
63. J. E. Smith. *Self-Adaptation in Evolutionary Algorithms*. PhD thesis, University of the West of England, Bristol, 1998.

64. J. E. Smith. Modelling GAs with self adaptive mutation rates. In L. Spector and et al., editors, *Proceedings of the Genetic and Evolutionary Computation Conference (GECCO-2001)*, pages 599–606, San Francisco, California, USA, 7-11 July 2001. Morgan Kaufmann.
65. J. E. Smith. Parameter perturbation mechanisms in binary coded gas with self-adaptive mutation. In K. DeJong, R. Poli, and J. Rowe, editors, *Foundations of Genetic Algorithms 7*, pages 329–346. Morgan Kauffman, 2004.
66. J. E. Smith and T. C. Fogarty. Operator and parameter adaptation in genetic algorithms. *Soft Computing*, 1(2):81–87, June 1997.
67. W. Spears. Adapting crossover in evolutionary algorithms. In *Proceedings of the Evolutionary Programming Conference*, pages 367–384, 1995.
68. W.M. Spears. *Evolutionary Algorithms: The Role of Mutation and Recombination.* Springer-Verlag, Heidelberg, 2000.
69. Ch. Stone and J.E. Smith. Strategy parameter variety in self-adaptation of mutation rates. In W.B. Langdon and et al., editors, *GECCO*, pages 586–593. Morgan Kaufmann, 2002.
70. H.-M. Voigt, H. Mühlenbein, and D. Cvetković. Fuzzy recombination for the breeder genetic algorithm. In L.J. Eshelman, editor, *Proc. 6th Int'l Conf. on Genetic Algorithms*, pages 104–111, San Francisco, CA, 1995. Morgan Kaufmann Publishers, Inc.
71. R. Weinberg. *Computer Simulation of a Living Cell.* PhD thesis, University of Michigan, 1970.
72. J. M. Won and J. S. Lee. Premature convergence avoidance of self-adaptive evolution strategy,. In *The 5th International Conference on Simulated Evolution And Learning*, Busan, Korea, Oct 2004.
73. X. Yao and Y. Liu. Fast evolutionary programming. In L. J. Fogel, P. J. Angeline, and T. Bäck, editors, *Proceedings of the Fifth Annual Conference on Evolutionary Programming*, pages 451–460. The MIT Press, Cambridge, MA, 1996.

Adaptive Strategies for Operator Allocation

Dirk Thierens

Department of Information and Computing Sciences
Universiteit Utrecht
The Netherlands
dirk.thierens@cs.uu.nl

Summary. Learning the optimal probabilities of applying an exploration operator from a set of alternatives can be done by self-adaptation or by adaptive allocation rules. In this chapter we discuss the latter approach. The allocation strategies considered in the literature usually belong to the class of probability matching algorithms. These strategies adapt the operator probabilities in such a way that they match the reward distribution. We will also discuss an alternative adaptive allocation strategy, called the adaptive pursuit method, and compare this method with the probability matching approach in a controlled, non-stationary environment. Calculations and experimental results show the performance differences between the two strategies. If the reward distributions stay stationary for some time, the adaptive pursuit method converges rapidly and accurately to an operator probability distribution that results in a much higher probability of selecting the current optimal operator and a much higher average reward than with the probability matching strategy. Importantly, the adaptive pursuit scheme also remains sensitive to changes in the reward distributions.

1 Introduction

Genetic algorithms usually apply their exploration operators with a fixed probability. There are however no general guidelines to help determine an optimal value for these probability values. In practice, the user simply searches for a reasonable set of values by running a series of trial-and-error experiments. Clearly, this is a computationally expensive procedure and since genetic algorithms are often applied to computational intensive problems, the number of probability values explored has to remain limited. To make matters worse, the problem is compounded by the fact that there is no single, fixed set of values that is optimal during the entire run. Depending on the current state of the search process the optimal probability values continuously change.

This problem has long been recognized and different adaptation methods have been proposed to solve it [4][5][11]. In general two classes of adaptation methods can be found:

D. Thierens: *Adaptive Strategies for Operator Allocation*, Studies in Computational Intelligence (SCI) **54**, 77–90 (2007)
www.springerlink.com

1. **Self-adaptation.** The values of the operator probabilities are directly encoded in the representation of the individual solutions. These values basically hitchhike with the solutions that are being evolved through the regular search process. The idea is that the operator probability values and problem solutions co-evolve to (near)-optimal settings.
2. **Adaptive allocation rule.** The values of the operator probabilities are adapted following an 'out-of-the-evolutionary-loop' learning rule according to the quality of new solutions created by the operators.

Self-adaptation is particularly applied within Evolutionary Strategies and Evolutionary Programming for numerical optimization problems. When applied to discrete optimization or adaptation problems using genetic algorithms, its success is somewhat limited as compared to the adaptive allocation rule method. In this chapter we will focus on the latter class. Looking at the literature it becomes clear that most adaptive allocation rules used belong to the probability matching type [2][3][6][7][8][9][10][15]. Here we also look at an alternative allocation rule, called the adaptive pursuit method. We compare both approaches on a controlled, non-stationary environment. Results indicate that the adaptive pursuit method possess a number of useful properties for adaptive operator allocation rule which are less present in the traditionally used probability matching algorithm.

The chapter is organized as follows. Section 2 describes the probability matching method and specifies an implementation particularly suited for non-stationary environments. Section 3 explains the adaptive pursuit algorithm. Section 4 shows experimental results of both adaptive allocation techniques. Finally, Section 5 concludes this contribution.

2 Probability Matching

An rule is an algorithm that iteratively chooses one of its operators to apply to an external environment [12]. The environment returns a reward - possibly zero - and the allocation rule uses this reward and its internal state to adapt the probabilities with which the operators are chosen. It is crucial to note that the environment we consider here is non-stationary, meaning that the probability distribution specifying the reward generated when applying some operator changes during the runtime of the allocation algorithm. Unfortunately, the non-stationarity requirement excludes the use of a large number of adaptive strategies that have been developed for the well-known multi-armed bandit problem [1].

Formally, we have a set of K operators $\mathcal{A} = \{a_1, \ldots, a_K\}$, and a probability vector $\mathcal{P}(t) = \{\mathcal{P}_1(t), \ldots, \mathcal{P}_K(t)\}$ $(\forall t : 0 \leq \mathcal{P}_i(t) \leq 1; \sum_{i=1}^{K} \mathcal{P}_i(t) = 1)$. The adaptive allocation rule selects an operator to be executed in proportion to the probability values specified in $\mathcal{P}(t)$. When an operator a is applied to the environment at time t, a reward $\mathcal{R}_a(t)$ is returned. Each operator has an associated reward, which is a non-stationary random variable. All rewards are collected in the reward vector $\mathcal{R}(t) = \{\mathcal{R}_1(t), \ldots, \mathcal{R}_K(t)\}$. In addition to the operator probability vector $\mathcal{P}(t)$ the adaptive allocation rule maintains a quality vector $\mathcal{Q}(t) = \{\mathcal{Q}_1(t), \ldots, \mathcal{Q}_K(t)\}$ that specifies a running estimate of the reward for each operator. Whenever an operator is executed his current estimate $\mathcal{Q}_a(t)$ is adapted. The allocation algorithm is run for a period of T time steps. The goal is to maximize the expected value of the cumulative reward $\mathcal{E}[\mathcal{R}] = \sum_{t=1}^{T} \mathcal{R}_a(t)$ received by the adaptive allocation rule. Since the environment is non-stationary, the estimate of the reward for each operator is only reliable when the rewards received are not too old. An elegant, iterative method to compute such a running estimate is the exponential, recency-weighted average that updates the current estimate with a fraction of the difference of the target value and the current estimate:

$$\mathcal{Q}_a(t+1) = \mathcal{Q}_a(t) + \alpha[\mathcal{R}_a(t) - \mathcal{Q}_a(t)] \tag{1}$$

with the adaptation rate $\alpha : 0 < \alpha \leq 1$. The basic rule computes each operator's selection probability $\mathcal{P}_a(t)$ as the proportion of the operator's reward estimate $\mathcal{Q}_a(t)$ to the sum of all reward estimates $\sum_{i=1}^{K} \mathcal{Q}_i(t)$. However, this may lead to the loss of some operators. Once a probability $\mathcal{P}_a(t)$ becomes equal to 0, the operator will no longer be selected and its reward estimate can no longer be updated. This is an unwanted property in a non-stationary environment because the operator might become valuable again in a future stage of the search process. We always need to be able to apply any operator and update its current value estimate. To ensure that no operator gets lost we enforce a minimal value $P_{min} : 0 < P_{min} < 1$ for each selection probability. As a result the maximum value any operator can achieve is $P_{max} = 1-(K-1)P_{min}$ where K is the number of operators. The rule for updating the probability vector $\mathcal{P}_a(t)$ now becomes:

$$\mathcal{P}_a(t+1) = P_{min} + (1 - K \cdot P_{min}) \frac{\mathcal{Q}_a(t+1)}{\sum_{i=1}^{K} \mathcal{Q}_i(t+1)}. \tag{2}$$

It is easy to see in the above equation that when an operator does not receive any reward for a long time its value estimate $\mathcal{Q}_a(t)$ will converge to 0 and its probability of being selected $\mathcal{P}_a(t)$ converges to P_{min}. It is also clear that when only one operator receives a reward during a long period of time - and all other operators get no reward - then its selection probability $\mathcal{P}_a(t)$ converges to $P_{min} + (1 - K \cdot P_{min}).1 = P_{max}$. Furthermore, $0 < \mathcal{P}_a(t) < 1$ and their sum equals 1:

$$\sum_{a=1}^{K} \mathcal{P}_a(t) = K \cdot P_{min} + \frac{1 - K \cdot P_{min}}{\sum_{i=1}^{K} \mathcal{Q}_i(t)} \sum_{a=1}^{K} \mathcal{Q}_a(t) = 1.$$

Finally, our probability matching algorithm is specified as:

ProbabilityMatching($\mathcal{P}, \mathcal{Q}, K, P_{min}, \alpha$)

for $i \leftarrow 1$ **to** K
do $\mathcal{P}(i) \leftarrow \frac{1}{K}$; $\mathcal{Q}(i) \leftarrow 1.0$
while NotTerminated?()
do $a^s \leftarrow$ ProportionalSelectOperator($\mathcal{P}$)
  $R_{a^s}(t) \leftarrow$ GetReward(a^s)
  $\mathcal{Q}_{a^s}(t+1) = \mathcal{Q}_{a^s}(t) + \alpha[R_{a^s}(t) - \mathcal{Q}_{a^s}(t)]$
  for $a \leftarrow 1$ **to** K
  do $\mathcal{P}_a(t+1) = P_{min} + (1 - K \cdot P_{min}) \frac{\mathcal{Q}_a(t+1)}{\sum_{i=1}^{K} \mathcal{Q}_i(t+1)}$

The probability matching allocation rule as specified in equation 2 is able to adapt to non-stationary environments. Unfortunately, it pays a heavy price for this in terms of reward maximization. This is most obvious when we consider a non-stationary environment. Suppose we have only two operators a_1 and a_2 with constant reward values $\mathcal{R}_1$ and $\mathcal{R}_2$. From equation 2 it follows that

$$\frac{\mathcal{P}_1(t) - P_{min}}{\mathcal{P}_2(t) - P_{min}} \rightarrow \frac{\mathcal{R}_1}{\mathcal{R}_2}.$$

Assume that $\mathcal{R}_1 > \mathcal{R}_2$. An ideal adaptive allocation rule should notice in this stationary environment that the operator a_1 has a higher reward than operator a_2. The allocation rule should therefore maximize the probability of applying operator a_1 and only apply a_2 with the minimal probability P_{min}. However, the closer the rewards $\mathcal{R}_1$ and $\mathcal{R}_2$ are, the less optimal the probability matching rule behaves. For instance, when $P_{min} = 0.1$, $\mathcal{R}_1 = 10$, and $\mathcal{R}_2 = 9$ then $\mathcal{P}_1 = 0.52$ and $\mathcal{P}_2 = 0.48$, which is far removed from the desired values of $\mathcal{P}_1 = 0.9$ and $\mathcal{P}_2 = 0.1$.

Matching the reward probabilities is not an optimal strategy for allocating operator probabilities in an optimizing algorithm. In the next section we discuss the as an alternative allocation method that is better suited to maximize the rewards received while still maintaining the ability to swiftly react to any changes in a non-stationary environment.

3 Adaptive Pursuit algorithm

Pursuit algorithms are a class of rapidly converging algorithms for learning automata proposed by Thathachar and Sastry [13]. They represent adaptive allocation rules that adapt the operator probability vector $\mathcal{P}(t)$ in such a way that the algorithm pursues the operator a^* that currently has the maximal estimated reward $\mathcal{Q}_{a^*}(t)$. To achieve this, the pursuit method increases the selection probability $\mathcal{P}_{a^*}(t)$ and decreases all other probabilities $\mathcal{P}_a(t), \forall a \neq a^*$. Pursuit algorithms originated from the field of learning automata. However, they are designed for stationary environments for which it can be proved that they are ϵ-optimal. The ϵ-optimality property means that in every stationary environment, there exists a learning rate $\beta^* > 0$ and time $t_0 > 0$, such that for all learning rates $0 < \beta \leq \beta^* \leq 1$ and for any $\delta \in [0 \ldots 1]$ and $\epsilon \in [0 \ldots 1]$:

$$\text{Prob}[\mathcal{P}_a^{optimal}(t) > 1 - \epsilon] > 1 - \delta, \qquad \forall t > t_0.$$

In practice this means that if the the learning rate β is small enough as a function of the reward distribution correct convergence is assured.

As in the probability matching allocation rule, the pursuit algorithm proportionally selects an operator to execute according to the probability vector $\mathcal{P}(t)$, and updates the corresponding operator's quality or estimated reward $\mathcal{Q}_a(t)$. Subsequently, the current best operator a^* is chosen ($a^* = \text{argmax}_a[\mathcal{Q}_a(t+1)]$), and its selection probability is increased

$$\mathcal{P}_{a^*}(t+1) = (1-\beta)\mathcal{P}_{a^*}(t) + \beta,$$

while the other operators have their selection probability decreased

$$\forall a \neq a^* : \mathcal{P}_a(t+1) = (1-\beta)\mathcal{P}_a(t).$$

It is clear from $(1-\beta)\mathcal{P}_{a^*}(t) + \beta = \mathcal{P}_{a^*}(t) + \beta(1 - \mathcal{P}_{a^*}(t))$ that if a particular operator is repeatedly the best operator its selection probability will converge to 1, while the selection probabilities of the other operators will converge to 0 and they will no longer be applied. Consequently, the pursuit algorithm cannot be used in a non-stationary environment. To make the method suitable for non-stationary environments the probability updating scheme needs to be adjusted [14]. The modified update rule ensures that the probability vector is still pursuing the current best operator at the same rate as the standard method, but now the exponential, recency-weighted averages of the operator probabilities are enforced to stay within the interval $[P_{min} \ldots P_{max}]$ with $0 < P_{min} < P_{max} < 1$. Calling $a^* = \text{argmax}_a[\mathcal{Q}_a(t+1)]$ the current best operator, we get:

$$\mathcal{P}_{a^*}(t+1) = \mathcal{P}_{a^*}(t) + \beta[P_{max} - \mathcal{P}_{a^*}(t)] \tag{3}$$

and

$$\forall a \neq a^* : \mathcal{P}_a(t+1) = \mathcal{P}_a(t) + \beta[P_{min} - \mathcal{P}_a(t)] \tag{4}$$

under the constraint:

$$P_{max} = 1 - (K-1)P_{min}. \tag{5}$$

The constraint ensures that if $\sum_{a=1}^{K} \mathcal{P}_a(t) = 1$ the sum of the updated probabilities remains equal to 1:

$$\begin{aligned}
&\sum_{a=1}^{K} \mathcal{P}_a(t+1) = 1 \\
&\Leftrightarrow \text{rhs. eqt.}(3) + \text{rhs. eqt.}(4) = 1 \\
&\Leftrightarrow (1-\beta)\sum_{a=1}^{K} \mathcal{P}_a(t) + \beta[P_{max} + (K-1)P_{min}] = 1 \\
&\Leftrightarrow P_{max} = 1 - (K-1)P_{min}.
\end{aligned}$$

Note that since $P_{min} < P_{max}$ the constraint can only be fulfilled[1] if $P_{min} < \frac{1}{K}$. An interesting value for the minimal probability is $P_{min} = \frac{1}{2K}$ which results in the maximum probability $P_{max} = \frac{1}{2} + \frac{1}{2K}$. An intuitive appealing way to look at these values is that the optimal operator will be selected half the time, while the other half of the time all operators have an equal probability of being selected.

Finally, we can now specify more formally the adaptive pursuit algorithm:

AdaptivePursuit($\mathcal{P}, \mathcal{Q}, K, P_{min}, \alpha, \beta$)

$P_{max} \leftarrow 1 - (K-1)P_{min}$
for $i \leftarrow 1$ **to** K
do $\mathcal{P}(i) \leftarrow \frac{1}{K}$; $\mathcal{Q}(i) \leftarrow 1.0$
while NotTerminated?()
do $a^s \leftarrow$ ProportionalSelectOperator($\mathcal{P}$)
    $R_{a^s}(t) \leftarrow$ GetReward(a^s)
    $\mathcal{Q}_{a^s}(t+1) = \mathcal{Q}_{a^s}(t) + \alpha[R_{a^s}(t) - \mathcal{Q}_{a^s}(t)]$
    $a^* \leftarrow \text{ArgMax}_a(\mathcal{Q}_a(t+1))$
    $\mathcal{P}_{a^*}(t+1) = \mathcal{P}_{a^*}(t) + \beta[P_{max} - \mathcal{P}_{a^*}(t)]$
    for $a \leftarrow 1$ **to** K
    do if $a \neq a^*$
        then $\mathcal{P}_a(t+1) = \mathcal{P}_a(t) + \beta[P_{min} - \mathcal{P}_a(t)]$

Consider again the 2-operator stationary environment at the end of the previous section with $P_{min} = 0.1$, $\mathcal{R}_1 = 10$, and $\mathcal{R}_2 = 9$. As opposed to the probability matching rule, the adaptive pursuit method will play the better operator a_1 with maximum probability $P_{max} = 0.9$. It also keeps playing the poorer operator a_2 with minimal probability $P_{min} = 0.1$ in order to maintain its ability to adapt to any change in the reward distribution.

4 Experimental results

To get an idea of the dynamic behavior of these adaptive allocation rules we compare the probability matching algorithm, the adaptive pursuit method, and the non-adaptive, equal-probability strategy, on the following non-stationary environment. We consider an environment with 5 operators (or arms in the multi-bandit problem terminology). Each operator a receives a uniformly distributed reward $\mathcal{R}_a$ between the respective boundaries $\mathcal{R}_5 = \mathcal{U}[4\ldots6], \mathcal{R}_4 = \mathcal{U}[3\ldots5], \mathcal{R}_3 = \mathcal{U}[2\ldots4], \mathcal{R}_2 = \mathcal{U}[1\ldots3]$, and $\mathcal{R}_1 = \mathcal{U}[0\ldots2]$. After a fixed time interval ΔT these reward distributions are randomly reassigned to the operators, under the constraint that the current best operator-reward association effectively has to change to a new couple. Specifically, the non-stationary

[1] Strictly speaking, P_{min} can be equal to $\frac{1}{K}$. This happens in the case of only 2 operators ($K = 2$) and a lower bound probability of $P_{min} = 0.5$. However, now P_{max} also equals 0.5 so there is no adaptation possible.

environment in the simulation switches 10 times with the following pattern: 01234 → 41203 → 24301 → 12043 → 41230 → 31420 → 04213 → 23104 → 14302 → 40213, where each sequence orders the operators in descending value of reward. For instance '41203' means that operator a_4 receives the highest reward $\mathcal{R}_5$, operator a_1 receives the second highest reward $\mathcal{R}_4$, operator a_2 receives reward $\mathcal{R}_3$, operator a_0 receives reward $\mathcal{R}_2$, and finally operator a_3 receives the lowest reward $\mathcal{R}_1$. If we had full knowledge of the reward distributions and their switching pattern we could always pick the optimal operator a^* and achieve an expected reward

$$\mathcal{E}[\mathcal{R}^{Opt}] = \mathcal{R}_{a^*} = 5 \ .$$

Clearly, this value can never be obtained by any adaptive allocation strategy since it always needs to pay a price for exploring the effects of alternative actions. Nevertheless, it does represent an upper bound of the expected reward. When operating in a stationary environment the allocation strategies converge to a fixed operator probability distribution. Using this distribution we can compute the maximum achievable expected reward for each allocation rule, which is preferably close to the theoretical upper bound. In a non-stationary environment, we aim to achieve this value as quick as possible, while still being able to react swiftly to any change in the reward distributions. In the experiments we have taken the value $P_{min} = \frac{1}{2K} = 0.1$ for the minimum probability each operator will be applied in the adaptive allocation schemes. For a stationary environment - this is, when the assignment of reward distributions to the arms are not switched - we can compute the expected reward and the probability of choosing the optimal operator once the operator probability vectors have converged.

1. **Non-adaptive, equal-probability allocation rule.**
   This strategy simply selects each operator with equal probability. The probability of choosing the optimal operator a^*_{Fixed} is

$$\begin{aligned}\text{Prob}[a^s = a^*_{Fixed}] &= \frac{1}{K} \\ &= 0.2 \ .\end{aligned}$$

   The expected reward becomes

$$\begin{aligned}\mathcal{E}[\mathcal{R}^{Fixed}] &= \sum_{a=1}^{K} \mathcal{E}[\mathcal{R}_a]\text{Prob}[a^s = a] \\ &= \frac{\sum_{a=1}^{K} \mathcal{E}[\mathcal{R}_a]}{K} \\ &= 3 \ .\end{aligned}$$

2. **Probability matching allocation rule.**
   For the probability matching updating scheme the probability of choosing the optimal operator $a^*_{ProbMatch}$ is

$$\begin{aligned}
&\text{Prob}[a^s = a^*_{ProbMatch}] \\
&= P_{min} + (1 - K.P_{min})\frac{\mathcal{E}[\mathcal{R}_{a^*}]}{\sum_{a=1}^{K}\mathcal{E}[\mathcal{R}_a]} \\
&= 0.2666\ldots .
\end{aligned}$$

The expected reward becomes

$$\begin{aligned}
&\mathcal{E}[\mathcal{R}^{ProbMatch}] \\
&= \sum_{a=1}^{K}\mathcal{E}[\mathcal{R}_a]\text{Prob}[a^s = a] \\
&= \sum_{a=1}^{K} a[P_{min} + (1 - K \cdot P_{min})\frac{\mathcal{E}[\mathcal{R}_a]}{\sum_{a=1}^{K}\mathcal{E}[\mathcal{R}_a]}] \\
&= 3.333\ldots .
\end{aligned}$$

3. **Adaptive pursuit allocation rule.**
For the adaptive pursuit updating scheme the probability of choosing the optimal operator $a^*_{AdaPursuit}$ is

$$\begin{aligned}
\text{Prob}[a^s = a^*_{AdaPursuit}] &= 1 - (K - 1) \cdot P_{min} \\
&= 0.6 .
\end{aligned}$$

The expected reward becomes

$$\begin{aligned}
&\mathcal{E}[\mathcal{R}^{AdaPursuit}] \\
&= \sum_{a=1}^{K}\mathcal{E}[\mathcal{R}_a]\text{Prob}[a^s = a] \\
&= P_{max}\ \mathcal{E}[\mathcal{R}_{a^*}] + P_{min} \sum_{a=1, a\neq a^*}^{K} \mathcal{E}[\mathcal{R}_a] \\
&= 4 .
\end{aligned}$$

The computed expected rewards and probabilities of applying the optimal operator show that both adaptive allocation rules have a better performance than the non-adaptive strategy that simply selects each operator with equal probability. More interesting, they also show that - after convergence - the adaptive pursuit algorithm has a significantly better performance than the probability matching algorithm in the stationary environment. The probability matching algorithm will apply the optimal operator in only 27% of the trials while the pursuit algorithm will be optimal in 60% of the cases. Similarly, the probability matching algorithm has an expected reward of 3.3 versus an expected reward of 4 for the pursuit method. Of course, this assumes that both adaptive strategies are able to converge correctly and rapidly. For non-stationary environments it is vital that the adaptive allocation techniques

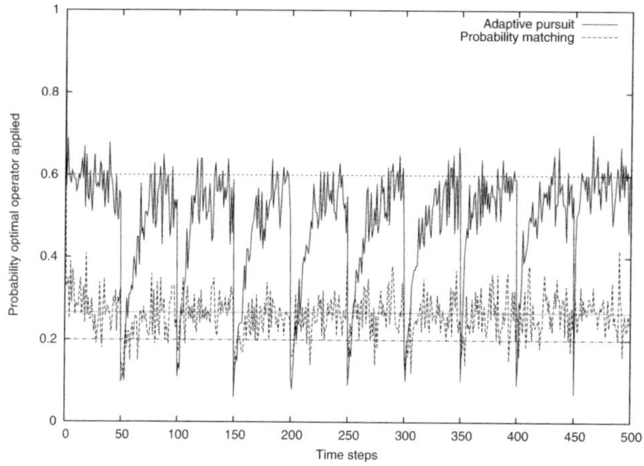

Fig. 1. The probability of selecting the optimal operator at each time step in the non-stationary environment with switching interval $\Delta T = 50$ time steps (learning rates $\alpha = 0.8$; $\beta = 0.8$; $Pmin = 0.1$; $K = 5$; results are averaged over 100 runs). The horizontal lines show the expected values for the non-switching, stationary environment for resp. adaptive pursuit (0.6), probability matching (0.27), and random selection (0.2).

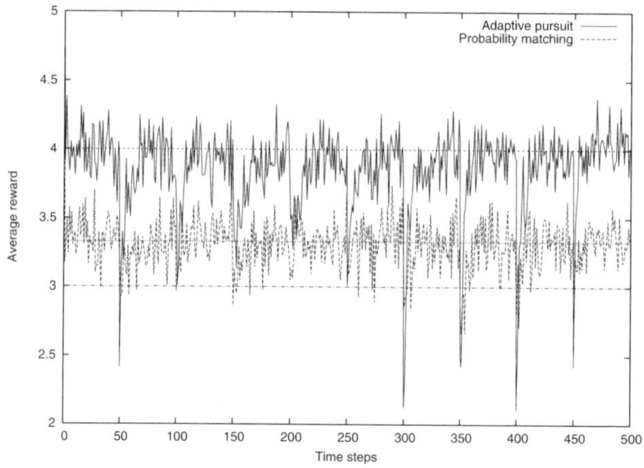

Fig. 2. The average reward received at each time step in the non-stationary environment with switching interval $\Delta T = 50$ time steps (learning rates $\alpha = 0.8$; $\beta = 0.8$; $Pmin = 0.1$; $K = 5$; results are averaged over 100 runs). The horizontal lines show the expected values for the non-switching, stationary environment for resp. adaptive pursuit (4), probability matching (3.33), and random selection (3).

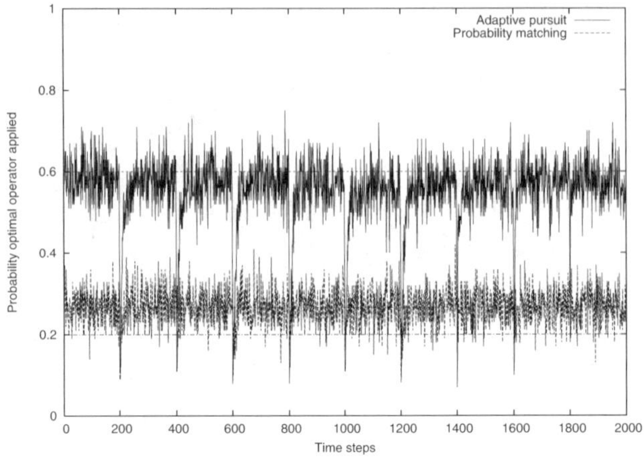

Fig. 3. The probability of selecting the optimal operator at each time step in the non-stationary environment with switching interval $\Delta T = 200$ time steps (learning rates $\alpha = 0.8$; $\beta = 0.8$; $Pmin = 0.1$; $K = 5$; results are averaged over 100 runs). The horizontal lines show the expected values for the non-switching, stationary environment for resp. adaptive pursuit (0.6), probability matching (0.27), and random selection (0.2).

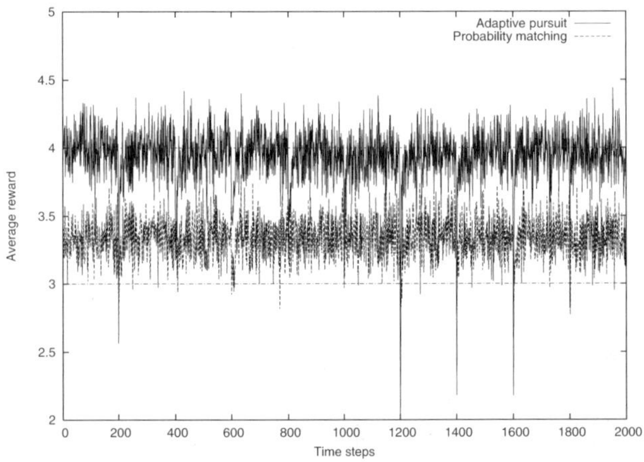

Fig. 4. The average reward received at each time step in the non-stationary environment with switching interval $\Delta T = 200$ time steps (learning rates $\alpha = 0.8$; $\beta = 0.8$; $Pmin = 0.1$; $K = 5$; results are averaged over 100 runs). The horizontal lines show the expected values for the non-switching, stationary environment for resp. adaptive pursuit (4), probability matching (3.33), and random selection (3).

converge quickly and accurately, and at the same time maintain the flexibility to swiftly track any changes in the reward distributions. Experimental results on the above specified non-stationary environment show that the adaptive pursuit method does indeed possess these capabilities. In our first simulation we have taken a switching interval $\Delta T = 50$ time steps. The results shown are all averaged over 100 independent runs. Figures 1 and 2 clearly show that the adaptive pursuit algorithm is capable of accurate and fast convergence. At the same time it is very responsive to changes in the reward distribution. Whenever the operator-reward associations are reassigned the performance of the adaptive pursuit algorithm plunges since it is now pursuing an operator that is no longer optimal. It does not take long though for the strategy to correct itself, and to pursue the current optimal operator again. This is in contrast with the probability matching algorithm where the differences between the operator selection probabilities are much smaller and the changes in the reward distributions cause only minor adaptations. Of course a more significant reaction would be observed for the probability matching method if the rewards would have a much large difference between them. The key point though is that in practice one usually will have to deal with reward differences of a few percent, not an order of magnitude.

In a second experiment we have increased the switching interval ΔT to 200 times steps (Figures 3 and 4). Given more time to adapt one can see that both adaptive allocation strategies approach the values that where computed above for the stationary environment.

The results in the Figures 1, 2, 3 and 4 were obtained for a learning rate $\alpha = 0.8$ when updating $Q_a(t)$ in Equation 1, and a learning rate $\beta = 0.8$ when updating $P_a(t)$ in Equations 3 and 4. These values gave the best performance for this particular problem instance. Tables 1, 2, 3, and 4 show the performance for different settings of the learning rates. For low values of the learning rates the adaptive schemes do not react swiftly enough to the rapidly changing reward distributions. Naturally, the high learning rates are only possible because at each time step an actual reward is given by the environment. If the rewards would only be given with a probability less than 1, the learning rates would necessarily be small to ensure meaningful running estimates that are exponentially, recency-weighted. It should be noted though that whatever the values of the learning rates the adaptive pursuit method keeps outperforming the probability matching scheme.

5 Conclusion

Adaptive allocation rules are often used for learning the optimal probability values of applying a fixed set of exploration operators. Traditionally, the allocation strategies adapt the operator probabilities in such a way that they match the distribution of the rewards. In this chapter, we have compared this probability matching algorithm with an adaptive pursuit allocation rule in

α	Probab. Match.	Adaptive Pursuit: (β) 0.1	0.2	0.3	0.4	0.5	0.6	0.7	0.8	0.9
0.10	0.218	0.248	0.272	0.281	0.276	0.290	0.284	0.289	0.288	0.287
0.20	0.229	0.281	0.298	0.315	0.321	0.327	0.329	0.332	0.329	0.340
0.30	0.243	0.313	0.356	0.373	0.381	0.388	0.392	0.386	0.393	0.397
0.40	0.249	0.352	0.401	0.411	0.429	0.427	0.434	0.439	0.436	0.445
0.50	0.254	0.381	0.423	0.443	0.451	0.456	0.448	0.459	0.467	0.471
0.60	0.259	0.392	0.447	0.461	0.474	0.477	0.484	0.484	0.492	0.488
0.70	0.259	0.404	0.448	0.477	0.480	0.490	0.492	0.496	0.496	0.493
0.80	0.257	0.408	0.462	0.478	0.482	0.491	0.495	0.499	0.507	0.502
0.90	0.262	0.404	0.455	0.468	0.476	0.492	0.497	0.493	0.502	0.495

Table 1. The average **probability** of selecting the **optimal operator** in the non-stationary environment with switching interval $\Delta T = 50$ time steps for different adaptation rates α and β ($Pmin = 0.1$; $K = 5$; results are averaged over 100 runs).

α	Probab. Match.	Adaptive Pursuit: (β) 0.1	0.2	0.3	0.4	0.5	0.6	0.7	0.8	0.9
0.10	3.115	3.269	3.392	3.434	3.442	3.472	3.478	3.489	3.478	3.474
0.20	3.166	3.428	3.515	3.568	3.597	3.607	3.621	3.612	3.625	3.635
0.30	3.221	3.489	3.619	3.669	3.689	3.703	3.713	3.710	3.717	3.730
0.40	3.243	3.553	3.685	3.715	3.746	3.751	3.767	3.777	3.778	3.788
0.50	3.270	3.589	3.715	3.765	3.787	3.791	3.787	3.803	3.812	3.825
0.60	3.276	3.612	3.742	3.791	3.807	3.822	3.831	3.839	3.848	3.844
0.70	3.286	3.634	3.740	3.808	3.815	3.840	3.839	3.856	3.846	3.842
0.80	3.288	3.627	3.758	3.808	3.829	3.830	3.853	3.859	3.871	3.862
0.90	3.308	3.627	3.743	3.789	3.815	3.844	3.845	3.840	3.861	3.851

Table 2. The **average reward** received in the non-stationary environment with switching interval $\Delta T = 50$ time steps for different adaptation rates α and β ($Pmin = 0.1$; $K = 5$; results are averaged over 100 runs).

a controlled, non-stationary environment. Calculations and experimental results show a better performance of the adaptive pursuit method. The adaptive pursuit strategy converges accurately and rapidly, yet remains able to swiftly react to any change in the reward distributions.

References

1. P. Auer, N. Cesa-Bianchi, Y. Freund, and R.E. Schapire (2002) The nonstochastic multiarmed bandit problem. *SIAM j. Computing* Vol.32, No.1, pp. 48–77
2. D.W. Corne, M.J. Oates, and D.B. Kell (2002) On fitness distributions and expected fitness gain of mutation rates in parallel evolutionary algorithms. *Proc. 7th Intern. Conf. on Parallel Problem Solving from Nature.* LNCS Vol. 2439, pp. 132–141

α	Probab. Match.	Adaptive Pursuit: (β)								
		0.1	0.2	0.3	0.4	0.5	0.6	0.7	0.8	0.9
0.10	0.247	0.399	0.414	0.416	0.422	0.423	0.427	0.422	0.423	0.429
0.20	0.257	0.491	0.498	0.508	0.508	0.509	0.515	0.514	0.511	0.516
0.30	0.260	0.520	0.530	0.537	0.537	0.538	0.542	0.540	0.543	0.547
0.40	0.264	0.534	0.546	0.550	0.551	0.554	0.556	0.555	0.559	0.558
0.50	0.265	0.539	0.553	0.557	0.557	0.559	0.559	0.561	0.561	0.562
0.60	0.264	0.537	0.552	0.556	0.558	0.561	0.562	0.565	0.564	0.563
0.70	0.264	0.538	0.552	0.555	0.556	0.560	0.560	0.561	0.560	0.561
0.80	0.267	0.528	0.541	0.549	0.550	0.552	0.557	0.554	0.556	0.560
0.90	0.266	0.521	0.537	0.538	0.546	0.547	0.547	0.549	0.550	0.553

Table 3. The average **probability** of selecting the **optimal operator** in the non-stationary environment with switching interval $\Delta T = 200$ time steps for different adaptation rates α and β ($Pmin = 0.1$; $K = 5$; results are averaged over 100 runs).

α	Probab. Match.	Adaptive Pursuit: (β)								
		0.1	0.2	0.3	0.4	0.5	0.6	0.7	0.8	0.9
0.10	3.233	3.719	3.757	3.767	3.768	3.775	3.778	3.780	3.776	3.789
0.20	3.287	3.834	3.853	3.877	3.879	3.879	3.893	3.891	3.887	3.892
0.30	3.302	3.873	3.896	3.916	3.912	3.914	3.922	3.921	3.923	3.934
0.40	3.315	3.886	3.915	3.926	3.932	3.933	3.939	3.942	3.948	3.938
0.50	3.320	3.891	3.925	3.940	3.939	3.945	3.940	3.946	3.946	3.950
0.60	3.323	3.890	3.926	3.936	3.941	3.949	3.947	3.956	3.955	3.951
0.70	3.322	3.894	3.928	3.936	3.943	3.948	3.948	3.947	3.947	3.951
0.80	3.333	3.878	3.912	3.934	3.937	3.934	3.946	3.940	3.945	3.951
0.90	3.329	3.881	3.916	3.913	3.933	3.933	3.933	3.938	3.936	3.944

Table 4. The **average reward** received in the non-stationary environment with switching interval $\Delta T = 200$ time steps for different adaptation rates α and β ($Pmin = 0.1$; $K = 5$; results are averaged over 100 runs).

3. L. Davis (1989) Adapting operator probabilities in genetic algorithms. *Proc. Third Intern. Conf. on Genetic Algorithms and their Applications.* pp. 61–69
4. A.E. Eiben, R. Hinterding, and Z. Michalewicz (1999) Parameter control in evolutionary algorithms. *IEEE Transactions on Evolutionary Computation*, 3(2):124-141
5. D.E. Goldberg (1990) Probability matching, the magnitude of reinforcement, and classifier system bidding. *Machine Learning.* Vol.5, pp. 407–425
6. T.P. Hong, H.S. Wang, and W.C. Chen (2000) Simultaneously applying multiple mutation operators in genetic algorithms. *Journal of Heuristics.* Vol.6, pp. 439–455
7. C. Igel, and M. Kreutz (2003) Operator adaptation in evolutionary computation and its application to structure optimization of neural networks. *Neurocomputing* Vol.55, pp. 347–361

8. B. Julstrom (1995) What have you done for me lately? Adapting operator probabilities in a steady-state genetic algorithm. *Proc. Sixth Intern. Conf. on Genetic Algorithms.* pp. 81–87
9. F. G. Lobo, and D. E. Goldberg (1997) Decision making in a hybrid genetic algorithm. *Proc. IEEE Intern. Conf. on Evolutionary Computation.* pp. 122–125
10. D. Schlierkamp-Voosen, and H. Mühlenbein (1994) Strategy adaptation by competing subpopulations. *Proc. Intern. Conf. on Parallel Problem Solving from Nature* pp. 199–208
11. J.E. Smith, and T.C. Fogarty (1997) Operator and parameter adaptation in genetic algorithms. *Soft Computing* No.1, pp. 81–87
12. R.S. Sutton, and A.G. Barto (1998) *Reinforcement Learning: an introduction.* MIT Press
13. M.A.L. Thathachar, and P.S. Sastry (1985) A Class of Rapidly Converging Algorithms for Learning Automata. *IEEE Transactions on Systems, Man and Cybernetics.* Vol.SMC-15, pp. 168-175
14. D. Thierens (2005) An Adaptive Pursuit Strategy for Allocating Operator Probabilities. *Proc. Intern. Conf. on Genetic and Evolutionary Computation.* pp. 1539–1546
15. A. Tuson, and P. Ross (1998) Adapting operator settings in genetic algorithms. *Evolutionary Computation* Vol.6, No.2, pp. 161–184

Sequential Parameter Optimization Applied to Self-Adaptation for Binary-Coded Evolutionary Algorithms

Mike Preuss and Thomas Bartz-Beielstein

Dortmund University, D-44221 Dortmund, Germany.
mike.preuss@cs.uni-dortmund.de, thomas.bartz-beielstein@udo.edu

Summary. Adjusting algorithm parameters to a given problem is of crucial importance for performance comparisons as well as for reliable (first) results on previously unknown problems, or with new algorithms. This also holds for parameters controlling adaptability features, as long as the optimization algorithm is not able to completely self-adapt itself to the posed problem and thereby get rid of all parameters. We present the recently developed *sequential parameter optimization* (SPO) technique that reliably finds good parameter sets for stochastically disturbed algorithm output. SPO combines classical regression techniques and modern statistical approaches for deterministic algorithms as Design and Analysis of Computer Experiments (DACE). Moreover, it is embedded in a twelve-step procedure that targets at doing optimization experiments in a statistically sound manner, focusing on answering scientific questions.

We apply SPO to a question that did not receive much attention yet: Is self-adaptation as known from real-coded evolution strategies useful when applied to binary-coded problems? Here, SPO enables obtaining parameters resulting in good performance of self-adaptive mutation operators. It thereby allows for reliable comparison of modified and traditional evolutionary algorithms, finally allowing for well founded conclusions concerning the usefulness of either technique.

1 Introduction

The evolutionary computation (EC) field currently seems to experience a state of flux, at least as far as experimental research methods are concerned. A purely theoretical approach is not reasonable for many optimization problems. In fact, there is a huge gap between theory and experiment in evolutionary computation. However, empiricism in EC cannot compensate this shortcoming due to a lack of standards. A broad spectrum of presentation techniques makes new results almost incomparable. At present, it is intensely discussed which experimental research methodologies should be used to improve the acceptance and quality of *evolutionary algorithms* (EA).

Several authors from related fields ([32], [57], and [39]) and from within EC ([81], [23]) have criticized usual experimental practice. Besides other topics, they ask for

M. Preuss and T. Bartz-Beielstein: *Sequential Parameter Optimization Applied to Self-Adaptation for Binary-Coded Evolutionary Algorithms*, Studies in Computational Intelligence (SCI) **54**, 91–119 (2007)
www.springerlink.com

increased thoughtfulness when selecting benchmark problems, a better structured process of experimentation, including presentation of results, and proper use of statistical techniques.

In the light of the *no free lunch* theorem (NFL, [82]), research aims are gradually changing from demonstration of superiority of new algorithms towards experimental analysis, addressing questions as: What makes an algorithm work well on a certain problem class? In spite of this development, the so-called *horse race* papers[1] still seem to prevail. [23] report about the situation in experimental EC and name the reasons for their discontentment with current practice. Put into positive formulation, their main demands are:

- assembly of concrete research questions and concrete answers to these, no claims that are not backed up by tests
- selection of test problems and instances motivated by these questions
- utilization of adequate performance measures for answering the questions, and
- reproducibility, which requires all necessary details and/or source code to repeat experiments

This line is carried further in [22]. Partly in response to these demands, [7] proposes SPO as a structured experimentation procedure based on modern statistic techniques (§4). In its heart, SPO contains a parameter tuning method that enables *adapting* parameters to the treated test problems.

Undoubtedly, comparing badly parametrized algorithms is rather useless. However, a good parameter set allows for near-optimal performance of an algorithm. As this statement suggests, parameter tuning is of course an optimization problem on its own. Following from that, some interesting questions arise when algorithms with parameter control operators are concerned, because their own adaptability is in turn guarded by control parameters, which can be tuned. The broadest of these questions may be where to put effort into when striving for best performance—tune the simple algorithm, or tune the adaptability of more complex algorithms—eventually resulting in recommendations when to apply adaptive operators.

1.1 Parameter Control or Parameter Tuning?

Parameter Control refers to parameter adaptation during the run of an optimization algorithm, whereas parameter tuning improves their setting before the run is started. These two different mechanisms are the roots of the two subtrees of parameter setting methods in the global taxonomy given by [21]. One may get the impression that they are contradictory and researchers should aim at using parameter control as much as possible. In our view, the picture is somewhat more complex, and the two are rather complementary.

In a production environment, an EA and any alternative stochastic optimization algorithm would be run several times, not once, possibly solving the next or trying the same problem again. This obliterates the separating time factor of above. What is more, the EA as well as the problem representation will most likely undergo structural changes during these iterations (new operators, criteria, etc.). These will entail changes in the "optimal" parameters, too.

[1] [39] uses this term for papers that aim at showing predominance of one algorithm over (all) others by reporting performance comparisons on (standard) test problems.

Furthermore, parameter control methods do not necessarily decrease the number of parameters. For example, the 1/5th adaptation rule by [63] provides at least three quantities where there has been one—mutation step size—before, namely the time window length used to measure success, the required success rate (1/5), and the rate of change applied. Even if their number remains constant, is it justified to expect that parameter tuning has become simpler now? We certainly hope so, but need more evidence to decide.

In principle, every EA parameter may be (self-)adapted. However, concurrent adaptation of many does not seem particularly successful and is probably limited to two or three at a time. We thus cannot simply displace parameter tuning with parameter control. On the other hand, manual tuning is surely not the answer, as well, and grid based tuning seems intractable for higher order design spaces. At this point, we have three possibilities, either to

- fall back on default values,
- abandon tuning alltogether and take random parameters, or
- to apply a tuning method that is more robust than doing it manually, but also more efficient than grid search.

We argue that SPO is such a method and thus of particular importance for experimental studies on non-trivial EAs, as such utilizing self-adaptation. As the experimental analysis in EC currently is a hot topic, other parameter tuning approaches are developed, too, e.g. [61]. But whatever method is used, there is hardly a way around parameter tuning, at least for scientific investigations. SPO will be discussed further in §4.

1.2 Self-adaptation for Binary Representations

When comparing recent literature, it is astounding that for real-coded EAs as e.g. *evolution strategies* (ES), self-adaptation is an almost ubiquitously applied feature, at least as far as mutation operators are concerned. In binary-coded EA however, it did not become a standard method. Optimization practitioners working with binary representations seem largely unconvinced that it may be of any use. What may be the reason for this huge difference? One could first think of traditional reasons. Of the three standard evolutionary methods for numerical optimization, genetic algorithms (GA), evolutionary programming (EP) and evolution strategies, the last one first adopted self-adaptation in various forms to improve mutation operators ([64, 71, 72]), followed by EP in [25]. These two mainly work with real-valued representations. Nevertheless, since then so much time has passed that it is hard to imagine that others would not employ a technique that clearly provides an advantage. Considering this, tradition appears not as a substantial reason for the divergent developments.

In fact, meanwhile, several authors have tried to transfer self-adaptation to EAs for binary encodings: [4, 3, 70, 74, 75, 77, 29]. If we accept the view of an evolutionary epistemology underlying the development of our scientific domain, so that mostly successful changes are inherited to the next stages, these studies have apparently not been very convincing. Either they were ignored by the rest of the field, or they failed to provide reason good enough for a change. Let us assume the latter case. In consequence, most EAs for binary representations published nowadays still do without self-adaptation.

Evolutionary algorithms employing different forms of self-adaptation are usually applied to continuous, or, more rarely, ordinal discrete or mixed search spaces. So the distinction may result from differences in representation. When dealing with real-valued numbers, it is quite easy to come by a good intuition why adapting mutation strengths (also called mutation step sizes) during optimization can speed up search. Features of a fitness landscape changing slowly compared to the fitness values themselves can be learned, like gradients or covariances ([31, 30]). In a binary space, this intuition is lacking. Definition of gradients is meaningless if a variable can only take two values. Consequently, research is trying different ways here, e.g. linkage learning or distribution estimation as in [59]. As much as the experiences from continuous search spaces cannot be simply transferred, neither can the methods. Self-adaptation of mutation rates, as straightforward continuation of existing techniques for real-valued variables, if working at all for binary representations, can be expected to work differently.

2 Aims and Methods

Our aims are twofold: To demonstrate usefulness of the SPO approach for experimental analysis, and to perform an experimental analysis of self-adaptation mechanisms.

Methodologically, we want to convey that SPO is a suitable tool for finding good parameter sets (designs) within a fixed, low budget of algorithm runs. It seems safe to assume that a parameter set exists in the parameter design space that is better than the best of a small (≈ 100) sample. We therefore require that the best configurations detected by SPO are significantly better than the best of a first sample. Furthermore, we want to suggest a flexible yet structured methodology for performing and documenting experimentation by employing a parameter tuning technique.

Concerning the experimental analysis of self-adaptation mechanisms on binary represented problems, our aim is to collect evidence for or against the following conjectures:

- Well parametrized self-adaptation significantly speeds up optimization when compared to well tuned constant mutation rates for many problems.
- Detecting good parameter sets for self-adaptive EAs is usually not significantly harder than doing so for non self-adaptive EAs.
- Problem knowledge, e.g., shared properties or structural similarities of problem classes, gives useful hints for selecting a self-adaptation mechanism.

Before these conjectures are detailed, we give an overview of existing parameter tuning methods.

3 Parameter Optimization Approaches

Modern search heuristics have proved to be very useful for solving complex real–world optimization problems that cannot be tackled through classical optimization techniques [73]. Many of these search heuristics involve a set of *exogenous parameters*, i.e., values are specified before the run is performed, that affect their convergence properties. The population size in EA is a typical example for an exogenous strategy parameter. The determination of an adequate population size is crucial for many optimization problems. Increasing the population size from 10 to 50 while keeping the number of function evaluations constant might improve the algorithm's performance—whereas a further increase might result in a performance decrease, if the number of function evaluations remains constant.

SPO is based on statistical *design of experiments* (DOE) which has its origins in agriculture and industry. However, DOE has to be adapted to the special requirements of computer programs. For example, computer program are per se deterministic, thus a different concepts of randomness has to be considered. Law & Kelton [46] and Kleijnen [43, 44] demonstrated how to apply DOE in simulation. Simulation is related to optimization (simulation models equipped with an objective function define a related optimization problem), therefore we can benefit from simulation studies.

DOE related parameter studies were performed to analyze EA: Schaffer et al. [67] proposed a complete factorial design experiment, Feldt & Nordin [24] use statistical techniques for designing and analyzing experiments to evaluate the individual and combined effects of genetic programming parameters. Myers & Hancock [58] presented an empirical modeling of genetic algorithms. François & Lavergne [26] demonstrate the applicability of *generalized linear models* to design evolutionary algorithms. These approaches require up to 100,000 program runs, whereas SPO is applicable even if a small amount of function evaluations are available only.

Because the search for useful parameter settings of algorithms itself is an optimization problem, meta-algorithms have been proposed. Bäck [2] and Kursawe [45] presented meta-algorithms for evolutionary algorithms. But these approaches do not solve the original problem completely, because they require the determination of additional parameter setting of the meta-algorithm. Furthermore, we argue that the experimenter's skill plays an important role in this analysis. It cannot be replaced by automatic "meta" rules.

SPO can also be run on auto-pilot without any user intervention. This convenience cannot be obtained for free: The user gains limited insight into the working mechanisms of the tuned algorithm and, which is even more severe, the validity and thus the predictive power of the regression model might be quite poor.

The reader should note that our approach is related to the discipline of *experimental algorithmics*, which offers methodologies for the design, implementation, and performance analysis of computer programs for solving algorithmic problems [18, 56]. Further valuable approaches have been proposed by McGeoch [50], Barr & Hickman [5], and Hooker [32].

Design and analysis of computer experiments (DACE) as introduced in Sacks et al. [65] models the deterministic output of a computer experiment as the realization of a stochastic process. The DACE approach focuses entirely on the correlation structure of the errors and makes simplistic assumptions about the regressors. It describes "how the function behaves," whereas regression as used in classical DOE describes "what the function is" [40, p. 14]. DACE requires other experimental designs than classical DOE, e.g., *Latin hypercube designs* (LHD) [51].

We claim that it is beneficial to combine some of these well-established ideas from DOE, DACE, and further statistical techniques to improve the acceptance and quality of evolutionary algorithms.

4 Sequential Parameter Optimization Methodology

How can optimization practitioners determine if concepts developed in theory work in practice? Hence, experiments are necessary. Experiment has a long tradition in science. To analyze experimental data, statistical methods can be applied. It is not a trivial task to answer the final question "Is algorithm A better than algorithm B?" Results that are statistically significant are not automatically scientifically meaningful.

Example 4.1 (Floor and ceiling effects). The statistical meaningful result "all algorithms perform equally" can be scientifically meaningless, because the problem instances are too hard for any algorithm. A similar effect occurs if the problem instances are too easy. The resulting effects are known as floor or ceiling effects, respectively. □

SPO is more than a simple combination of existing statistical approaches. It is based on the new experimentalism, a development in the modern philosophy of science, which considers that an experiment can have a life of its own. SPO provides a statistical methodology to learn from experiments, where the experimenter should distinguish between statistical significance and scientific meaning.

An optimization run is considered as an experiment. An optimal parameter setting, or statistically speaking, an optimal *algorithm design*, depends on the problem at hand as well as on the restrictions posed by the environment (i.e., time and hardware constraints). Algorithm designs are usually either determined empirically or set equal to widely used default values. SPO is a methodology for the experimental analysis of optimization algorithms to determine improved algorithm designs and to learn, how the algorithm works. The proposed technique employs computational statistic methods to investigate the interactions among optimization problems, algorithms, and environments.

An optimization practitioner is interested in robust solutions, i.e., solutions independent from the random seeds that are used to generate the random numbers during the optimization run. The proposed statistical methodology provides guidelines to design robust algorithms under restrictions, such as a limited number of function evaluations and processing units. These restrictions can be modeled by considering the performance of the algorithm in terms of the (expected) best function value for a limited number of function evaluations.

To justify the usefulness of our approach, we analyze the properties of several algorithms from the viewpoint of a researcher who wants to develop and understand self-adaptation mechanisms for evolutionary algorithms. SPO provides numerical and graphical tools to test if the statistical results are really relevant or have been caused by the experimental setup only. It is based on a framework that permits a delinearization of the complex steps from raw data to scientific hypotheses. Substantive scientific questions are broken down into several local hypotheses, that can be tested experimentally. The optimization process can be regarded as a process that enables learning. SPO consists of the twelve steps that are reported in Table 1. These steps and the necessary statistical techniques will be presented in the following. SPO has been applied on search heuristics in the following domains:

1. machine engineering: design of mold temperature control [53, 80, 52]
2. aerospace industry: airfoil design optimization [11]

3. simulation and optimization: elevator group control [13, 49]
4. technical thermodynamics: nonsharp separation [10]
5. economy: agri-environmental policy-switchings [17]

Other fields of application are in fundamental research:

1. algorithm engineering: graph drawing [79]
2. statistics: selection under uncertainty (optimal computational budget allocation) for PSO [8]
3. evolution strategies: threshold selection and step-size adaptation [6]
4. other evolutionary algorithms: genetic chromodynamics [76]
5. computational intelligence: algorithmic chemistry [10]
6. particle swarm optimization: analysis and application [9]
7. numerics: comparison and analysis of classical and modern optimization algorithms [12]

Further projects, e.g., vehicle routing and door-assignment problems and the application of methods from computational intelligence to problems from bioinformatics are subject of current research. An SPO-toolbox is freely available under the following link: `http://www.springer.com/3-540-32026-1`.

4.1 Tuning

In order to find an optimal algorithm design, or to tune the algorithm, it is necessary to define a performance measure. Effectivity (robustness) and efficiency can guide

Table 1. Sequential parameter optimization (SPO). This approach combines methods from computational statistics and exploratory data analysis to improve (tune) the performance of direct search algorithms.

Step	Action
(S-1)	Preexperimental planning
(S-2)	Scientific claim
(S-3)	Statistical hypothesis
(S-4)	Specification of the
	(a) optimization problem
	(b) constraints
	(c) initialization method
	(d) termination method
	(e) algorithm (important factors)
	(f) initial experimental design
	(g) performance measure
(S-5)	Experimentation
(S-6)	Statistical modeling of data and prediction
(S-7)	Evaluation and visualization
(S-8)	Optimization
(S-9)	Termination: If the obtained solution is good enough, or the maximum number of iterations has been reached, go to step (S-11)
(S-10)	Design update and go to step (S-5)
(S-11)	Rejection/acceptance of the statistical hypothesis
(S-12)	Objective interpretation of the results from step (S-11)

the choice of an adequate performance measure. Note that optimization practitioners do not always choose the absolute best algorithm. Sometimes a robust algorithm or an algorithm that provides insight into the structure of the optimization problem is preferred. From the viewpoint of an experimenter, design variables (factors) are the parameters that can be changed during an experiment. Generally, there are two different types of factors that influence the behavior of an optimization algorithm:

- problem specific factors, e.g., the objective function
- algorithm specific factors, i.e., the population size or other exogenous parameters

We will consider experimental designs that comprise problem specific factors and exogenous algorithm specific factors. Algorithm specific factors will be considered first. *Endogenous* can be distinguished from *exogenous parameters* [16]. The former are kept constant during the optimization run, whereas the latter, e.g., standard deviations, are modified by the algorithms during the run. Consider $\mathcal{D}_A$, the set of all parameterizations for one algorithm. An *algorithm design* X_A is a set of vectors, each representing one specific setting of the design variables of an algorithm. A design can be specified by defining ranges of values for the design variables. A *design point* $x_a \in \mathcal{D}_A$ presents exactly one parameterization of an algorithm. Note that a design can contain none, one, several or even infinitely many design points. The *optimal algorithm design* is denoted as X_A^*. The term "optimal design" can refer to the best design point x_a^* as well as the most informative design points [60, 66].

Let $\mathcal{D}_P$ denote the set of all problem instances for one optimization problem. *Problem designs* X_P provide information related to the optimization problem, such as the available resources (number of function evaluations) or the problem's dimension.

An *experimental design* $X_E \in \mathcal{D}$ consists of a problem design X_P and an algorithm design X_A. The run of a stochastic search algorithm can be treated as an experiment with a stochastic output $Y(x_a, x_p)$, with $x_a \in \mathcal{D}_A$ and $x_p \in \mathcal{D}_P$. If the random seed is specified, the output would be deterministic. This case will not be considered further, because it is not a common practice to specify the seed that is used in an optimization run. Our goals of the experimental approach can be stated as follows:

(G-1) *Efficiency.* To find a design point $x_a^* \in \mathcal{D}_A$ that improves the performance of an optimization algorithm for one specific problem design point $x_p \in \mathcal{D}_P$.
(G-2) *Robustness.* To find a design point $x_a^* \in \mathcal{D}_A$ that improves the performance of an optimization algorithm for several problem design points $x_p \in \mathcal{D}_P$.

Performance can be measured in many ways, e.g., as the best or the average function value for n runs. Statistical techniques to attain these goals will be presented next.

4.2 Stochastic Process Models as Extensions of Classical Regression Models

The classical DOE approach consists of three steps: Screening, modeling, and optimization. Each step requires different experimental designs. Linear regression models are central elements of the classical design of experiments approach [19, 55]. We propose an approach that extends the classical regression techniques, because the assumption of a linear model for the analysis of computer programs and the implicit model assumption that observation errors are independent of one another are highly speculative [7]. To keep the number of experiments low, a sequential procedure has been developed. Our approach relies on a stochastic process model, that will be presented next.

We consider each algorithm design with associated output as a realization of a stochastic process. *Kriging* is an interpolation method to predict unknown values of a stochastic process and can be applied to interpolate observations from computationally expensive simulations. Our presentation follows concepts introduced in Sacks et al. [65], Jones et al. [40], and Lophaven et al. [48].

Consider a set of m design points $x = (x^{(1)}, \ldots, x^{(m)})^T$ with $x^{(i)} \in \mathbb{R}^d$. In the *design and analysis of computer experiments* (DACE) *stochastic process model*, a deterministic function is evaluated at the m design points x. The vector of the m responses is denoted as $y = (y^{(1)}, \ldots, y^{(m)})^T$ with $y^{(i)} \in \mathbb{R}$. The process model proposed in Sacks et al. [65] expresses the deterministic response $y(x^{(i)})$ for a d-dimensional input $x^{(i)}$ as a realization of a regression model $\mathcal{F}$ and a stochastic process Z,

$$Y(x) = \mathcal{F}(\beta, x) + Z(x). \tag{1}$$

DACE Regression Models

We use q functions $f_j : \mathbb{R}^d \to \mathbb{R}$ to define the regression model

$$\mathcal{F}(\beta, x) = \sum_{j=1}^{q} \beta_j f_j(x) = f(x)^T \beta. \tag{2}$$

Regression models with polynomials of orders 0, 1, and 2 have been used in our experiments. Regression models with a constant term only, i.e., $f_1 = 1$, have been applied successfully to model the data and to predict new data points in the sequential approach.

DACE Correlation Models

The random process $Z(\cdot)$ (Equation 1) is assumed to have mean zero and covariance $V(w, x) = \sigma^2 \mathcal{R}(\theta, w, x)$ with process variance σ^2 and correlation model $\mathcal{R}(\theta, w, x)$. Consider an algorithm with d factors (parameters). Correlations of the form

$$\mathcal{R}(\theta, w, x) = \prod_{j=1}^{d} \mathcal{R}_j(\theta, w_j - x_j)$$

will be used in our experiments. The correlation function should be chosen with respect to the underlying process [34]. Lophaven et al. [47] discuss seven different models. The Gaussian correlation function is a well-known example. It is defined as

$$\text{GAUSS} : \qquad \mathcal{R}_j(\theta, h_j) = \exp(-\theta_j h_j^2), \tag{3}$$

with $h_j = w_j - x_j$, and for $\theta_j > 0$. The regression matrix R is the matrix with elements

$$R_{ij} = R(x_i, x_j) \tag{4}$$

that represent the correlations between $Z(x_i)$ and $Z(x_j)$. The vector with correlations between $Z(x_i)$ and a new design point $Z(x)$ is

$$r(x) = (R(x_1, x), \ldots, R(x_m, x)) . \tag{5}$$

Large θ_j's indicate that variable j is active: function values at points in the vicinity of a point are correlated with Y at that point, whereas small θ_j's indicate that also distant data points influence the prediction at that point. The *empirical best unbiased linear predictor* (EBLUP) can be shown to be

$$\hat{y}(x) = f^T(x)\hat{\beta} + r^T(x)R^{-1}(y - F\hat{\beta}), \tag{6}$$

where

$$\hat{\beta} = \left(F^T R^{-1} F\right)^{-1} F^T R^{-1} y \tag{7}$$

is the generalized least-squares estimate of β in Equation 1, $f(x)$ are the q regression functions in Equation 2, and F represents the values of the regression functions in the m design points.

Maximum likelihood estimation methods to estimate the parameters θ_j of the correlation functions from Equation 3 are discussed in Lophaven et al. [47]. DACE methods provide an estimation of the prediction error on an untried point x, the *mean squared error* (MSE) of the predictor

$$\mathrm{MSE}(x) = E\left(\hat{y}(x) - y(x)\right) . \tag{8}$$

The stochastic process model, which was introduced as an extension of the classical regression model will be used in our experiments. Next, we have to decide how to generate design points, i.e., which parameter settings should be used to test the algorithm's performance.

Space Filling Designs and Expected Improvement

Often, designs that use sequential sampling are more efficient than designs with fixed sample sizes. Therefore, we specify an initial design $X_A^{(0)} \in \mathcal{D}_A^{(0)}$ first. *Latin hypercube sampling* (LHS) was used to generate the initial algorithm designs. Consider n number of levels being examined and d design variables. A Latin hypercube is a matrix of n rows and d columns. The d columns contain the levels $1, 2, \ldots, n$, randomly permuted, and the d columns are matched at random to form the Latin hypercube. The resulting *Latin hypercube designs* are space-filling designs. McKay et al. [51] introduced LHDs for computer experiments, Santner et al. [66] give a comprehensive overview.

Information obtained in the first runs can be used for the determination of the second design $X_A^{(1)}$ in order to choose new design points sequentially and thus more efficiently.

Sequential sampling approaches have been proposed for DACE. For example, in Sacks et al. [65] sequential sampling approaches were classified to the existing meta-model. We will present a sequential approach that is based on the expected improvement. In Santner et al. [66, p. 178] a heuristic algorithm for unconstrained global minimization problems is presented. Consider one problem design point x_p. Let $y_{\min}^{(t)}$ denote the smallest known minimum value after t runs of the algorithm,

SEQUENTIALPARAMETEROPTIMIZATION($\mathcal{D}_A, \mathcal{D}_P$)
1 Select $p \in \mathcal{D}_P$ and set $t = 0$ /* Select problem instance */
2 $X_A^{(t)} = \{x_1, x_2, \ldots, x_k\}$ /* Sample k initial points, e.g., LHS */
3 **repeat**
4 $y_{ij} = Y_j(x_i, p) \forall x_i \in X_A^{(t)}$ and $j = 1, \ldots, r^{(t)}$ /* Fitness evaluation */
5 $\overline{Y}_i^{(t)} = \sum_{j=1}^{r^{(t)}} y_{ij}^{(t)} / r^{(t)}$ /* Sample statistic for the ith design point */
6 x_b with $b = \arg\min_i(\overline{y}_i)$ /* Determine best point */
7 $Y(x) = \mathcal{F}(\beta, x) + Z(x)$ /* DACE model from Eq. 1 */
8 $X_S = \{x_{k+1}, \ldots, x_{k+s}\}$ /* Generate s sample points, $s \gg k$ */
9 $y(x_i), i = 1, \ldots, k+s$ /* Predict fitness from DACE model */
10 $I(x_i)$ for $i = 1, \ldots, s+k$ /* Expected improvement (Eq. 9) */
11 $X_A^{(t+1)} = X_A^{(t)} \cup \{x_{k+i}\}_{i=1}^{m} \notin X_A^{(t)}$ /* Add m promising points */
12 **if** $x_b^{(t)} = x_b^{(t+1)}$
13 **then** $r^{(t+1)} = 2r^{(t)}$ /* Increase number of repeats */
14 $t = t + 1$ /* Increment iteration counter */
15 **until** BUDGETEXAUSTED?()

Fig. 1. Sequential parameter optimization

$y(x)$ be the algorithm's response, i.e., the realization of $Y(x)$ in Equation (1), and let x_a represent a specific design point from the algorithm design X_A. Then the improvement is defined as

$$I(x_a) = \begin{cases} y_{\min}^{(t)} - y(x_a), & y_{\min}^{(t)} - y(x_a) > 0 \\ 0, & \text{otherwise} \end{cases} \tag{9}$$

for $x_a \in \mathcal{D}_A$. As $Y(\cdot)$ is a random variable, its exact value is unknown. The goal is to optimize its expectation, the so-called *expected improvement*. New design points, which are added sequentially to the existing design, are attractive "if either there is a high probability that their predicted output is below [minimization] the current observed minimum and/or there is a large uncertainty in the predicted output." This leads to the *expected improvement heuristic* [7]. Based on theorems from Schonlau [68, p. 22] we implemented a program to estimate and plot the main factor effects. Furthermore, three dimensional visualizations produced with the DACE toolbox [48] can be used to illustrate the interaction between two design variables and the associated mean squared error of the predictor.

Figure 1 describes the SPO in a formal manner. The selection of a suitable problem instance is done in the pre-experimental planning phase to avoid floor and ceiling effects (l.2). Latin hypercube sampling can be used to determine an initial set of design points (l.3). After the algorithm has been run with these k initial parameter settings (l.5), the DACE process model is used to discover promising design points (l.10). Note that other sample statistics than the mean, e.g., the median, can be used in l.6. The m points with the highest expected improvement are added to the set of design points, where m should be small compared to s. The update rule for the number of reevaluations $r^{(t)}$ (l.13–15) guarantees that the new best design point $x_b^{(t+1)}$ has been evaluated at least as many times as the previous best design point $x_b^{(t)}$. Obviously, this is a very simple update rule and more elaborate rules are possible. Other termination criteria exist besides the budget based termination (1.17).

4.3 Experimental Reports

Surprisingly, despite around 40 years of empirical tradition, in EC a standardized scheme for reporting experimental results never developed. The natural sciences, e.g. physics, possess such schemes as de-facto standards. We argue that for both groups, readers and writers, an improved report structure is beneficial: As with the common overall paper structure (introduction, conclusions, etc.), a standard provides guidelines for readers, what to expect, and where. Writers are steadily reminded to describe the important details needed to understand and possibly replicate their experiments. For the structured documentation of experiments, we propose organizing their presentation into 7 parts, as follows.

ER-1: **Focus/Title**
Briefly names the matter dealt with, the (possibly very general) objective, preferably in one sentence.

ER-2: **Pre-experimental planning**
Reports the first—possibly explorative—program runs, leading to task and setup (steps ER-3 and ER-4) . Decisions on used benchmark problems or performance measures may often be influenced by the outcome of preliminary runs. This may also include negative results, e.g. modifications to an algorithm that did not work, or a test problem that turned out to be too hard, if they provide new insight.

ER-3: **Task**
Concretizes the question in focus and states scientific and derived statistical hypotheses to test. Note that one scientific hypothesis may require several, sometimes hundreds of statistical hypotheses. In case of a purely explorative study, as with the first test of a new algorithm, statistical tests may be not applicable. Still, the task should be formulated as precise as possible.

ER-4: **Setup**
Specifies problem design and algorithm design, containing fixed and variable parameters and criteria of tackled problem, investigated algorithm and chosen performance measuring. The information in this part should be sufficient to replicate an experiment.

ER-5: **Experimentation/Visualization**
Gives raw or produced (filtered) data on the experimental outcome, additionally provides basic visualizations where meaningful.

ER-6: **Observations**
Describes exceptions from the expected, or unusual patterns noticed, without subjective assessment or explanation. As an example, it may be worthwhile to look at parameter interactions. Additional visualizations may help to clarify what happens.

ER-7: **Discussion**
Decides about the hypotheses specified in step 4.3, and provides necessarily subjective interpretations for the recorded observations.

This scheme is tightly linked to the 12 steps of experimentation suggested in [7] and depicted in Table 1, but on a slightly more abstract level. The scientific and statistical hypothesis steps are treated together in part ER-3, and the SPO core (parameter tuning) procedure, much of which may be automated, is included in part ER-5. In our view, it is especially important to divide parts ER-6 and ER-7, to facilitate different conclusions drawn by others.

5 Self-Adaptation Mechanisms and Test Problems

In order to prepare a meaningful experiment, we want to take up the previously cited warning words from [39] and [23] concerning the proper selection of ingredients for a good setup and carefully select mechanisms and problems to test.

5.1 Mechanisms: Self-Adaptive, Constant, Asymmetrical

[54], to be found in this volume, give an extensive account of the history and current state of adaptation mechanisms in evolutionary algorithms. However, in this work, we consider only self-adaptive mechanisms for binary represented (often combinatorial) problems. Self-adaptiveness basically means to introduce unbiased deviations into control parameters and let the algorithm chose the value that apparently works best. Another possibility would be to use a rule set as in the previously stated 1/5th rule by [63], which grants adaptation, but not self-adaptation. The mechanism suggested by [29] is of this type. Others aim for establishing a feedback loop between the probability of using one operator and the fitness advancement this operator is responsible for, e.g. [42, 33, 78].

Thus, whereas several adaptation techniques have been proposed for binary representations, few are purely self-adaptive. The mechanisms proposed in [3] and [75] have this property, though they origin from very different sources. The former is a variant of a self-adaptation scheme designed for real-valued search spaces; the latter has been shaped especially for binary search spaces and to overcome premature convergence. Mutation of the mutation rate in [3] is accomplished according to the formula:

$$p'_k = \frac{1}{1 + \frac{1-p_k}{p_k} \exp(-\gamma N_k(0,1))}, \tag{10}$$

where $N_k(0,1))$ is a standard normally distributed random variable. p'_k stands for the new, and p_k the current mutation rate, which must be prevented from becoming 0 as this is a fixpoint of the iteration. The approach of [75] is completely different; it employs a discrete, small number q of mutation rates, one of which is selected according to the *innovation rate* z, so that the probability for alteration in one generation respects $p_a = z \cdot (q-1)/q$. The q values are given as $p_m \in \{0.0005, 0.001, 0.0025, 0.005, 0.0075, 0.01, 0.025, 0.05, 0.075, 0.1\}$ and thus unbalanced in the interval $[0, 1]$. We employ a different set to facilitate comparison with the first mechanism that produces balanced mutation rates. Thereby, the differences are reduced to continuity or discreteness with and without causal relationship between new and old values (the second mechanism is stateless). In the following experiment, we also set q to ten, with $p_m \in \{0.001, 0.01, 0.05, 0.1, 0.3, 0.7, 0.9, 0.95, 0.99, 0.999\}$.

Depending on the treated problem, it sometimes pays to use asymmetric mutation, that is, different mutation rates for ones and zeroes, as has been shown e.g.

for certain instances of the Subset Sum problem by [37]. In [38] a meta-algorithm—itself an EA—was used to learn good mutation rate pairs. In this work, we also want to enable comparison between constant and self-adaptive asymmetric mutation rates by simply extending the two named symmetric self-adaptive mechanisms to two independently adapted mutation rates in the fashion of the asymmetric constant rate operator. It shall not be concealed that recent theoretical investigations also deal with asymmetric mutation operators, e.g. [36], even though their notion of asymmetry is slightly different.

5.2 Problems and Expectations

We chose a set of six test problems that maybe divided into three pairs, according to similar properties. Our expectation is that shared problem attributes lead to comparable performance and parametrization of algorithms. However, we are currently not able to express problem similarity quantitatively. Thus, our experimental study may be seen as a first explorative attempt in this direction.

wP-PEAKS and SUFSAMP

wP-PEAKS stands for *weighted P-PEAKS* generator, a modification of the original problem by [41], which employed random N-bit strings to represent the location of P peaks in search space. A small/large number of peaks results in weakly/strongly epistatic problems. Originally, each peak represented a global optimum. We employ a modified version ([27]) by adding weights $w_i \in \mathbb{R}_+$ with only $w_1 = 1.0$ and $w_{[2...P]} < 1.0$, thereby requiring the optimization algorithm to find the one peak bearing the global optimum instead of just any peak. In our experiments, P and N were 100, and $w_i \in [0.9, 0.99]$ for local optima.

The SUFSAMP problem has been introduced by [35] as test for the fraction of neighborhoods actually visited by an EA when building the offspring generation. The key point is that only one of the direct neighbours, the one that provides the highest fitness gain, leads towards the global optimum. Thus, it is essential to perform sufficient sampling (high selection pressure) to maintain the chance of moving in the right direction. We used an instance with bitlength 30 for our tests, this is just beyond the point where the problem is solved easily.

Both problems are by far not similar, but still somewhat related because they are extreme in requiring global and local exploration. In case of the wP-PEAKS, the whole search space is to cover to find the best peak, in SUFSAMP the local neighborhood must be covered well to get hold of the path to the global optimum.

MMDP and COUNTSAT

The Massively Multimodal Deceptive Problem (MMDP) as suggested by [28] is a multiple of a 6 bit deceptive subproblem which has its maximum for 6 or 0 bits set, its minimum for 1 or 5 bits set, and a deceptive local peak at 3 set bits. The fitness of a solution is simply the sum over all subproblems. From the mode of construction, it becomes clear that mutation is deliberately mislead here whether recombination is essential. We use an instance with 40 blocks of 6 bit, each.

The COUNTSAT problem is an instance of the MAXSAT problem suggested by [20]. Its fitness only depends on the number of set bits, all set to 1 in the global optimum. Our test instance is of length 40.

These two problems share the property that their fitness values are invariant to permutations of the whole (COUNTSAT) or separate parts (MMDP) of the genome. Consequently, [27] empirically demonstrate the general aptitude of self-adaptive mutation mechanisms on these problems.

Number Partitioning and Subset Sum

The (Min) Number Partitioning Problem (MNP) requires arranging a set of long numbers, here 35 with 10 digits length each, into two groups adding up to the same sum. Fitness of a solution candidate is measured as the remaining difference in sums, meaning this is a minimization problem. It was treated e.g. in [15, 14] by means of a *memetic algorithm*. We used a randomly determined instance for each optimization run.

The Subset Sum problem is related to the MNP in that from a given collection of positive integers, a subset must be chosen to achieve a predetermined sum. This time, the target sum does not directly depend on the overall sum of the given numbers but can be any attainable number. Additionally, all solution candidates approaching it from above are counted as infeasible. We use the same setup as in [38], made up of 100 numbers from within $[0, 2^{100}]$ and a density of 0.1, thus constructing the target sum from 10 of the original numbers. For each optimization run, an instance of this form is randomly created.

As stated, both are set selection problems. However, the expected fraction of ones in a good solution is different: Around 0.5 for the MNP, and 0.1 for the Subset Sum. Therefore, me may expect asymmetric mutation rates (self-adpative or not) to perform well on the latter, and symmetric on the former.

6 Assessment via Parameter Optimization

Within this section, we perform and discuss an SPO-based experiment to collect evidence for or against the claims made in §2. However, allowing for multiple starts during an optimization run on one and simultaneously employing algorithms with enormous differences in speed and quality on the other hand complicates assessing performance by means of standard measures like the MBF (mean best fitness), AES (average evaluations to solution), or SR (success rate). This is also confirmed by other views, e.g. [62], who express that dealing with aggregated measures, especially for multistart algorithms, requires some creativity. Additionally, we want to investigate the "tunability" of an algorithm-problem pair, for which no common measures exist. We therefore resort to introducing the needed, prior to describing the experiment itself.

6.1 Adequate Measures

LHS Average and Best

For assessing tunability of an algorithm towards a problem, the best performance found by a parameter optimization method shall be related to a base performance that is easily obtained without, or by manual tuning. Additionally, comparison to a simple hillclimber makes sense, as the more complex EAs should be able to attain better solutions than these to claim any importance.

The performance achieved by manual tuning surely depends on the expertise of the experimenter. As a substitute for this hardly quantifiable measure we propose to employ the best performance contained in an LHS of size 10 × #parameters, which in our case resembles the result of an initial SPO step. In contrast to the best, the average performance of all LHS design points is an approximation of the expected quality of a random configuration. Moreover, the variance of this quantity hints to low large the differences are. Large variances may indicate two things at once: a) There are very bad configurations the user must avoid, and b) there exist very good configurations that cover only a small amount of the parameter space and are thus hard to find. Low variances, however, mean that the algorithm is neither tunable nor mis-configurable. In consequence, we suggest to observe the measures f_{hc}, the MBF of a standard hillclimber, f_{LHSa}, the average fitness of an LHS design, σ_{LHSa}, its standard deviation, f_{LHSb}, fitness of the best configuration of an LHS, and f_{SPO}, the best performance achieved when tuning is finished.

AEB: Average Evaluations to Best

The AES needs a definition of success and is thus a problem-centric measure. If the specified success rarely occurs, its meaning is questionable. Lowering the quality of what counts as success may lead from floor to ceiling effect if performance differences of the algorithms in scope are large and the meaning of success is not carefully balanced. We turn this measure into an algorithm-centric one by taking the average of the number of evaluations needed to obtain the peak performance in each single optimization run. Stated differently, this is the average time an algorithm continues to make progress. This quantity still implies dependency of the time horizon of the experiment (the total number of evaluations allowed), but in contrast to the AES, it is always defined. The AEB alone is of limited value, as fast converging algorithms are prefered, regardless of the quality of their final best solution. However, it is useful for defining the following measure.

RFA: Robustness Favoring Aggregate

Assessing robustness and efficiency of an optimization algorithm at the same time is difficult, especially in situations where the optimum is seldomly reached, or even worse, is unknown. Common fitness measures as the MBF, AES, or SR, apply only to one of the two aspects. Combined measures as the success performance suggested by [1] require definition of success, as well. Two possibilities remain to resolve this dilemma: Multicriterial assessment, or aggregation. We choose the latter and use a robustness favoring aggregate (RFA), defined as:

$$RFA = \Delta f \cdot \left| \frac{\Delta f}{AEB} \right| \; , \; \Delta f = \begin{cases} MBF - f_{hc} & : \; minimization \\ f_{hc} - MBF & : \; maximization \end{cases} \tag{11}$$

The RFA consists of fitness progress, relative to the hillclimber, multiplied by the linear relative progress rate, which depends on the average number of evaluations to reach the best fitness value (AEB). It weights robustness higher than efficiency, but takes both into account. Using solely the linear relative progress rate is no alternative; it rates fast algorithms achieving poor quality equal to slow algorithms able to attain good quality. If an algorithm performs better than the hillclimber (in quality), its RFA is negative, positive otherwise. Note that this is only one of a virtually unlimited number of ways to define an aggregate.

6.2 Experiment

Apply SPO to EAs including different self-adaptive mechanisms and fixed rate mutation, compare tunability and achievable quality.

Table 2. Algorithm design for the six EA variants. Only asymmetric mutation operators need a second mutation rate and its initialization range (*). The learning rate (**) is only applicable to adaptive mutation operators, hence not used for variants with constant mutation rates. Algorithm designs thus have 7, 8, 9, or 10 parameters.

Parameter name	$\mathbb{N}/\mathbb{R}$	Min	Max	Parameter name	$\mathbb{N}/\mathbb{R}$	Min	Max
Population size μ	$\mathbb{N}$	1	500	Maximum age κ	$\mathbb{N}$	1	50
Selection pressure λ/μ	$\mathbb{R}_+$	1	10	Recombination prob. p_r	$\mathbb{R}$	0	1
Stagnation restart r_s	$\mathbb{N}$	5	30	Learning rate** τ	$\mathbb{R}$	0.0	1.0
Mutation rate p_m	$\mathbb{R}_+$	10^{-3}	1.0	Mut. rate range $r(p_m)$	$\mathbb{R}$	0	0.5
Mutation rate 2* p_{m2}	$\mathbb{R}_+$	10^{-3}	1.0	Mut. rate 2 range* $r(p_{m2})$	$\mathbb{R}$	0	0.5

Pre-experimental planning:
The originally intended performance measure (MBF) has been exchanged with the RFA fitness measure, due to first experiences with certain EA configurations on the MMDP problem, namely with asymmetric mutation operators. RFA provides a gradient even between the best found configurations that always solved the problem to optimality before, even for increased problem sizes (100 or more 6-bit groups).

Task:
Based on the aims named in §2, we investigate two scientific claims:

1. Tuned self-adaptative mutation allows for better performance than tuned constant mutation for many problems.
2. The tuning process is not significantly harder for self-adaptive mutation.

Additionally, the third aim shall be achieved by searching for patterns in the virtual relation between problem class properties and performance results. We perform statistical hypothesis testing by employing bootstrap permutation tests with the commonly used significance level of 5%. For accepting the first hypothesis, we demand a significant difference between the fastest self-adaptive and the fastest non-adaptive EA variant for at least half of the tested problems. The second claim is more difficult to test; in absence of a well-defined measure, we have to resort to just expressing the impression obtained from evaluating data and visualizations.

Setup:
The optimization algorithm used is a panmictic (μ, κ, λ)-ES with different mutation operators and 2-point crossover for recombination. As κ is one of the factors the parameter optimization is allowed to change, the degree of elitism (complete for $\kappa \geq \max(\#\text{generations})$, none for $kappa = 1$), that is the time any individual may

survive within the population, maybe varied. As common for binary representations, mutation works with bit flip probabilities (rates). We employ 6 different modes of performing mutation, namely constant rates, the self-adaptative method of [70], and the self-adaptive method of [75], each in turn symmetrical and asymmetrical (2 mutation rates, one for 0s and 1s, respectively). Table 2 shows the algorithm design for all variants. Note that not every parameter applies to all mutation operators. Parameter ranges are chosen relatively freehanded, bounded only by resource limits (μ, λ) or to enable a reasonable gradient (r_s, κ) in configuration qualities. Two features of the used algorithm variants may appear unusual, namely the recombination probability, allowing for gradually switching recombination on and off, and the stagnation restart, enabling restarts after a given number of generations without improvement.

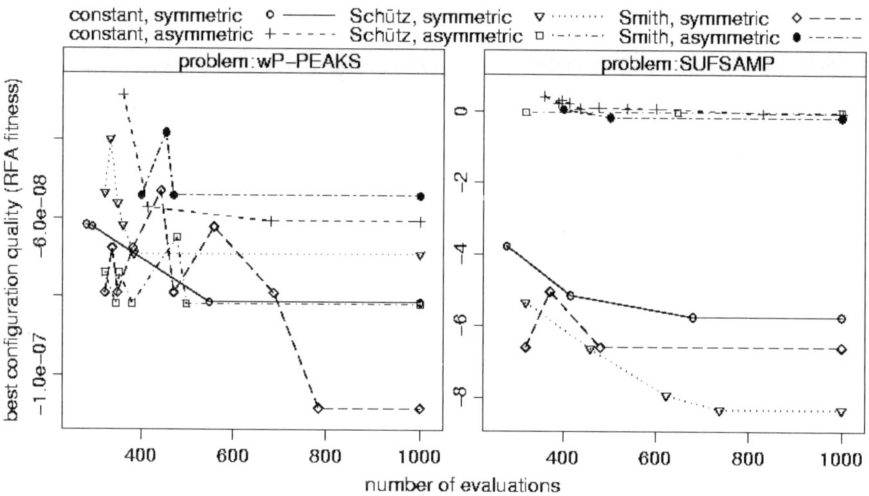

Fig. 2. SPO performance on weighted P-PEAKS (left) and SUFSAMP (right) problems. Points denote the best configurations found up to the corresponding number of evaluations (runs), using all repeats available after 1000 runs to reduce noise. Except the last one, all points stand for changed configurations regarded as better by SPO. The ones tried in between are estimated as worse than the current best. The graphs start on the left with the best configuration of the initial LHS (f_{LHSb}).

SPO parameters are kept at default values where possible. However, the total budget of allowed runs always is a design choice. We set this to 1000 as compromise between effectivity and operability. The initial LHS size is set to 10 · #parameters —following practical experience, also suggested by [69]— with 4 repeats per configuration, the maximum number of repeats to 64. Testing is performed at significance level 5%.

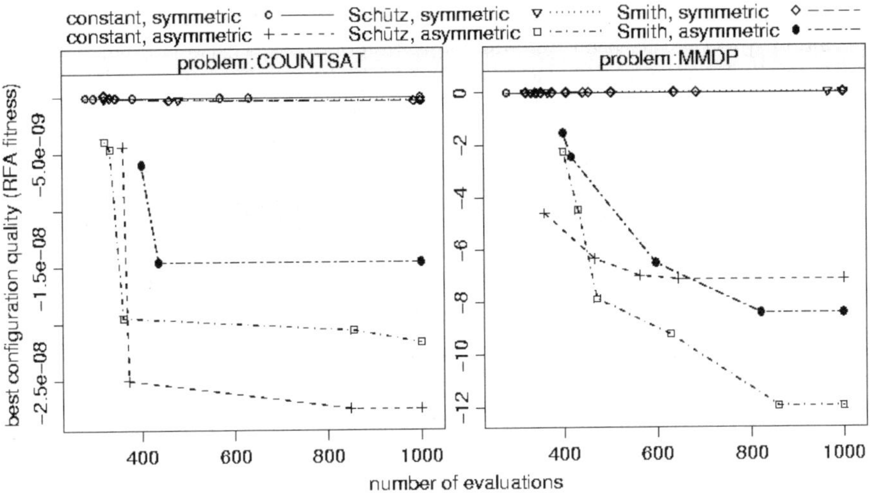

Fig. 3. SPO performance on COUNTSAT (left) and MMDP (right) problems. Remarks of Fig. 2 apply here as well.

The tackled problems are the 6 named in §5. For each problem and mutation operator variant ($6 \times 6 = 36$), we perform a separate SPO run (1000 algorithm runs each).

Experimentation/Visualization:
The development of the SPO runs on either of the 6 problems is presented in figures 2, 3, and 5. Note that for the depicted configurations, we use information gained after a SPO is finished, which is more than that available at runtime, due to possible re-evaluations (additional runs) of good configurations. Restriction to runtime information leads to heavily increased noise that would render the figures almost useless. Numerical results for the weighted P-PEAKS, Subset Sum, and MMDP problems are given in Table 4, the last column giving error probabilities for rejecting the hypothesis that f_{LHSb} and f_{SPO} are equal (p-values), based on all measures of this configuration obtained during the SPO run.

Observations:
As is evident from figures 2 to 5, there are two different behaviors of the six algorithm variants on on the considered test problems: Either a clear distinction into two groups of three can be recognized (SUFSAMP, COUNTSAT, and MMDP), or the curves are much more intermingled. A closer look at the constituents of these groups reveals that the separating feature is the symmetry type. For the COUNTSAT and MMDP problems, the better group is formed by all three asymmetric mutation variants, whereas all three symmetric variants are performing better on the SUFSAMP problem.

For the wP-PEAKS problem, we obtain two surprising observations:

1. At both ends of the performance spectrum, we find the self-adaptation variant of Smith, the symmetric being the best, the asymmetric the worst.

2. Ordering of the symmetric/asymmetric variants of one type is not the same for all, for the variants using the self-adaptation of Schütz, it is reversed.

As the SUFSAMP problem was constructed to favor enumerating the local neighborhood, it can be expected that SPO adjusts the selection pressure to high values. A look at the parameter distributions of the best variant, figure 4, reveals that SPO surprisingly circumvented doing so by more frequently using higher maximum age values; at the same time, the mean stagnation restart time was also increased.

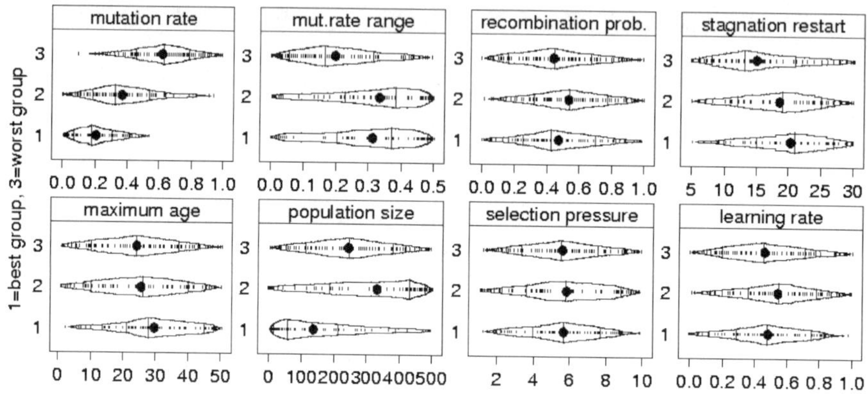

Fig. 4. Parameter distributions of configurations chosen by SPO (including the initial LHS) on the SUFSAMP problem with symmetric self-adaptation after Schütz, separated into three equally sized groups according to measured fitness.

The number partitioning problem results surprisingly indicate constant asymmetric mutation as the best and self-adaptation after Schütz as the worst variant, the latter being obviously very hard to improve. The tuning process on this as well as on the Subset Sum problem (figure 5), looks much more disturbed than e.g. on the MMDP, containing many huge jumps.

Discussion:
A look at Table 4 reveals that for the chosen, discrete test problems, averaged performances and standard deviations are most often very similar. In contrast to the situation usually found for real-valued test problems treated with SPO, the noise level here makes further improvement very difficult. It obviously cannot be easily lowered by increasing the number of repeats as the best configurations typically already reached the imposed maximum of 64 runs. This hardship is probably due to the discrete value set of the objective functions which hinder achieving narrow result distributions. It does not render SPO useless, but its performance seems to be limited tighter than for the real-valued case.

Another unexpected conclusion is that asymmetric mutation operators often perform very well on functions usually treated with symmetric constant mutation rates. COUNTSAT and MMDP are best approximated with asymmetric mutation operators, the former with the constant, the latter with the self-adaptive variant by Schütz. Table 3 lists the 2 best variants for each problem and the p-Values obtained from the comparison. Concerning aim 1 of §2, we cannot verify the original claim.

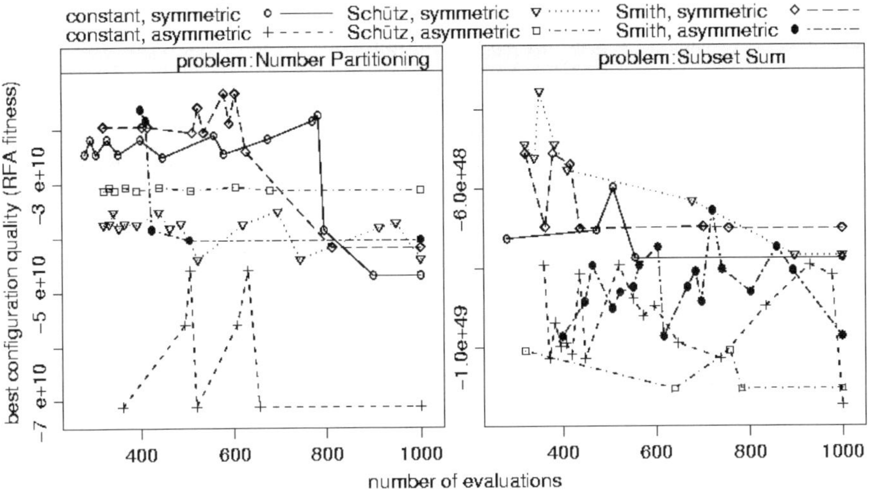

Fig. 5. SPO performance on Number Partitioning (left) and Subset Sum (right).

There are problems for which self-adaptation is very useful, and there are some where it is not. Only in 2 of the six test cases, they have significant advantage.

Table 3. Bootstrap Permutation hypothesis tests between performances of the best constant and the best self-adaptive variant for each problem. The tests use all best configuration runs available after SPO is finished, this is in most cases 64.

Problem	Best Variant	Second Variant	p-Value	Significant?
wP-PEAKS	Smith, symmetric	constant, symmetric	0.125	No
SUFSAMP	Schütz, symmetric	constant, symmetric	2e-05	Yes
COUNTSAT	constant, asymmetric	Schütz, asymmetric	0.261	No
MMDP	Schütz, asymmetric	constant, asymmetric	0.008	Yes
Number Part.	constant, asymmetric	Schütz, symmetric	0.489	No
Subset Sum	constant, asymmetric	Schütz, asymmetric	0.439	No

Concerning the effort needed for tuning constant and self-adaptive mutation operators, we cannot account for a clear difference, thereby adhering to the original claim that tuning efforts are similar. However, there is neither statistical evidence in favor nor against this statement.

Nevertheless, when comparing the final best configurations to the average LHS results, it becomes clear that without any tuning, algorithms may easily be misconfigured, leading to performance values much worse than that of a simple EA-based hillclimber. Moreover, the best point of an initial LHS appears as a reasonable estimate for the quality achievable by tuning. Within this experiment, the LHS best fitness f_{LHSb} has always been superior to that of the (single-start) hillclimber.

Is it possible to derive any connection between the properties of the tackled problems and the success of different self-adaptation variants? We can detect similarities in the progression of the SPO runs, e.g. between COUNTSAT and MMDP, and Number Partitioning and Subset Sum, like the huge jumps found in case of the latter; these probably indicate result distributions much more complex than for the former. However, it appears difficult to foresee if self-adaptation enhanced EAs will perform better or worse. In case of the MMDP, recombination is likely to preserve existing good blocks that may be optimized individually so that mutation rate schedules, once learned, can be successfully reapplied. But without further tests, this is rather speculation.

7 Conclusions

We motivated and explained the SPO procedure and, as a test case, applied it to self-adaptive EA variants for binary coded problems. This study revealed why self-adaptation is rarely applied for these problem types: It is hard to predict whether it will increase performance, and which variant to choose. However, if properly parametrized, it can be significantly faster than well tuned standard mutation operators. Modeling only a rough problem-mechanism interaction would require a much more thoroughly conducted study than presented here.

Surprisingly, specialized and seldomly used operators like the (constant and self-adaptive) asymmetric mutation performed very well when parametrized accordingly. This probably leads the way to superior operators still to develop—with or without self-adaptation. Mutation rate learning seems to provide rather limited potential.

The invented performance measures, RFA and AEB, were found capable of leading the meta-search into the right direction. Unfortunately, the absence of well-defined measures for the tuning process itself currently prevents effectively using SPO for answering questions concerning the tunability of an algorithm-problem combination. This problem we want to tackle in future work.

Recapitulating, SPO works on binary problems, and proves to be a valuable tool for experimental analysis. However, there is room for improvement, first and foremost by taking measures to reduce the variances within the results of a single configuration.

Acknowledgment

The research leading to this paper has been supported by the DFG (Deutsche Forschungsgemeinschaft) as project no. 252441, "Mehrkriterielle Struktur- und Parameteroptimierung verfahrenstechnischer Prozesse mit evolutionären Algorithmen am Beispiel gewinnorientierter unscharfer destillativer Trennprozesse". T. Bartz–Beielstein's research was supported by the DFG as part of the collaborative research center "Computational Intelligence" (531).

The autors want to thank Thomas Jansen for providing his code of the SUF-SAMP problem, and Silja Meyer-Nieberg and Nikolaus Hansen for sharing their opinions with us.

Table 4. Best found configuration performances, mean value of LHS, best of LHS, and best after SPO is finished, on the weighted P-PEAKS, Subset Sum, and MMDP problems. Besides the RFA fitness measure used within SPO (to minimize, relative to hillclimber), we also give the MBF and AEB performance measures. AEB values are meant as multiples of 10^3. Note that wP-PEAKS and MMDP(40) are to be maximized with optimal values at 1.0 and 40.0, respectively, and Subset Sum is to be minimized to 0.0.

	Hillclimber, f_{hc}		LHS mean, f_{LHSa}				LHS best, f_{LHSb}				SPO best, f_{SPO}				
Adaptation	MBF	AEB	RFA	stddev	MBF	AEB	RFA	stddev	MBF	AEB	RFA	stddev	MBF	AEB	p-val
Problem: wP-PEAKS															
none, sym	0.945	<5	4.7e-07	9.5e-07	0.852	61.9	-6.2e-08	6.7e-08	0.984	52.1	-8.1e-08	8.2e-08	0.993	50.4	0.29
none, asym	0.945	<5	4.2e-07	8.8e-07	0.851	63.4	-2.9e-08	8.3e-08	0.964	48.8	-6.1e-08	2.5e-08	0.983	30.4	0.00
Schütz, sym	0.945	<5	5.0e-07	9.1e-07	0.844	60.5	-5.7e-08	3.1e-08	0.995	57.3	-6.9e-08	1.1e-07	0.980	46.2	0.46
Schütz, asym	0.945	<5	4.1e-07	1.1e-06	0.858	62.0	-7.4e-08	8.2e-08	0.984	48.3	-8.2-08	4.4e-08	0.984	24.2	0.35
Smith, sym	0.945	<5	4.9e-07	1.0e-06	0.846	58.5	-7.9e-08	1.2e-07	0.990	54.2	-1.1e-07	1.6e-07	0.987	46.1	0.12
Smith, asym	0.945	<5	4.0e-07	8.6e-07	0.858	63.1	-5.4e-08	3.4e-08	0.984	41.6	-5.4e-08	3.4e-08	0.984	41.6	0.50
Problem: Subset Sum															
none, sym	5.5e+26	36.0	3.4e+59	3.9e+60	4.7e+31	54.7	-7.3e+48	1.2e+49	1.6e+26	49.3	-7.8e+48	6.8e+48	8.1e+25	49.2	0.44
none, asym	5.5e+26	36.0	1.4e+59	6.4e+59	3.2e+31	52.5	-7.9e+48	1.0e+49	7.7e+25	55.6	-1.1e+49	1.6e+49	7.7e+25	38.0	0.18
Schütz, sym	5.5e+26	36.0	1.9e+59	1.1e+60	5.0e+31	53.9	-4.9e+48	6.6e+48	2.0e+26	51.7	-7.7e+48	5.6e+48	4.3e+25	51.2	0.04
Schütz, asym	5.5e+26	36.0	2.5e+59	1.9e+60	2.9e+31	49.4	-1.0e+49	1.2e+49	4.2e+25	51.0	-1.1e+49	1.4e+49	7.1e+25	48.1	0.42
Smith, sym	5.5e+26	36.0	2.6e+59	1.7e+60	5.0e+31	55.7	-5.1e+48	5.2e+48	1.3e+26	58.3	-7.0e+48	9.0e+48	8.8e+25	52.6	0.21
Smith, asym	5.5e+26	36.0	3.2e+59	4.5e+60	2.8e+31	51.1	-9.7e+48	1.3e+49	7.3e+25	52.5	-9.7e+48	1.3e+49	7.3e+25	52.5	0.50
Problem: MMDP															
none, sym	23.80	58.1	-1.1e-03	1.4e-03	29.3	62.0	-4.3e-03	2.3e-03	37.0	54.3	-4.3e-03	4.3e-04	37.8	45.7	0.48
none, asym	23.80	58.1	-0.225	0.783	39.9	7.4	-4.60	4.83	40	<5	-7.13	11.7	40	<5	0.23
Schütz, sym	23.80	58.1	-1.3e-03	1.7e-03	30.0	63.0	-4.3e-03	2.8e-03	36.0	48.3	-4.6e-03	2.3e-03	37.6	54.3	0.37
Schütz, asym	23.80	58.1	-0.155	0.274	39.8	8.1	-2.28	0.894	40	<5	-12.0	10.9	40	<5	0.00
Smith, sym	23.80	58.1	-1.4e-03	1.8e-03	30.0	62.1	-4.1e-03	2.4e-03	37.3	59.6	-4.6e-03	2.1e-03	36.8	46.1	0.18
Smith, asym	23.80	58.1	-0.143	0.234	39.8	7.4	-1.56	0.774	40	<5	-8.41	6.59	40	<5	0.00

References

1. Anne Auger and Nikolaus Hansen. Performance Evaluation of an Advanced Local Search Evolutionary Algorithm. In B. McKay et al., editors, *Proc. 2005 Congress on Evolutionary Computation (CEC'05)*, Piscataway NJ, 2005. IEEE Press.
2. T. Bäck. *Evolutionary Algorithms in Theory and Practice.* Oxford University Press, New York NY, 1996.
3. T. Bäck and M. Schütz. Evolution strategies for mixed-integer optimization of optical multilayer systems. In J. R. McDonnell, R. G. Reynolds, and D. B. Fogel, editors, *Evolutionary Programming IV: Proceedings of the Fourth Annual Conference on Evolutionary Programming*, pages 33–51. MIT Press, Cambridge, MA, 1995.
4. Th. Bäck. Self-adaptation in genetic algorithms. In FJ Varela and P Bourgine, editors, *Toward a Practice of Autonomous Systems: proceedings of the first European conference on Artificial Life*, pages 263–271. MIT Press, 1992.
5. R. Barr and B. Hickman. Reporting computational experiments with parallel algorithms: Issues, measures, and experts' opinions. *ORSA Journal on Computing*, 5(1):2–18, 1993.
6. Thomas Bartz-Beielstein. Evolution strategies and threshold selection. In M. J. Blesa Aguilera, C. Blum, A. Roli, and M. Sampels, editors, *Proceedings Second International Workshop Hybrid Metaheuristics (HM'05)*, volume 3636 of *Lecture Notes in Computer Science*, pages 104–115, Berlin, Heidelberg, New York, 2005. Springer.
7. Thomas Bartz-Beielstein. *Experimental Research in Evolutionary Computation—The New Experimentalism.* Springer, Berlin, Heidelberg, New York, 2006.
8. Thomas Bartz-Beielstein, Daniel Blum, and Jürgen Branke. Particle swarm optimization and sequential sampling in noisy environments. In Richard Hartl and Karl Doerner, editors, *Proceedings 6th Metaheuristics International Conference (MIC2005)*, pages 89–94, Vienna, Austria, 2005.
9. Thomas Bartz-Beielstein, Marcel de Vegt, Konstantinos E. Parsopoulos, and Michael N. Vrahatis. Designing particle swarm optimization with regression trees. Interner Bericht des Sonderforschungsbereichs 531 Computational Intelligence CI–173/04, Universität Dortmund, Germany, Mai 2004.
10. Thomas Bartz-Beielstein, Christian Lasarczyk, and Mike Preuß. Sequential parameter optimization. In B. McKay et al., editors, *Proceedings 2005 Congress on Evolutionary Computation (CEC'05), Edinburgh, Scotland*, volume 1, pages 773–780, Piscataway NJ, 2005. IEEE Press.
11. Thomas Bartz-Beielstein and Boris Naujoks. Tuning multicriteria evolutionary algorithms for airfoil design optimization. Interner Bericht des Sonderforschungsbereichs 531 Computational Intelligence CI–159/04, Universität Dortmund, Germany, Februar 2004.
12. Thomas Bartz-Beielstein, Konstantinos E. Parsopoulos, and Michael N. Vrahatis. Design and analysis of optimization algorithms using computational statistics. *Applied Numerical Analysis & Computational Mathematics (ANACM)*, 1(2):413–433, 2004.
13. Thomas Bartz-Beielstein, Mike Preuß, and Sandor Markon. Validation and optimization of an elevator simulation model with modern search heuristics. In

T. Ibaraki, K. Nonobe, and M. Yagiura, editors, *Metaheuristics: Progress as Real Problem Solvers*, Operations Research/Computer Science Interfaces, pages 109–128. Springer, Berlin, Heidelberg, New York, 2005.
14. Regina Berretta, Carlos Cotta, and Pablo Moscato. Enhancing the performance of memetic algorithms by using a matching-based recombination algorithm. In *Metaheuristics: computer decision-making*, pages 65–90. Kluwer Academic Publishers, Norwell, MA, 2004.
15. Regina Berretta and Pablo Moscato. The number partitioning problem: an open challenge for evolutionary computation? In *New ideas in optimization*, pages 261–278. McGraw-Hill Ltd., Maidenhead, UK, 1999.
16. H.-G. Beyer and H.-P. Schwefel. Evolution strategies—A comprehensive introduction. *Natural Computing*, 1:3–52, 2002.
17. Marcel de Vegt. Einfluss verschiedener Parametrisierungen auf die Dynamik des Partikel-Schwarm-Verfahrens: Eine empirische Analyse. Interner Bericht der Systems Analysis Research Group SYS–3/05, Universität Dortmund, Fachbereich Informatik, Germany, Dezember 2005.
18. Camil Demetrescu and Giuseppe F. Italiano. What do we learn from experimental algorithmics? In *MFCS '00: Proceedings of the 25th International Symposium on Mathematical Foundations of Computer Science*, pages 36–51, Berlin, Heidelberg, New York, 2000. Springer.
19. N. R. Draper and H. Smith. *Applied Regression Analysis.* Wiley, New York NY, 3rd edition, 1998.
20. S. Droste, T. Jansen, and I. Wegener. A natural and simple function which is hard for all evolutionary algorithms. In *IEEE International Conference on Industrial Electronics, Control, and Instrumentation (IECON 2000)*, pages 2704–2709, Piscataway, NJ, 2000. IEEE Press.
21. A. E. Eiben, R. Hinterding, and Z. Michalewicz. Parameter control in evolutionary algorithms. *IEEE Transactions on Evolutionary Computation*, 3(2):124–141, 1999.
22. A. E. Eiben and J. E. Smith. *Introduction to Evolutionary Computing.* Springer, Berlin, Heidelberg, New York, 2003.
23. A.E. Eiben and M. Jelasity. A critical note on experimental research methodology in EC. In *Proceedings of the 2002 Congress on Evolutionary Computation (CEC'2002)*, pages 582–587, Piscataway NJ, 2002. IEEE.
24. Robert Feldt and Peter Nordin. Using factorial experiments to evaluate the effect of genetic programming parameters. In Riccardo Poli et al., editors, *Genetic Programming, Proceedings of EuroGP'2000*, volume 1802 of *Lecture Notes in Computer Science*, pages 271–282, Berlin, Heidelberg, New York, 15-16 April 2000. Springer.
25. D. B. Fogel. *Evolving Artificial Intelligence.* PhD thesis, University of California, San Diego, 1992.
26. O. François and C. Lavergne. Design of evolutionary algorithms—a statistical perspective. *IEEE Transactions on Evolutionary Computation*, 5(2):129–148, April 2001.
27. Mario Giacobini, Mike Preuß, and Marco Tomassini. Effects of scale-free and small-world topologies on binary coded self-adaptive CEA. In *6th European Conf. Evolutionary Computation in Combinatorial Optimization, Proc. (EvoCOP'06)*, Lecture Notes in Computer Science, pages 86–98, Berlin, 2006. Springer.

28. D. E. Goldberg, K. Deb, and J. Horn. Massively multimodality, deception and genetic algorithms. In R. Männer and B. Manderick, editors, *Parallel Prob. Solving from Nature II*, pages 37–46. North-Holland, 1992.
29. Garrison W. Greenwood. Adapting mutations in genetic algorithms using gene flow principles. In R. Sarker et al., editors, *Proc. 2003 Congress on Evolutionary Computation (CEC'03), Canberra*, pages 1392–1397, Piscataway NJ, 2003. IEEE Press.
30. N. Hansen and A. Ostermeier. Completely derandomized self-adaptation in evolution strategies. *Evolutionary Computation*, 9(2):159–195, 2001.
31. N. Hansen, A. Ostermeier, and A. Gawelczyk. On the adaptation of arbitrary normal mutation distributions in evolution strategies: The generating set adaptation. In L. J. Eshelman, editor, *Proc. 6th Int'l Conf. on Genetic Algorithms*, pages 57–64, San Francisco, CA, 1995. Morgan Kaufmann Publishers, Inc.
32. J. Hooker. Testing heuristics: We have it all wrong. *Journal of Heuristics*, 1(1):33–42, 1996.
33. Christian Igel and Martin Kreutz. Operator adaptation in evolutionary computation and its application to structure optimization of neural networks. *Neurocomputing*, 55(1-2):347–361, 2003.
34. E. H. Isaaks and R. M. Srivastava. *An Introduction to Applied Geostatistics.* Oxford University Press, Oxford, U.K., 1989.
35. Thomas Jansen, Kenneth A. De Jong, and Ingo Wegener. On the choice of the offspring population size in evolutionary algorithms. *Evolutionary Computation*, 13(4):413–440, 2005.
36. Thomas Jansen and Dirk Sudholt. Design and analysis of an asymmetric mutation operator. In B. McKay et al., editors, *Proc. 2005 Congress on Evolutionary Computation (CEC'05), Edinburgh, Scotland*, volume 1, pages 190–197, Piscataway NJ, 2005. IEEE Press.
37. Mark Jelasity. A wave analysis of the subset sum problem. In Thomas Bäck, editor, *Proceedings of the Seventh International Conference on Genetic Algorithms (ICGA97)*, pages 89–96, San Francisco, CA, 1997. Morgan Kaufmann.
38. Márk Jelasity, Mike Preuß, and Agoston E. Eiben. Operator learning for a problem class in a distributed peer-to-peer environment. In J. J. Merelo Guervós, P. Adamidis, H.-G. Beyer, J.L. Fernández-Villacañas, and H.-P. Schwefel, editors, *Parallel Problem Solving from Nature – PPSN VII, Proc. Seventh Int'l Conf., Granada*, pages 172–183, Berlin, 2002. Springer.
39. David S. Johnson. A theoretician's guide to the experimental analysis of algorithms. In M. H. Goldwasser, D. S. Johnson, and C. C. McGeoch, editors, *Data Structures, Near Neighbor Searches, and Methodology: Fifth and Sixth DIMACS Implementation Challenges*, pages 215–250, Providence, 2002. American Mathematical Society.
40. D.R. Jones, M. Schonlau, and W.J. Welch. Efficient global optimization of expensive black-box functions. *Journal of Global Optimization*, 13:455–492, 1998.
41. Kenneth A. De Jong, Michael A. Potter, and William M. Spears. Using problem generators to explore the effects of epistasis. In T. Bäck, editor, *Proceedings of the Seventh ICGA*, pages 338–345. Morgan Kaufmann, 1997.
42. Bryant A. Julstrom. Adaptive operator probabilities in a genetic algorithm that applies three operators. In *SAC '97: Proceedings of the 1997 ACM symposium on Applied computing*, pages 233–238, New York, NY, 1997. ACM Press.
43. J. P. C. Kleijnen. *Statistical Tools for Simulation Practitioners.* Marcel Dekker, New York NY, 1987.

44. J. P. C. Kleijnen. Experimental design for sensitivity analysis, optimization, and validation of simulation models. In J. Banks, editor, *Handbook of Simulation*. Wiley, New York NY, 1997.
45. F. Kursawe. *Grundlegende empirische Untersuchungen der Parameter von Evolutionsstrategien – Metastrategien*. Dissertation, Fachbereich Informatik, Universität Dortmund, Germany, 1999.
46. A.M. Law and W.D. Kelton. *Simulation Modeling and Analysis*. McGraw-Hill, New York NY, 3rd edition, 2000.
47. S.N. Lophaven, H.B. Nielsen, and J. Søndergaard. Aspects of the Matlab Toolbox DACE. Technical Report IMM-REP-2002-13, Informatics and Mathematical Modelling, Technical University of Denmark, Copenhagen, Denmark, 2002.
48. S.N. Lophaven, H.B. Nielsen, and J. Søndergaard. DACE—A Matlab Kriging Toolbox. Technical Report IMM-REP-2002-12, Informatics and Mathematical Modelling, Technical University of Denmark, Copenhagen, Denmark, 2002.
49. Sandor Markon, Hajime Kita, Hiroshi Kise, and Thomas Bartz-Beielstein, editors. *Modern Supervisory and Optimal Control with Applications in the Control of Passenger Traffic Systems in Buildings*. Springer, Berlin, Heidelberg, New York, 2006.
50. Catherine C. McGeoch. *Experimental Analysis of Algorithms*. PhD thesis, Carnegie Mellon University, Pittsburgh PA, 1986.
51. M. D. McKay, R. J. Beckman, and W. J. Conover. A comparison of three methods for selecting values of input variables in the analysis of output from a computer code. *Technometrics*, 21(2):239–245, 1979.
52. Jörn Mehnen, Thomas Michelitsch, Thomas Bartz-Beielstein, and Nadine Henkenjohann. Systematic analyses of multi-objective evolutionary algorithms applied to real-world problems using statistical design of experiments. In R. Teti, editor, *Proceedings Fourth International Seminar Intelligent Computation in Manufacturing Engineering (CIRP ICME'04)*, volume 4, pages 171–178, Naples, Italy, 2004.
53. Jörn Mehnen, Thomas Michelitsch, Thomas Bartz-Beielstein, and Christian W. G. Lasarczyk. Multiobjective evolutionary design of mold temperature control using DACE for parameter optimization. In H. Pfützner and E. Leiss, editors, *Proceedings Twelfth International Symposium Interdisciplinary Electromagnetics, Mechanics, and Biomedical Problems (ISEM 2005)*, volume L11-1, pages 464–465, Vienna, Austria, 2005. Vienna Magnetics Group Reports.
54. Silja Meyer-Nieberg and Hans-Georg Beyer. Self-adaptation in evolutionary algorithms. In Fernando Lobo, Claudio Lima, and Zbigniew Michalewicz, editors, *Parameter Setting in Evolutionary Algorithms*, Studies in Computational Intelligence. Springer, Berlin Heidelberg New York, 2006.
55. D. C. Montgomery. *Design and Analysis of Experiments*. Wiley, New York NY, 5th edition, 2001.
56. B. M. E. Moret. Towards a discipline of experimental algorithmics. In M.H. Goldwasser, D.S. Johnson, and C.C. McGeoch, editors, *Data Structures, Near Neighbor Searches, and Methodology: Fifth and Sixth DIMACS Implementation Challenges, DIMACS Monographs 59*, pages 197–213, Providence RI, 2002. American Mathematical Society.
57. Bernard M. Moret and Henry D. Shapiro. Algorithms and experiments: The new (and old) methodology. *Journal of Universal Computer Science*, 7(5):434–446, 2001.

58. R. Myers and E.R. Hancock. Empirical modelling of genetic algorithms. *Evolutionary Computation*, 9(4):461–493, 2001.
59. Martin Pelikan, David E. Goldberg, and Erick Cantú-Paz. Linkage problem, distribution estimation, and bayesian networks. *Evolutionary Computation*, 8(3):311–340, 2000.
60. F. Pukelsheim. *Optimal Design of Experiments.* Wiley, New York NY, 1993.
61. Iloneide C. O. Ramos, Marco C. Goldbarg, Elizabeth G. Goldbarg, and Adrião D. Dória Neto. Logistic Regression for Parameter Tuning on an Evolutionary Algorithm. In B. McKay et al., editors, *Proc. 2005 Congress on Evolutionary Computation (CEC'05)*, volume 2, pages 1061–1068, Piscataway NJ, 2005. IEEE Press.
62. Ronald L. Rardin and Reha Uzsoy. Experimental evaluation of heuristic optimization algorithms: A tutorial. *J. Heuristics*, 7(3):261–304, 2001.
63. I. Rechenberg. *Evolutionsstrategie: Optimierung technischer Systeme nach Prinzipien der biologischen Evolution.* Frommann-Holzboog Verlag, Stuttgart, 1973.
64. I. Rechenberg. Evolutionsstrategien. In B. Schneider and U. Ranft, editors, *Simulationsmethoden in der Medizin und Biologie*, pages 83–114. Springer-Verlag, Berlin, 1978.
65. J. Sacks, W. J. Welch, T. J. Mitchell, and H. P. Wynn. Design and analysis of computer experiments. *Statistical Science*, 4(4):409–435, 1989.
66. T. J. Santner, B. J. Williams, and W. I. Notz. *The Design and Analysis of Computer Experiments.* Springer, Berlin, Heidelberg, New York, 2003.
67. J. D. Schaffer, R. A. Caruana, L. Eshelman, and R. Das. A study of control parameters affecting online performance of genetic algorithms for function optimization. In J. D. Schaffer, editor, *Proceedings of the Third International Conference on Genetic Algorithms*, pages 51–60, San Mateo CA, 1989. Morgan Kaufman.
68. M. Schonlau. *Computer Experiments and Global Optimization.* PhD thesis, University of Waterloo, Ontario, Canada, 1997.
69. Matthias Schonlau, William J. Welch, and Ronald R. Jones. Global versus local search in constrained optimization of computer models. In *New Development and Applications in Experimental Design*, number 34 in IMS Lecture Notes, pages 11–25. Institute of Mathematical Statistics, Beachwood, OH, 1998.
70. Martin Schütz. Eine Evolutionsstrategie für gemischt-ganzzahlige Optimierprobleme mit variabler Dimension. Interner Bericht der Systems Analysis Research Group SYS–1/96, Universität Dortmund, Fachbereich Informatik, August 1996.
71. H.-P. Schwefel. Adaptive Mechanismen in der biologischen Evolution und ihr Einfluß auf die Evolutionsgeschwindigkeit. Technical report, Technical University of Berlin, 1974. Abschlußbericht zum DFG-Vorhaben Re 215/2.
72. H.-P. Schwefel. *Numerical Optimization of Computer Models.* Wiley, Chichester, 1981.
73. H.-P. Schwefel, I. Wegener, and K. Weinert, editors. *Advances in Computational Intelligence—Theory and Practice.* Springer, Berlin, Heidelberg, New York, 2003.
74. J. Smith and T. C. Fogarty. Self-adaptation of mutation rates in a steady state genetic algorithm. In *Proceedings of 1996 IEEE Int'l Conf. on Evolutionary Computation (ICEC '96)*, pages 318–323. IEEE Press, NY, 1996.

75. J. E. Smith. Modelling gas with self adaptive mutation rates. In L. Spector and et al., editors, *Proceedings of the Genetic and Evolutionary Computation Conference (GECCO-2001)*, pages 599–606, San Francisco, California, USA, 7-11 July 2001. Morgan Kaufmann.
76. Catalin Stoean, Mike Preuss, Ruxandra Gorunescu, and Dan Dumitrescu. Elitist Generational Genetic Chromodynamics - a New Radii-Based Evolutionary Algorithm for Multimodal Optimization. In B. McKay et al., editors, *Proc. 2005 Congress on Evolutionary Computation (CEC'05)*, volume 2, pages 1839 – 1846, Piscataway NJ, 2005. IEEE Press.
77. Ch. Stone and J. E. Smith. Strategy parameter variety in self-adaptation of mutation rates. In W. B. Langdon and et al., editors, *GECCO*, pages 586–593. Morgan Kaufmann, 2002.
78. Dirk Thierens. An adaptive pursuit strategy for allocating operator probabilities. In *GECCO '05: Proceedings of the 2005 conference on Genetic and evolutionary computation*, pages 1539–1546, New York, NY, 2005. ACM Press.
79. Marko Tosic. Evolutionäre Kreuzungsminimierung. Diploma thesis, University of Dortmund, Germany, January 2006.
80. Klaus Weinert, Jörn Mehnen, Thomas Michelitsch, Karlheinz Schmitt, and Thomas Bartz-Beielstein. A multiobjective approach to optimize temperature control systems of moulding tools. *Production Engineering Research and Development, Annals of the German Academic Society for Production Engineering*, XI(1):77–80, 2004.
81. Darrell Whitley, Soraya B. Rana, John Dzubera, and Keith E. Mathias. Evaluating evolutionary algorithms. *Artificial Intelligence*, 85(1-2):245–276, 1996.
82. D.H. Wolpert and W.G. Macready. No free lunch theorems for optimization. *IEEE Transactions on Evolutionary Computation*, 1(1):67–82, 1997.

Combining Meta-EAs and Racing for Difficult EA Parameter Tuning Tasks

Bo Yuan and Marcus Gallagher

School of Information Technology and Electrical Engineering
University of Queensland, Qld. 4072, Australia
boyuan@itee.uq.edu.au, marcusg@itee.uq.edu.au

Summary. This chapter presents a novel framework for tuning the parameters of Evolutionary Algorithms. A hybrid technique combining Meta-EAs and statistical Racing approaches is developed, which is not only capable of effectively exploring the search space of numerical parameters but also suitable for tuning symbolic parameters where it is generally difficult to define any sensible distance metric.

1 Introduction

One of the major issues in applying Evolutionary Algorithms (EAs) is how to choose an appropriate parameter setting. The importance of parameter tuning is at least threefold. Firstly, EAs are not completely parameter robust and may not be able to solve challenging problems effectively with inappropriate values. Secondly, when two or more EAs are compared, arbitrarily specified parameter values may make the comparison unfair and conclusions misleading. Finally, finding the optimal setting may be also helpful for better understanding the behavior of EAs.

There has been a long history of work on finding the optimal parameter values of EAs [11]. However, it has shown that this kind of optimal setting does not exist in general and there are different optimal values for different problems [17]. To better understand the relationship between parameter setting and performance, for each pair of algorithm and problem, a performance landscape could be defined with one dimension for each parameter and an extra dimension for the performance. As a result, the global optimum of this landscape corresponds to the optimal parameter setting of the EA on that particular problem. Note that, the structure of such landscapes is unlikely to be always trivial (e.g., monotonic or separable).

One principled approach to parameter tuning is to exhaustively explore the performance of an EA with systematically varied parameter values. Statistical techniques such as ANalysis Of VAriance (ANOVA) could then be applied on the results to formally investigate the influence and contribution of each parameter towards the algorithm's performance as well as the interaction among parameters [4, 5, 16], which may be used as heuristic knowledge in choosing the appropriate parameter setting. However, a large number of experiments are usually required to collect enough data for such statistical analysis, which is inefficient for the purpose of parameter tuning.

B. Yuan and M. Gallagher: *Combining Meta-EAs and Racing for Difficult EA Parameter Tuning Tasks*, Studies in Computational Intelligence (SCI) **54**, 121–142 (2007)
www.springerlink.com

Alternatively, mathematical modeling techniques such as Response Surfaces could be employed to search for the optimum parameter setting by only testing a relatively small number of parameter combinations [2].

If the performance landscape is suspected to be multimodal and there is no prior knowledge to come up with a good starting position, global optimization methods could be applied to conduct parameter tuning. A typical example is to use an external EA, called a Meta-EA [1, 8] as the high-level optimization procedure to conduct searching on the performance landscape. In these search-based methods, all parameters to be tuned are usually encoded into a vector and each instantiation of this vector/individual/algorithm instance corresponds to a fully-specified EA. Consequently, the process of parameter tuning is to find the individual or algorithm instance that yields the best performance.

It is worth mentioning that, in typical parameter tuning, algorithm parameters are chosen in advance and remain fixed during evolution. Another branch of research is parameter *control* in which algorithm parameters are allowed to vary with time as the solution vectors of the objective function are optimized [7]. The advantage is that different parameter values may be needed in different stages of evolution in order to achieve optimal performance. However, there are still some exogenous parameters used to control those self-adaptive parameters. Furthermore, not all algorithm parameters can be conveniently adapted in this manner. After all, finding the optimal settings in different situations may also provide valuable information for designing better parameter control strategies.

The major contributions of this chapter are listed as follows. Firstly, the properties of traditional search-based methods are given a critical analysis. We point out that, in addition to being very time-consuming in evaluating each candidate, they have inherent difficulty in tuning nominal or symbolic parameters where it is very hard to define any sensible distance metric over the parameter space. For these parameters, no search-based methods could be expected to be better than the exhaustive or random search.

Secondly, a novel statistical technique called Racing [12], which is originally proposed to solve the model selection problem in Machine Learning, is introduced to reliably choose the best parameter setting. The advantage of Racing is that it does not require a well-defined parameter space but is still much more efficient than the exhaustive search.

Finally, with a good understanding of the properties of the two distinct classes of tuning techniques, a framework of hybridization is proposed, which is not only able to handle symbolic parameters but also to search and explore the parameter space efficiently. Two case studies on tuning the parameters of a Genetic Algorithm (GA) are also conducted to highlight the effectiveness of the proposed techniques.

2 The Challenge of Symbolic Parameters

As mentioned above, in search-based methods, the EA to be tuned is often encoded as a vector and each individual/algorithm instance corresponds to a unique point in the parameter/search space. The major reason that these search-based methods can be expected to outperform random or exhaustive search is largely due to the assumption of the existence of some regular structure in the performance landscape.

In other words, there should be some kind of "path" in the landscape that could be exploited by the search operators to effectively drive the overall tuning process.

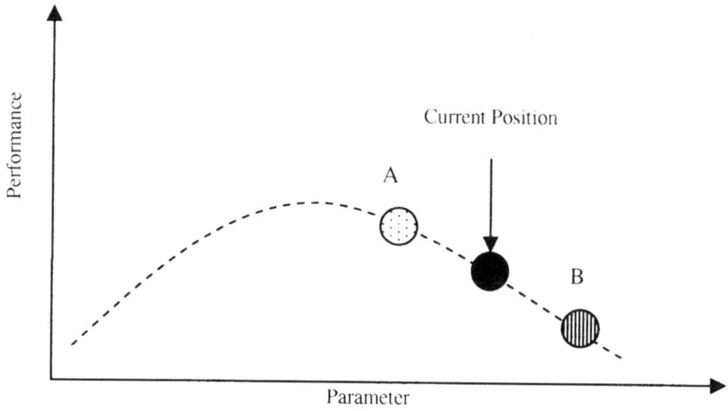

Fig. 1. An example of the tuning process conducted by a hill-climbing method.

Figure 1 shows the tuning process conducted by a hill-climbing method working on a single parameter. After evaluating the two neighbors A and B of the current individual, it turns out that A is better than B as well as the current individual. As a result, in the next step, the hill-climbing method will move its current position to A (i.e., towards the optimum).

Although this example is very simple, it does reveal a key factor of all non-trivial search-based methods: a meaningful distance metric among individuals, which could create a performance landscape compatible with the search operators. In general, the distance metric in the search space (i.e., the spatial relationship among parameter settings) is determined by the way that parameters are encoded. For example, if the continuous crossover rate and mutation rate parameters are directly encoded using their original data type, it would be reasonable to say that two algorithm instances similar to each other in terms of these two parameters are also close to each other in the search space defined by the Euclidean distance. However, if they are encoded into binary strings, these two algorithm instances may be quite distant from each other in terms of the Hamming distance.

The major issue here is that EAs with similar parameter settings should be represented by individuals that are also close to each other in the search space. Note that the actual distance between two individuals in terms of searching is also influenced by the specific search operator in use [10]. If the distance metric does not match the similarity among algorithm instances, the fitness values/performance of individuals close to each other in the corresponding search space may not be closely correlated.

From the landscape point of view, this means that the performance landscape may present a significant level of noisy structure, which could reduce the effectiveness of almost all search-based methods. In fact, finding the appropriate encoding scheme is also a well-known common issue in the field of Evolutionary Computation.

Unfortunately, unlike those numerical parameters such as population size and mutation rate, some parameters are nominal and only have symbolic meanings, which cannot be encoded in a natural way. For instance, an EA could adopt one of many different selection strategies but it is hard to design a meaningful encoding scheme for this parameter. The reason is that there is usually no clear knowledge about the relationship among these selection strategies and it is difficult to quantify the distance between each pair of them (e.g., the distance between the Tournament selection and the Truncation selection). In fact, it is even difficult to say which one is more similar to which one in some general sense. Certainly, it is always possible to apply any arbitrary encoding scheme on this parameter but the shape of the performance landscape could also be arbitrary, which can hardly be guaranteed to be solved efficiently by a given search-based method.

Suppose that there are 10 candidate selection strategies and the performance of the EA with each strategy is shown in Figure 2 (top), labeled according to a certain encoding scheme compatible with their inherent similarity. Note that this is an idealized situation and if an arbitrary encoding scheme is applied, selection strategies that produce similar performance could be mapped into quite different individuals (i.e., large distances along the horizontal axis). As a result, the original smooth landscape may turn into something pretty nasty as shown in Figure 2 (bottom).

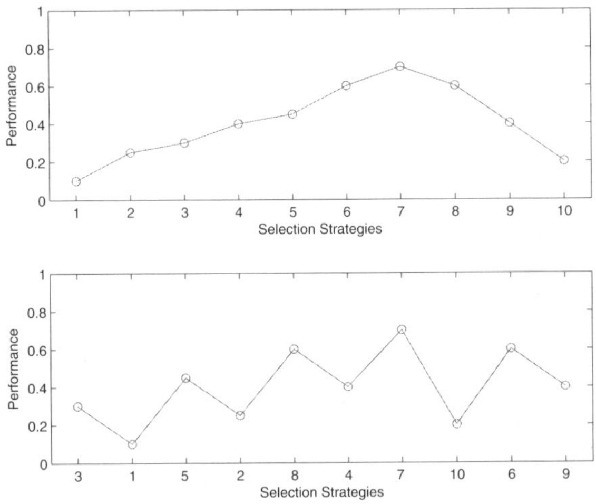

Fig. 2. The influence of the encoding scheme on the performance landscape: A smooth landscape defined in consistence with the similarity of selection strategies (top) and a nasty landscape defined by reordering the sequence (bottom).

More specifically, in Figure 2 (top), selection strategies close to each other usually result in similar algorithm performance and the global optimum could be easily reached by the hill-climbing method starting from any position. By contrast, in Figure 2 (bottom), although the algorithm's performance with each selection strategy is kept unchanged, the spatial relationship has been altered. As a result, the previous smoothness is broken and the new landscape contains multiple local optima.

There are further challenges associated with applying search-based methods to parameter tuning as shown in previous research. Firstly, it is possible that some parameters only play a role in certain situations. As a result, the encoding scheme may have to contain some redundancies and the tuning algorithms also need to be specifically customized to handle these parameters. Secondly, in order to evaluate each individual, the corresponding EA must be run for several trials in order to get reliable performance. However, in practice, it would often require an enormous amount of computational time and may become impractical in many cases.

3 Racing and Model Selection

Racing was originally proposed to efficiently solve the model selection problem in Machine Learning [12, 13]. Typically, the performance of a group of M candidate supervised learning models is evaluated on a data set of interest containing N test points using a validation approach. The general mechanism of the leave-one-out cross validation (LOOCV) is to use all but a single test point to train the model. After training, the model is tested on the test point absent from the training. This process is repeated N times with each time a unique test point left out and the model with the lowest average error is regarded as the best model. In other words, the performance of each model is determined by the mean of its performance distribution, which consists of the outputs from a series of experiments.

The most straightforward approach would be to exhaustively test each model on all available test points. With this brute force method, after finishing N experiments, the true performance E_{true} of a certain model could be found and after totally $M \times N$ experiments, it will guarantee to find the best model according to the LOOCV measure. However, the complexity of this method is $O(M \cdot N)$, which may become quite time consuming for large M and/or N values. After all, each experiment may involve quite heavy computation as the model needs to be trained on $N - 1$ test points.

By contrast, Racing techniques approach this model selection problem from a different angle: whether it is necessary to test a model for N times? Certainly, in principle, each model must undertake all tests to reveal its E_{true}. However, the ultimate goal here is *not* to examine E_{true} but to find the best model. From a statistical point of view, the errors of a model on all N test points create a population and it is possible to estimate the population mean E_{true} based on E_{sam}, which is the mean of a random sample of size N' (i.e., $N' < N$). Usually, it is more helpful to estimate the true difference between two models based on the difference of their sample means.

The basic idea behind Racing (Table 1) is to test all available models in parallel on a single unseen test point at each step. The errors of each model on all test points that have been selected are maintained and used in some statistical test. If there is enough evidence, as indicated by the result of the statistical test, that model i is significantly worse than another model, it will be removed from the model selection process and undertake no further testing because it is unlikely to be the best model. By doing so, models are forced to race against each other and only promising ones could survive through to the next iteration. As a result, the overall computational cost could be reduced compared to the brute force approach described above by avoiding running unnecessary experiments on inferior models. Figure 3 gives an illustration of a virtual Racing experiment with 50 starting models

Table 1. The framework of Racing for model selection problems.

Repeat steps 1-4 until there is no more unseen data point or only one model is remaining:

1. Randomly select a new data point
2. Calculate the LOOCV error of each remaining model on it
3. Apply some statistical test to conduct pair-wise comparison
4. Remove models that are significantly worse than others
5. Return the remaining model with the lowest average error

and totally 200 test points where the length of each bar indicates when a model was removed from the Racing process. It is clear that after being tested on 20 test points quite a few models had already been removed and only 12 models could pass through 60 iterations. There were only two models (i.e., No. 34 and No. 47) left after 102 iterations and model No. 34 won the competition against No. 47 after being tested on additional 24 test points and thus the Racing process was terminated. Note that, in practice, it is also possible that some models are indistinguishable even at the end of Racing.

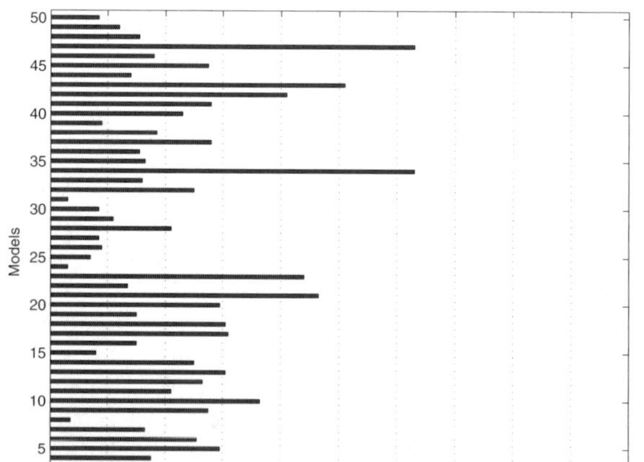

Fig. 3. An illustration of a virtual Racing experiment with 50 candidate models and 200 test points in the dataset. It shows the number of iterations that each model has gone through before being removed from the Racing process.

The advantage of Racing is obvious from the above example: there is no need to test all models thoroughly on 200 test points and most of the models were actually tested on far less number of test points before being discarded. The cost of Racing is indicated by the area of the bars (i.e., 2,264 tests), compared to the cost of the brute force method, which is equal to the entire box area (i.e., 50 × 200=10,000 tests). The heart of the Racing technique, as shown in Table 1, is the statistical test used to judge whether one model is significantly worse than another, which directly determines the efficiency and reliability of Racing. Depending on the properties of the performance distributions, there are two major tests that have been used in previous work: the non-parametric Hoeffding's bounds [9] and the parametric Student-t test [14]. For a single model, suppose its errors are bounded within $[b, a]$. Then, for $a - b > \xi > 0$, the probability of E_{sam} being more than ξ away from E_{true} after n iterations is:

$$P(|E_{\text{sam}} - E_{\text{true}}| \geq \xi) \leq 2e^{\frac{-2n\xi^2}{(a-b)^2}} \tag{1}$$

Based on Eq. 1, it is easy to derive the value of ξ for a given significance level α:

$$\xi = \sqrt{\frac{(a-b)^2 \log(2/\alpha)}{2n}} \tag{2}$$

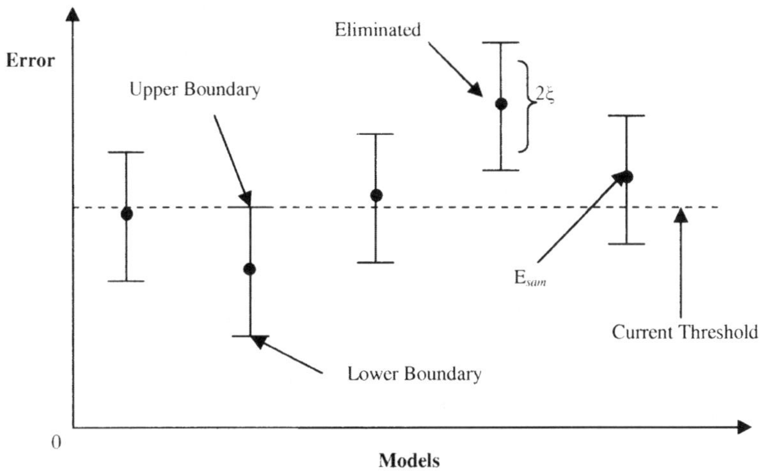

Fig. 4. An example of Hoeffding Racing at a specific iteration with 5 remaining models left. The current threshold is indicated by the dashed line.

According to Eq. 2, at each step, the upper and lower boundary of E_{true} is estimated for each remaining model: $[E_{\text{sam}} - \xi, E_{\text{sam}} + \xi]$. The eliminating rule is to remove models with lower boundaries (i.e., best possible errors) still higher than the upper boundary (i.e., worst possible error) of the best model [12]. Figure 4 shows the competition among five models at a certain stage of Racing where the dashed line is the threshold above which all models are to be eliminated (e.g., the second rightmost model).

In addition to Eq. 2, it would also be possible to utilize another Hoeffding's bound for the difference of two sample means with regard to the difference of the true means [9]:

$$P(|E^1_{\text{sam}} - E^2_{\text{sam}} - (E^1_{\text{true}} - E^2_{\text{true}})| \geq \xi) \leq 2e^{\frac{-n\xi^2}{(a-b)^2}} \tag{3}$$

The major advantage of Hoeffding's bound is that it has no assumption on the underlying probability distribution and can be applied in a variety of situations. However, the bounds acquired are typically not very tight and consequently quite a lot of test points may be needed in order to distinguish models [18].

If it is reasonable to assume that the errors are approximately normally distributed, the parametric Student-t test is usually more powerful. Under the null hypothesis that there is no difference between the true means and the assumption of equal variances, the test statistic is given by:

$$z = \frac{E^1_{\text{sam}} - E^2_{\text{sam}}}{s\sqrt{2/n}} \tag{4}$$

In Eq. 4, s is the pooled estimator of the standard deviation σ and the test statistic z follows the Student-t distribution with $2n - 2$ degrees of freedom. The null hypothesis is to be rejected if the value of z falls into the critical region based on the desired significance level.

If there is reason to doubt the assumption of equal variances, the test statistic is then given by (i.e., s_1 and s_2 are the estimators of σ_1 and σ_2):

$$z = \frac{E^1_{\text{sam}} - E^2_{\text{sam}}}{\sqrt{(s_1^2 + s_2^2)/n}} \tag{5}$$

The corresponding degree of freedom is approximated by:

$$\text{d.f.} \approx \frac{(s_1^2 + s_2^2)^2}{\frac{s_1^4 + s_2^4}{n-1}} \tag{6}$$

Figure 5 shows an example with five remaining models at a certain stage of Racing where the performance distributions are approximately normal but have different variances. In this case, based on the Student-t test, the model in the middle is likely to be eliminated.

Note that in the Hoeffding Racing and the Student-t Racing with equal variances, a model only needs to be tested against the currently best model because if it is not significantly worse than the best model, it cannot be eliminated by any other models. However, in the Student-t Racing with unequal variances, a model that cannot be eliminated by the best model with a large variance might be eliminated by a good model with a small variance. As a result, each model needs to be compared against all models that have better sample errors.

Furthermore, in Table 1, all remaining models are tested on the same data point in each step. However, both statistical tests above assume that samples are independently drawn from the populations and take no advantage of this paired experiment. Nevertheless, for statistical tests with blocking design, this feature could

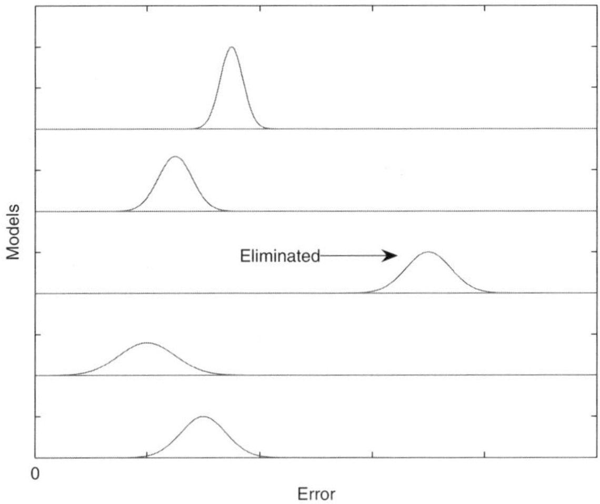

Fig. 5. An example of Racing based on the Student-t test (5 remaining models).

be explicitly exploited to reduce the variance among test points and increase the power of statistical tests.

In summary, the major advantage of Racing is that it only works on the statistics of the experimental results instead of the structure of models and could be applied in a much wider situation than many other methods.

4 Racing Evolutionary Algorithms

4.1 Overview

There is a clear similarity between the model selection problem in Machine Learning and the task of parameter tuning in EAs. In each case, there is a meta-optimization problem: to find values for all of the adjustable parameters of the model or algorithm in order to produce the best results when applied to the problem of interest. More importantly, the performance of a model or an algorithm is often defined as the mean value of the outputs from a series of random experiments or trials, which provides the basis for applying statistical tests. In fact, the set of EAs to be raced do not necessarily need to be different instances of the same EA. Hence, Racing can be used in quite a general sense in an attempt to reduce the experimental effort required whenever a comparison is needed over a range of experimental configurations provided that the evaluation of each configuration involves iterating over a number of independent trials (restarts).

In addition to some early attempts in applying Racing in the field of EAs [3, 19], there are a few important differences between model selection and evaluation of EAs that must be made clear.

Firstly, the number of test points in model selection problems is fixed, which means that it is always possible to find the true performance of each model (i.e., assume that its behavior is not stochastic). However, for EAs, the number of trials that could be conducted for each pair of algorithm and problem is unlimited, which means that it is impossible to know the true performance of EAs. Since, in practice, 50 to 100 trials are usually believed to be sufficient to produce trustable results, it is reasonable to put a similar upper boundary on the maximum number of trials in Racing experiments. Otherwise, it may need an extremely long period to terminate.

For example, in scientific studies, it is preferable to compare the best model or algorithm found by the brute force method and that found by Racing to show its reliability (i.e., whether the best model/algorithm was eliminated incorrectly at some stage). Although it is very straightforward in model selection as the performance is deterministic, in EAs, the only claim that can be made is to say whether the top EAs found by the brute force method (i.e., certainly it is based on a limited number of trials and the results are still stochastic) are also available at the end of Racing.

Secondly, in model selection, all models could be tested on the same test point at each time and the blocking techniques may improve the power of statistical tests. However, in the experiments of EAs, the output from the ith trial of one EA typically has nothing to do with that of another EA, which means that there is no explicit correspondence/relationship between the results. Consequently, it is not beneficial to use any paired statistical tests in the comparison of EAs with regard to random trials.

Thirdly, in some model selection problems, the error range is known in advance. For example, in classification problems, the model's error at each test point is either 0 or 1, representing correct or incorrect classification. As a contrast, it is usually not easy to give a tight estimation of the range of the performance of EAs on test problems, which may make it difficult to apply the Hoeffding's method.

Finally, the number of test points in a typical dataset could easily reach a few hundreds or even thousands while the number of trials to be conducted in an EA experiment is usually no more than 100. A major implication is that, in model selection, it is acceptable to start Racing after a number of, say 30, test points have been selected and this would not have a significant impact on the efficiency of Racing methods. The advantage is that, although it is usually not known in advance whether the errors of those models are normally distributed, it is still reasonable to apply those statistical tests based on the assumption of normality such as the Student-t test as long as the sample size is relatively large. However, for EAs, this would make Racing methods require at least 60% cost of the brute force method if the maximum number of trials is 50.

4.2 Statistical Tests

Based on the above analysis, a good statistical test for racing EAs should be able to work with small sample sizes and not rely on the assumption of normality. Traditionally, the nonparametric Wilcoxon rank-sum test is used as the alternative of the Student-t test. However it still assumes that the two underlying populations follow distributions with similar shapes, which is not always reasonable.

To overcome the difficulty faced by many existing statistical tests, a hypothesis test based on the bootstrap method is explained next. The bootstrap method is a computationally intensive statistical technique based on the resampling of data [6]. It could be used to calculate many important statistics such as standard errors, bias and confidence intervals without the need of using complicated mathematical models. When used in hypothesis testing, it could avoid the danger of making various assumptions such as normality and equal variances required by other statistical tests.

The basic procedure of the bootstrap method is described in Figure 6. Given a sample X containing n data points, B bootstrap samples of the same size are generated by sampling from X with replacement. This means that X is regarded as the whole population and each X^* is a sample from it (i.e., each data point in X has equal probability to be selected). For each bootstrap sample $X^*(i)$, the bootstrap replication of the statistic of interest θ denoted by $\theta^*(i)$ is calculated from a certain function S such as the mean or the median of the sample. Finally, all such replications θ^* are combined together to, for example, calculate the standard error of θ or create a bootstrap table like the Student-t table for estimating confidence intervals.

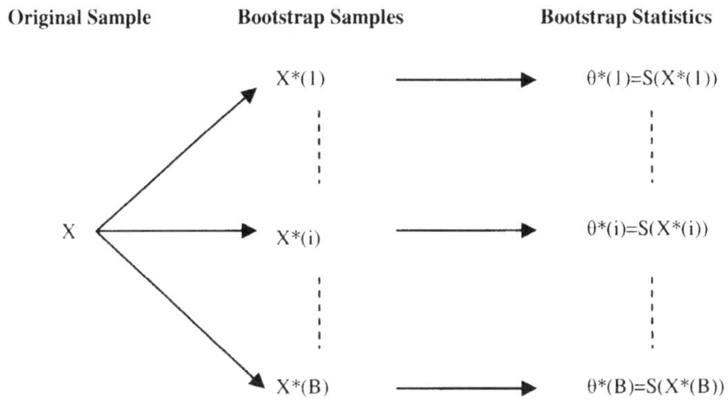

Fig. 6. The basic procedure of the bootstrap method.

In general, better estimation could be achieved with larger B values such as $B = 2000$. Obviously, this requires much more computational resource compared to classical statistical tests. As a result, in the following experiments, the bootstrap test is applied on samples with less than 30 data points combined with the Student-t test in other situations to try to find a balance between reliability and computational cost.

4.3 Racing Experiments

In this section, Racing is applied to the situation where researchers are interested in finding out which algorithm out of a set of candidates performs best on a particular benchmark problem. For this purpose, we chose the well-known Rastrigin function, which is a multimodal problem (i.e., to be minimized) with the global optimum at the origin:

$$f(X) = 10n + \sum_{i=1}^{n} (X_i^2 - 10\cos(2\pi X_i)), X_i \in [-5, 10] \tag{7}$$

The algorithm to be tuned is a continuous EDA (Estimation of Distribution Algorithm) based on Gaussian distributions, which is similar to RECEDA [15] but with some additional features.

Table 2. The framework of the continuous EDA based on Gaussian distributions.

	Initialize and evaluate the starting population P
	`while` stopping criteria not met
1.	Select top n individuals P_{sel} (Truncation selection)
2.	Fit a multivariate Gaussian $G(\mu, \Sigma)$ to P_{sel}
	$\mu = \frac{1}{n}\sum_{i=1}^{n} P_{\text{sel}}$
	$\Sigma = \frac{1}{n}(P_{\text{sel}} - \mu)^T \cdot (P_{\text{sel}} - \mu)$
3.	Update μ towards the best individual X^{best}
	$\mu' = (1 - \alpha) \cdot \mu + \alpha \cdot X^{\text{best}}$
4.	Update Σ by an amplification scalar γ
	$\Sigma' = \gamma^2 \cdot \Sigma$
5.	Sample a set of new individuals P' from $G(\mu', \Sigma')$
6.	Evaluate individuals in P'
7.	Choose the best individuals from the union of P and P' to form the new population P
	`end while`

A set of 5 × 4 × 3=60 candidate algorithms were created by systematically varying the following three parameters described in Table 2 (i.e., each of these algorithm instances corresponds to a unique parameter setting):

- Selection ratio (step 1): [0.1, 0.3, 0.5, 0.7, 0.9]
- Learning rate (step 3): [0.0, 0.2, 0.5, 1.0]
- Amplification scalar (step 4): [1.0, 1.5, 2.0]

Other experimental configurations were fixed as: dimensionality = 5, population size = 100, number of generations = 100 and maximum number of trials = 100. The fitness value of the best individual found in each trial was recorded as the performance criterion.

In order to provide a benchmark for the performance of Racing, an exhaustive experiment was conducted by running all 60 algorithms for 100 trials (i.e., 6,000 trials in total) and the overall performance (i.e., sorted based on the mean values) is summarized in Figure 7 with error bars (i.e., one standard deviation above and below the mean). It is possible to inspect these results visually and determine, for example, which algorithms are most likely to find good solutions. Note that there is clear evidence that the variances tend to be different, which violates the widely adopted assumption of equal variances.

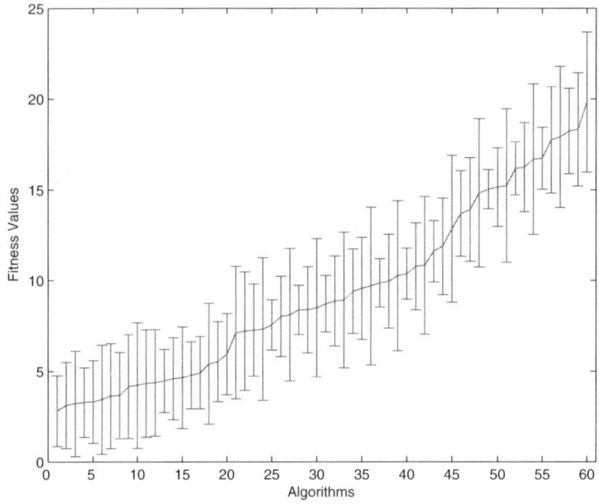

Fig. 7. The performance distributions of 60 candidates on Rastrigin's function.

Using Racing techniques, we aim to make a similar observation at a fraction of the computational effort required for the above exhaustive experiment. The procedure operates here by eliminating poorly performing algorithms on the basis of a small number of restarts. The significance level α was 0.05 and the Racing algorithm was started after 5 trials, taking into account the limited maximum number of trials (100).

Note that since the brute force search has already been conducted, in the Racing experiment, the performance of an algorithm instance in each trial was retrieved as a sample from the corresponding performance population without actually re-running the EA (i.e., certainly this is only applicable for comparison purpose). Since the sequence of such samples may have more or less impact on the performance of Racing, the Racing experiment was repeated by 10 times with random sequences of samples.

The average number of algorithm instances remaining at each step is shown in Figure 8 from which it is clear that Racing eliminated a large portion of algorithms in the early stage and usually there were only around 5 algorithm instances remaining after 20 restarts. The average number of candidate algorithms left after 100 restarts was 1.6 and sometimes Racing was terminated even without the need of fully testing any candidate for 100 trials. The efficiency of Racing can be visually examined by comparing the area under the curve and the area of the whole rectangular region.

From a quantitative point of view, the average cost of Racing is 10.89% of the cost of the brute force method in terms of the number of trials that have actually been conducted. In other words, Racing reduced the cost by almost 90%, which could be very significant for expensive experiments. Certainly, the actual efficiency of Racing depends on the properties of the specific performance distributions encountered.

Another important performance metric of Racing is reliability: whether the top algorithms could be preserved and avoid being removed incorrectly. In the above

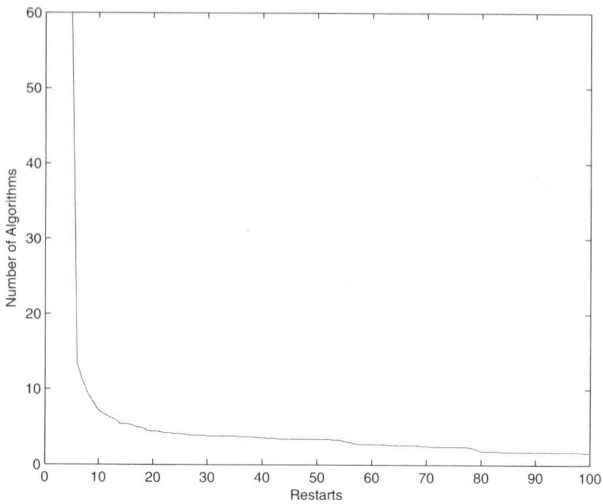

Fig. 8. The average number of remaining algorithm instances during Racing. The cost of Racing is indicated by the area below the curve.

Racing experiments, the best algorithm based on the exhaustive experiment was remaining till the end of Racing in 9 out of 10 times, showing pretty impressive robustness. After all, for the single Racing experiment in which the best problem was eliminated, three other top algorithms were still able to survive through to the end of Racing. If greater reliability is desired, a higher significance level could be used to make Racing more conservative in eliminating models. Furthermore, Racing could be allowed to start after more samples have been collected in order to improve the accuracy of estimation.

5 Hybrid Techniques

5.1 Racing vs. Meta-EA

As shown in the last section, Racing does not work on the parameters of EAs in the way that a search-based method might. Instead, Racing is based on the statistical comparison of different algorithm instances, which are regarded as atomic black boxes (i.e., only the performance evaluation of the instances is required). By doing so, the requirement of defining an appropriate encoding scheme or distance metric is removed. However, a closer look at Racing reveals that there is an inherent lack of any mechanism of exploration. In fact, Racing always starts with a fixed set of candidates and no other candidates can be reached. If Racing is to be applied to the task of parameter tuning, the optimal parameter setting must be included in the initial set. Otherwise, there is no hope to find it by Racing.

Unfortunately, there is usually no such guarantee in practice. One solution would be to start with an exhaustive set containing all possible combinations of parameter values. However, for continuous parameters, the candidate set might be very large assuming certain discretization of these parameters, which makes this approach impractical. Also, the accuracy of searching may also suffer accordingly.

In summary, the major challenge faced by all search-based methods is the difficulty in defining a distance metric over those nominal/symbolic parameters. Without a sensible distance metric, the resulting landscape may present high level irregularity and no search algorithm is expected to do well on it. For this reason, these parameters are generally referred as non-searchable parameters. Another practical issue is that in search-based methods each individual is to be evaluated independently of others and need to be tested thoroughly on the benchmark problem, just like the brute force method. If the size of the search space is relatively small, the advantage of search-based methods may not be significant as they still need to evaluate a large portion of all possible candidates (i.e., this is particularly true for those population-based methods such as Meta-EAs).

By contrast, the difficulty in Racing is mainly due to the fixed initial set of algorithm instances. If the number of all possible candidates is not too large, say a few hundreds, Racing is expected to work reasonably well. If the number is far beyond this figure, it would be difficult to apply Racing directly on such a huge set of candidates (e.g., even testing one million candidates for a single trial would require an enormous amount of time). In this situation, it is only practical to require Racing to work on a small number of candidates at each time.

5.2 Hybridization

It is easy to see that the properties of Racing and search-based methods are complimentary to each other. In order to efficiently handle situations where the search space is potentially very large and/or contains symbolic parameters, it is useful to consider a hybrid tuning framework combining Racing and search-based methods. In this section, two examples are given to demonstrate how this hybrid approach works.

In the first scheme, a relatively simple situation is considered where a $(1+\lambda)$ ES, which is a special case of EAs, is combined with Racing and it is assumed that all parameters to be tuned are numerical and thus searchable. At each step, a set of λ new individuals/algorithm instances are sampled from a Gaussian distribution centered at the current solution.

The major computational cost in this method is on finding the best individual from a set of candidates, which is exactly what Racing is designed for. In the proposed hybrid approach, instead of sequentially evaluating each individual, all individuals are brought into Racing to find out the best one and as shown in previous experiments, it is expected that many of them may be eliminated after just a few trials, saving a large portion of computational resource (Figure 9).

The second scheme is proposed to handle the more general and complicated situation where parameters to be tuned include symbolic ones, which are generally not searchable. Also, it is assumed that a more general and population-based Meta-EA is in use.

The general idea is to treat these two classes of parameters separately: only searchable parameters are encoded into individuals in the population of the Meta-EA and each individual no longer corresponds to a fully specified algorithm instance. Instead, it represents a set of complete EAs sharing the same searchable parameters and different from each other on those non-searchable parameters. Note that

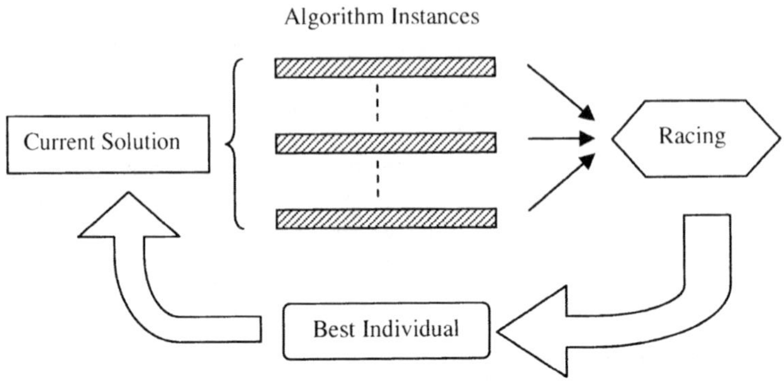

Fig. 9. The framework of the hybrid (1+λ) ES + Racing scheme.

the size of the set depends on the number of non-searchable parameters and their cardinalities. In order to evaluate each individual, Racing is applied to find the best algorithm instance from the corresponding set of EAs and its performance is returned as the fitness of the individual (Figure 10). An alternative approach would be to apply the brute force method to evaluate those algorithm instances, which may be very time-consuming.

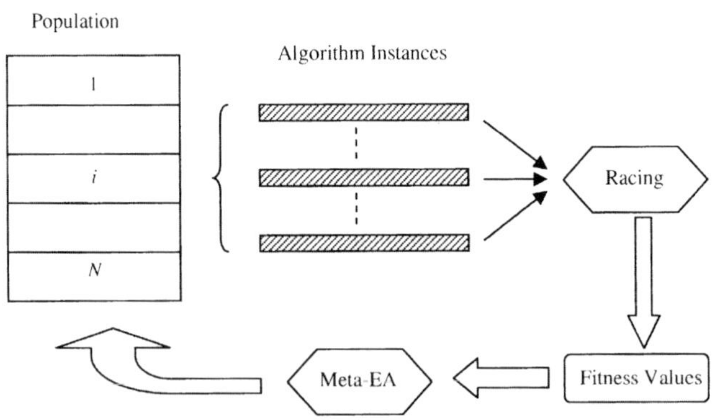

Fig. 10. The framework of the hybrid Meta-EA + Racing scheme.

5.3 Case Studies

In this section, a standard GA with binary representation is used in the experiments to illustrate the hybrid approach. Two parents from the current population are selected at each time and two offspring are generated through recombination with probability P_c (the crossover rate). Otherwise, these two parents are kept unchanged and copied to the mating pool. When the mating pool is full, mutation is applied, which changes the value of each bit by flipping it from 0 to 1 and vice versa with probability P_m. Finally, new individuals are evaluated and replace the old population. If elitism is applied, the best individual found so far in previous generations will be copied to the new population, replacing a randomly chosen individual.

There are a number of parameters to be specified. The population size (P) is a discrete parameter of high cardinality (i.e., it should be a positive even number because individuals are generated in pairs). It is well-known that a very small population may cause premature convergence while a very large one may result in a slow convergence rate. The crossover rate P_c and mutation rate P_m are both continuous parameters within [0, 1], which are important for controlling the balance between exploration and exploitation. The selection strategy (S) is a symbolic parameter, which contains various candidates (e.g., Tournament Selection) combined with their corresponding parameters (e.g., Tournament size). Also, a symbolic parameter (C) is required to specify the type of crossover (e.g., one-point or two-point). At last, a binary parameter (E) is used to indicate whether elitism is used or not. The feasible values of S, C & E are given in Table 3.

Table 3. Feasible values of S, C & E.

Parameters	Values
S	"Truncation" 0.1, 0.2, 0.3, 0.4, 0.5 "Tournament" 2, 4, 6 "Linear Ranking" 1.5, 1.8, 2.0
C	"One-Point", "Two-Point", "Uniform"
E	"0" (without elitism), "1" (with elitism)

According to the above discussion, the GA could be fully specified by a six-element tuple: $< P, P_c, P_m, S, C, E >$. Note that some parameters are continuous (i.e., the population size could be regarded as a continuous parameter during optimization due to its high cardinality and then rounded up to the nearest feasible value) while others are symbolic, which created a mixed-value optimization problem.

In the first case study, a simple situation was considered in which only P, P_c and P_m were to be tuned while the rest of the parameters were set to some fixed values. In this experiment, although all parameters were searchable, evaluating each individual using the brute force method could still be very time-consuming. As a result, the hybrid (1+λ) ES+Racing approach was applied to help reduce this computational burden.

The (1+λ) ES in use was based on a fixed Gaussian distribution with a diagonal covariance matrix. The boundaries of the search space and the standard deviations are given in Table 4, which were chosen based on some general knowledge without any specific tuning. Note that for some values of the population size, it is possible that, given the fixed number of fitness evaluations (i.e., 500), the number of generations may not be an integer. The solution adopted was to increase the population size in the final generation to accommodate those extra individuals to maintain the same number of fitness evaluations among all candidates.

Table 4. Definition of search space and standard deviations.

	Range	Standard Deviation
P	[20, 100]	5
P_c	[0, 1]	0.1
P_m	[0, 0.2]	0.02

The 100-bit One-Max problem was used as the benchmark problem. The (1+λ) ES with $\lambda = 20$ started from the middle of the search space while other GA parameters were arbitrarily set to (i.e., not necessarily a good choice) $\langle$"Truncation", 0.5, "One-Point", 1$\rangle$. Note that, in addition to identifying the best candidate, its performance/fitness value is usually needed in search-based methods. As a result, the starting individual was evaluated exhaustively. The best candidate found by Racing (i.e., significance level 0.10) was also required to go through all trials (100) to find out its mostly reliable performance for the comparison against the currently best one even if it was the only one left.

The parameter settings found during evolution in a typical trial (i.e., 20 generations) and the corresponding performance of the GA are shown in Figure 11. It is clear that this approach could find much better parameter settings than the initial one and the average performance of the GA increased from 71.51 to 83.97 within 15 generations.

In the meantime, all three parameters under tuning showed a clear pattern. The population size dropped from 60 to 20 while the crossover rate increased from 0.5 to 1.0. The mutation rate also quickly settled down to a very small value around 0.02. Since there is no dependence among the variables of the One-Max problem and it has a simple structure with only one optimum, it seems unnecessary to employ a large population to maintain the genetic diversity and a small mutation rate is well justified in this case. As for the crossover rate, it plays a major role in GAs in combining building blocks and a quite large value is often recommended.

Obviously, the hybrid approach did a good job in this case study and found some very good parameter settings for the GA. After all, it only took around 13.4% of the cost of an ordinary (1+λ) ES (i.e., individuals are exhaustively evaluated) in terms of the number of trials conducted. Note that the lower limit of the cost of Racing in this experiment is around 9.75% as all 20 candidates needed to be tested for at least 5 trials and the best one needed to be fully tested for 100 trials.

Next, we will show how the GA's performance can be further improved through the simultaneous tuning of all six parameters.

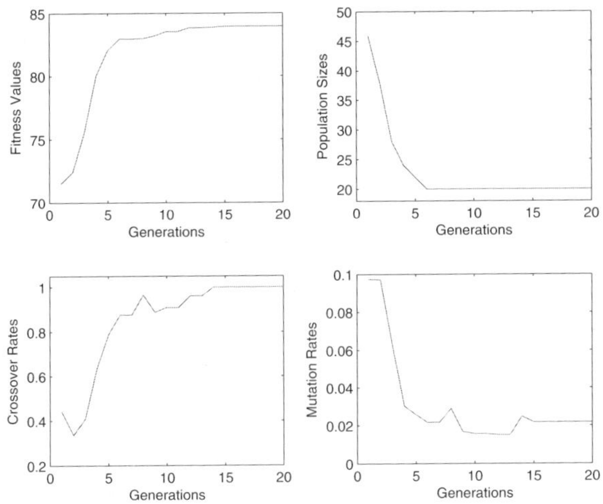

Fig. 11. The evolution process of the hybrid (1+λ) ES + Racing scheme.

In the second case study, a $(\mu+\lambda)$ ES with $\mu=\lambda=20$ was employed as the Meta-EA (i.e., the initial population was randomly generated) in which each individual consisted of three parameters P, P_c and P_m and represented a total of 66 algorithm instances (i.e., 11 selection strategies × 3 crossover operators × 2 options of elitism). Note that although E is a binary parameter, it was still included in the Racing so that the Meta-EA only needed to handle a continuous optimization problem. As a result, the role of Racing here was to, for each individual, find out the best algorithm instance out of a set of 66 candidates as efficiently as possible and return its performance as the fitness value of that individual. Also, the best algorithm was to be fully tested even if it was the only one left.

Figure 12 presents the evolution process of this hybrid scheme in which all results were averaged over the population. It is clear that much better parameter settings were found while the superiority of small population sizes, large crossover rates and small mutation rates was again evident. Furthermore, the uniform crossover dominated the final population while a strong selection pressure was created by either the Tournament selection with tournament size 6 or the Truncation selection with selection ratio 0.2. As to Elitism, it was not shown to be beneficial in this case. The cost of Racing was only around 7.4% of the cost of the brute force method, had it been applied to find the best algorithm instances.

6 Summary

In this Chapter, the major issue addressed is how to efficiently find the best parameter settings for EAs through experimental studies, assuming no specific knowledge on those parameters. Two classes of tuning techniques: search-based methods and

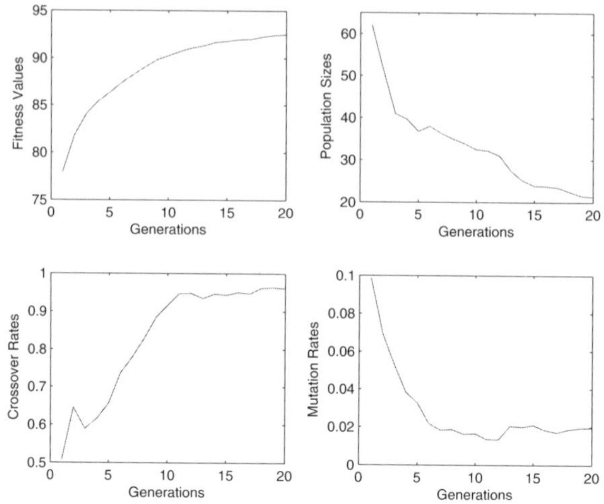

Fig. 12. The evolution process of the hybrid Meta-EA + Racing scheme.

statistical Racing techniques, which are based on distinctly different mechanism, are given a critical review on their strengths and weaknesses.

We point out that the applicability of Racing is inherently restricted by its lack of exploration capability while it could be very difficult for typical search-based methods such as Meta-EAs to efficiently handle nominal parameters that only have symbolic meanings. In order to take the advantages of both methods, two hybrid schemes are proposed in which search-based methods are responsible for exploring the parameter space while Racing is applied, in the place of the brute force method, to significantly reduce the cost associated with finding the best performing algorithm instance from a moderate size of candidates.

In the case studies, Racing has shown to be able to reduce the cost of the brute force method by around 90% while maintaining quite high reliability. The major direction for future work would be to establish some theoretical framework for Racing in order to conduct principled analysis of the influence of various experimental factors. It is also an important topic to investigate the possibility of combining more features into Racing from other tuning techniques.

Finally, some rules of thumb are given as follows to assist in choosing appropriate parameter tuning techniques:

1. If the size of the parameter space is small (e.g., ten candidates or even less), it is usually straightforward and reliable to conduct an exhaustive search to find the best parameter setting.
2. If the size of the parameter space is moderate (e.g., up to a few hundreds of candidates), Racing could be applied to significantly reduce the computational time required, regardless of whether the parameter space is searchable or not.
3. If the sizes of the searchable parameter space and non-searchable parameter space are both large, it is worthwhile to consider the proposed scheme of Meta-EA + Racing.

4. If the size of the searchable parameter space is large while there is no nominal parameter or the size of the non-searchable parameter space is small, it is possible to use Meta-EAs only (i.e., apply random/exhaustive search on nominal parameters if necessary).
5. An exception of rule No. 4: if the Meta-EA in use is similar to the $(1 + \lambda)$ or $(1, \lambda)$ ES, which requires choosing the best individual from a set of candidates, it is still possible to consider the scheme of Meta-EA + Racing.

References

1. T. Bäck. *Evolutionary Algorithms in Theory and Practice.* Oxford University Press, New York, 1996.
2. T. Bartz-Beielstein. Experimental analysis of evolution strategies - overview and comprehensive introduction. Technical Report Reihe CI 157/03, SFB 531, Universität Dortmund, 2003.
3. M. Birattari, T. Stutzle, L. Paquete, and K. Varrentrapp. A racing algorithm for configuring metaheuristics. In *Genetic and Evolutionary Computation Conference (GECCO 2002)*, pages 11–18, 2002.
4. P. A. Castillo-Valdivieso, J. J. Merelo, and A. Prieto. Statistical analysis of the parameters of a neuro-genetic algorithm. *IEEE Transactions on Neural Networks*, 13(6):1374–1393, 2002.
5. A. Czarn, C. MacNish, K. Vijayan, B. Turlach, and R. Gupta. Statistical exploratory analysis of genetic algorithms. *IEEE Transactions on Evolutionary Computation*, 8(4):405–421, 2004.
6. B. Efron and R. J. Tibshirani. *An Introduction to the Bootstrap.* Chapman and Hall, 1993.
7. A. E. Eiben, R. Hinterding, and Z. Michalewicz. Parameter control in evolutionary algorithms. *IEEE Transactions on Evolutionary Computation*, 3(2):124–141, 1999.
8. J. J. Grefenstette. Optimization of control parameters for genetic algorithms. *IEEE Transactions on System, Man, And Cybernetics*, 16(1):122–128, 1986.
9. W. Hoeffding. Probability inequalities for sums of bounded random variables. *Journal of the American Statistical Association*, 58(301):13–30, 1963.
10. T. Jones. *Evolutionary Algorithms, Fitness Landscapes and Search.* PhD thesis, University of New Mexico, 1995.
11. K. De Jong. *The Analysis of the Behavior of a Class of Genetic Adaptive Systems.* PhD thesis, University of Michigan, 1975.
12. O. Maron and A. W. Moore. Hoeffding races: accelerating model selection search for classification and function approximation. In *Advances in Neural Information Processing Systems 6*, pages 59–66, 1994.
13. O. Maron and A. W. Moore. The racing algorithm: model selection for lazy learners. *Artificial Intelligence Review*, 11(1):193–225, 1997.
14. D. S. Moore. *Introduction to the practice of statistics.* W. H. Freeman, New York, 4th edition, 2003.
15. T. K. Paul and H. Iba. Real-coded estimation of distribution algorithm. In *Proceedings of the 5th Metaheuristics International Conference (MIC2003)*, pages 61–66, 2003.

16. I. Rojas, J. Gonzalez, H. Pomares, J. J. Merelo, P.A. Castillo, and G. Romero. Statistical analysis of the main parameters involved in the design of a genetic algorithm. *IEEE Transactions on Systems, Man, And Cybernetics-Part C*, 32(1):31–37, 2002.
17. D. H. Wolpert and W. G. Macready. No free lunch theorems for optimization. *IEEE Transactions on Evolutionary Computation*, 1(1):67–82, 1997.
18. F. Y-H. Yeh and M. Gallagher. An empirical study of heoffding racing for model selection in k-nearest neighbor classification. In J. Hogan M. Gallagher and F. Maire, editors, *Sixth International Conference on Intelligent Data Engineering and Automated Learning (IDEAL'05)*, volume 3578 of *Lecture Notes in Computer Science*, pages 220–227. Springer, 2005.
19. B. Yuan and M. Gallagher. Statistical racing techniques for improved empirical evaluation of evolutionary algorithms. In X. Yao et al., editor, *Proc. Parallel Problem Solving from Nature (PPSN VIII)*, volume 3242 of *Lecture Notes in Computer Science*, pages 172–181. Springer, 2004.

Genetic Programming: Parametric Analysis of Structure Altering Mutation Techniques

Alan Piszcz and Terence Soule

Department of Computer Science
University of Idaho, Moscow, ID, 83844-1010
apiszcz@acm.org, tsoule@cs.uidaho.edu

Summary. We hypothesize that the relationship between parameter settings, specifically parameters controlling mutation, and performance is non-linear in genetic programs. Genetic programming environments have few means for a priori determination of appropriate parameters values. The hypothesized nonlinear behavior of genetic programming creates difficulty in selecting parameter values for many problems. In this paper we study three structure altering mutation techniques using parametric analysis on a problem with scalable complexity. We find through parameter analysis that two of the three mutation types tested exhibit nonlinear behavior. Higher mutation rates cause a larger degree of nonlinear behavior as measured by fitness and computational effort. Characterization of the mutation techniques using parametric analysis confirms the nonlinear behavior. In addition, we propose an extension to the existing parameter setting taxonomy to include commonly used structure altering mutation attributes. Finally we show that the proportion of mutations applied to internal nodes, instead of leaf nodes, has a significant affect on performance.

1 Introduction

Parameter setting includes specification of values for population size, selection rate, crossover probability, mutation probability, reproduction rate, the number of generations and other environment attributes. In many problems, the parameters of genetic programming (GP) algorithms interact, making optimal value selection difficult [3]. This interaction leads to unpredictable behavior with respect to computational effort and may limit the ability to evolve a solution. If linear changes at the input parameter space translate into linear changes in performance, a transfer function or model would enable appropriate parameter selection. Luke and Spector studied crossover and mutation noting interesting nonlinearities in the results, they suggest that further analysis is needed [12]. Identification of specific nonlinearity in GP parameter settings will improve the researchers probability of choosing appropriate values. The GP community typically treats mutation as a secondary operator, when implemented it often has less emphasis than crossover and reproduction. Optimizing

A. Piszcz and T. Soule: *Genetic Programming: Parametric Analysis of Structure Altering Mutation Techniques*, Studies in Computational Intelligence (SCI) **54**, 143–160 (2007)
www.springerlink.com

population size and crossover rates appear frequently as methods to improve GP solutions, whereas mutation rates are rarely considered. Feldt and Nordin identified the importance of population size and mutation in [4]. In the following experiments, we examine the affects of a single parameter related to mutation across three structure altering mutation techniques, and several mutation selection methods. We observe that nonlinearity is the behavior for the majority of the structure altering mutation technique (SAMT) cases examined.

2 Background

Mutation is an operator for genetic programming that introduces diversity in the building blocks created during evolution. Due to a lack of research on mutation in GP, choosing mutation parameters is difficult, which is likely to lead to suboptimal parameter choices. Additionally, improved characterization of mutation parameters could further our understanding on the affects of mutation on the evolutionary process. It is typical to use $\frac{1}{size\ of\ individual}$ for mutation probability when it is possible to estimate or control the problem size.

Koza's view of mutation and crossover considers the value of mutation versus crossover. Koza introduces a framework for comparing mutation with crossover in GP [8] [9]. The framework describes the difference of the source of the new subtree introduced by mutation or crossover. Koza advises that mutation is unlikely to be beneficial when compared to a large population and crossover. Few of Koza's early problems include mutation, and he states two reasons for the omission of mutation in problems investigated in his first book [6]. First, it is unlikely to lose diversity when using a sufficient population size, and therefore mutation is simply not needed in GP. Second, when the crossover operation occurs using endpoints in both trees the affect is "very similar to point mutation" [6].

Feldt and Nordin applied the design of experiments techniques from mathematical statistics to evaluate the affects of genetic programming parameters on three problems [4]. The findings show the population size was the most important parameter (with an affect score of 11.51), followed by mutation (5.21) and the number of generations (5.14). Mutation and large crossover probabilities had a positive affect on one of the three screening experiments. We attempt to confirm mutation has a similar positive influence on our problem with the SAMT.

In addition to affecting performance there is research suggesting that mutation can also affect rates of code growth. Luke considers modification point depth as a new model for genome growth in GP [11]. The findings suggest correlation of code growth with node depth bias. The survivability of the child improves with deep mutation and crossover points. Luke's meaning of deep node depth bias refers to locations near the terminal nodes. In this paper, we attempt to confirm that the relationship between the depth and size of mutation replacement trees and population performance is nonlinear.

In addition to the rate of mutation, the ratio of internal/external mutations may have a significant affect on performance. Few researchers report this ratio. Exceptions do exist, Luke, Ross, and Burke et al. reference the internal/external mutation parameter in descriptions for the genetic programming environment problems in their work [12] [13] [1]. However none of the research describes or reports the affect of this parameter value. In this work we report preliminary results regarding the importance of this ratio.

3 Parameter Setting Taxonomy

To understand the affects of mutation we first need to define the different parameters that can be used to control mutation. Eiben et al. studied parameter settings for evolutionary algorithms (EA) and introduces a parameter setting taxonomy that defines attributes related to EA environments [3]. Eiben et al. examines parameter settings for mutation, crossover, evaluation functions, replacement operator or survivor selection, and population size in evolutionary algorithms. Eiben et al. divide the research into parameter tuning and parameter control. Figure 3 shows the parameter setting taxonomy. Eiben et al. prefaces the discussion of mutation operators and their probabilities with "There have been several efforts to tune the probability of mutation in genetic algorithms. Unfortunately, the results (and hence the recommended values) vary, leaving practitioners in the dark." [3].

We propose an extension to Eiben et al. global taxonomy for parameter setting, starting at the "Parameter tuning" node shown in Figure 3. The proposed extensions follow, the underlined attributes indicate the three SAMTs examined in this paper with parameter analysis.

- *Parameter tuning*: The extended taxonomy further refines relationships for mutation behavior.
- *Population proportion* determines the number of individuals selected for mutation.
- *Individual proportion* determines the techniques applied to each individual.s structure. Individual structure modifications include terminals versus non-terminals, or introns versus exons.
- *Selection bias* determines what members of the population are chosen for mutation, examples include the *best*, *worst*, or *random* individuals.
- *Individual proportion* is further subdivided into *Intron and exon* ratios that specifically mutate introns or exons.
- *Depth ratios* that specifically mutate nodes based on sub-tree depth.
- *Variable shallow node* mutation modifies 1 or more nodes by a controlled depth. An example would be to randomly modify 5 nodes throughout an individual. Shallow indicates a depth of 0 or a single node mutated, instead of the normal variable size replacement tree depth.
- *Replacement tree depth* controls the size of the replacement subtree randomly created and inserted by the mutation operation.
- *Internal, external ratio* controls which nodes within a tree for mutation based on their position in the tree hierarchy.

4 Structure Altering Mutation Techniques (SAMT)

The possible permutations, and interaction with other parameters make rationalization of the mutation rate parameter difficult to optimize for many problems. To

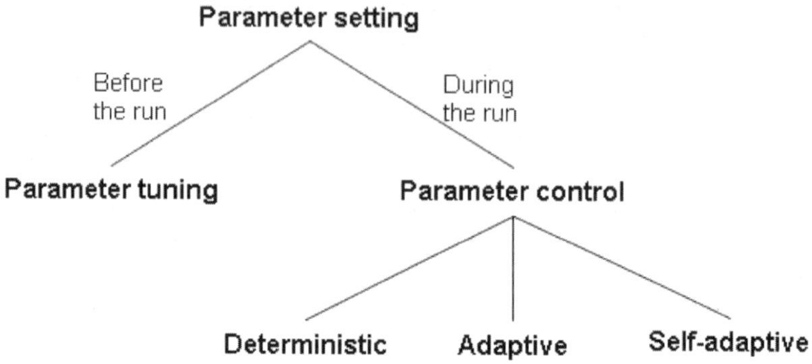

Fig. 1. Parameter Setting Taxonomy. Eiben original.

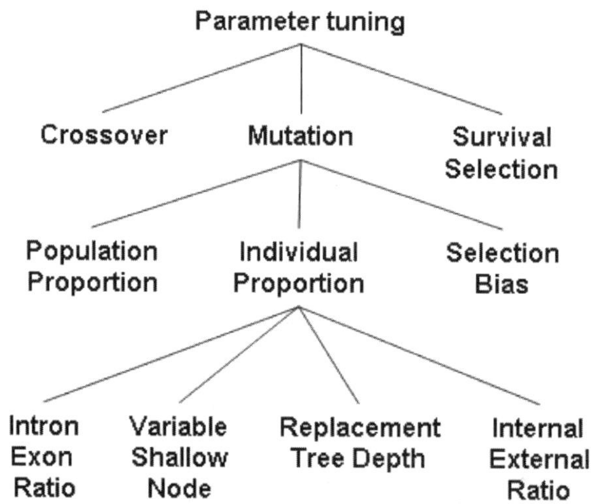

Fig. 2. Parameter Setting Taxonomy. The original taxonomy on the left is extended on the right starting at the 'Parameter tuning' node.

address the evaluation of structure altering mutation we introduce three SAMT techniques, each modifies the genetic program tree in the mutation phase. Figure 3 depicts visual examples for the descriptions of SAMT 1, 2, and 3.

SAMT-1 implements shallow variable node count mutation (SVNCM) and operates by selecting 1 to n random locations in a tree, mutating a single node at each of the locations.

SAMT-2 controls the depth of replacement tree for mutation. Genetic programming algorithms often consider the replacement tree depth for mutation a tunable parameter. SAMT-2 mutates the tree structure using increasing mutation replacement tree depth.

SAMT-3 controls selection frequency for internal versus external nodes or non terminal nodes to mutate. Researchers often omit defining the internal/external (INEX) mutation rate parameter, instead relying on the default values.

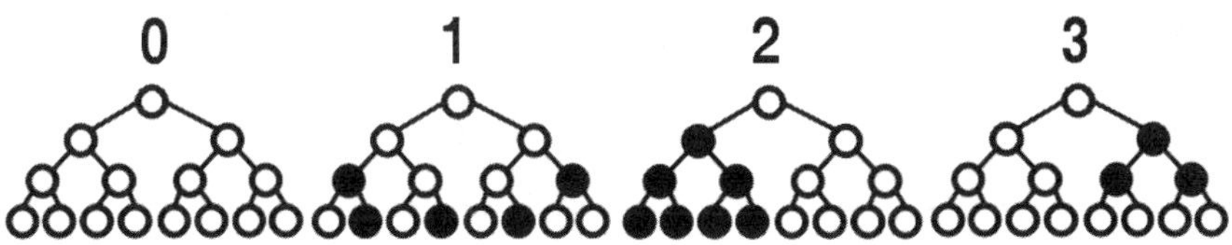

Fig. 3. SAMT Examples. The SAMT technique number is above each tree. A white node indicates no mutation, a black node indicates a mutation. **0** is the original tree. **1** depicts 5 single node locations mutated. **2** depicts a mutation replacement tree depth of 2. **3** depicts mutation of the internal nodes, the INEX ratio is 1.0.

5 The MAX Binary Tree Problem

The goal of MAX binary tree problem (BTP) is to create a full tree of a given depth that produces the maximum value. This problem has several advantages: it is easy to understand, fitness evaluation is fast and simple, and the problem complexity is easy to influence by changing the tree depth. The MAX BTP problem is similar to the MAX problem introduced by Gathercole for researching the interaction of crossover and tree depth [5]. Langdon and Poli extended the analysis of the MAX problem by Gathercole and Ross and developed a quantitative model that indicates the rate of improvement and shows that solution time grows exponentially with depth [10]. Langdon and Poli cite two reasons why GP finds the MAX problem difficult. (1) the tendency for GP populations to converge in the first few generations to suboptimal solutions from which they can never escape. (2) convergence to suboptima from which escape can only be made by slow search similar to randomized hill climbing [10].

In the MAX BTP the root node is at level (or depth) 0. All internal nodes have a degree of three, supporting one parent and two child nodes, and the terminal nodes have a degree of one (the parent). The problem operators are +, − and the terminals are ephemeral random constants 0 and 1. The maximum value for the tree is obtained when all internal nodes are + and all terminals are 1 in which case the $maxvalue = 2^{depth}$. Tree evaluation only occurs once for each individual. Thus, fitness evaluation is fast compared to many other GP problems, such as symbolic regression that may require tens to hundreds of tree evaluations to measure the fitness of a single individual.

If we consider a full tree with binary values as follow: +, − for internal node's, and 0, 1 for leaf nodes, the number of permutations for a given depth is:

$$Permutations = 2^{(Maximum\ Tree\ Size)}$$

For example, a depth 3 problem with 15 vertices has 256 possible permutations. Structural complexity is defined by Koza as a count of the number of points in a solution, this is the sum of functions and terminals. Structural complexity is one characteristic of a tree representation that can be measured [7]. The permutations nearly double with each level of depth and correlates with the structural complexity. The ratio of permutations to structural complexity increases exponentially. Through control of the tree depth, we can control the number of permutations, which determines the size of the search space, which determines the problem complexity. Thus, increasing the depth significantly increases problem complexity.

6 Experiment Approach

We present three mutation selection techniques to test for nonlinear performance by presenting an increasing linear input that mutates the tree structure. Each experiment performs a sweep of the parameter setting under study. Each problem run has a unique random seed assigned. We test three mutation selection modes: *none* (n), *random* (r) and *best* (b). *none* indicates no mutation mode or mutation phase. *random*, P percent of the population is randomly selected to undergo mutation. *best*, the best P percent of the population is selected for mutation. Table 1 lists the experiment parameters. The table has a few differences in parameter values, this does not affect analysis which takes place within an SAMT scope and not across SAMTs.

6.1 SAMT-1, Shallow Variable Node Count Mutation

The goal of this experiment is to determine if the relationship between the number of nodes mutated and the mean performance is linear or non-linear. Knowing the optimal mutation function will be useful for researchers using this type of mutation, and inform the community on optimal parameter settings for other types of mutation. This experiment consists of three trials, a control run with no mutation or *none*, then 5 repetitions for *random*, and *best* selection mutation mode (abbreviated n, r, and b in the figures). Each of the 5 repetitions consists of mutating 1, 2, 3, 4, and 5 nodes while iterating 100 separate runs differentiated by the random seed.

6.2 SAMT-2 and SAMT-2b, Mutation Replacement Tree Depth

The goal of this experiment is to determine if changes in the mutation replacement tree depth cause nonlinear changes in the mean fitness of the population. In this experiment, we test for nonlinearity by measuring the mean fitness of the population while the depth of the mutation replacement tree is increased linearly. We test three mutation selection modes including *none*. The replacement tree depth is tested with a range of values from 0 to the maximum tree depth (total replacement). The subtree created is the result of the same method used for the initial population at generation 0. We have chosen to continuously generate new replacement trees until one meets the maximum depth criteria. The new tree is tested for compliance with the maximum tree depth limit. If a new individual exceeds the depth limit another subtree be will be generated. For example in lilgp, a common GP engine which is used in these experiments the default behavior omits the mutation if the total tree depth exceeds the maximum. If the keep_trying flag is enabled lilgp will continuously search for a replacement until one meets the overall depth requirement. By default, lilgp also generates the new random subtree with a default depth range of 0 to 4 [2]. This depth range represents 1 to 15 new nodes for the subtree in the binary tree problem. For example depth 0 results in a single node replacement, depth 1 results in 3 nodes undergoing replacement, etc. For each of the replacement tree sizes we measure the resulting mean population fitness. Experiment SAMT-2b evaluates the use of the *keep_trying* flag at the default setting of false. *keep_trying* determines if the genetic programming simulation tree replacement method, should give up if a replacement depth exceeds the maximum tree depth, or continue to generate new subtrees until one complies with the maximum tree depth specification. This frequently overlooked parameter setting significantly changes the behavior of the algorithm and subsequent results.

6.3 SAMT-3, Internal/External Mutation Selection Ratios

The goal of this experiment is to understand if a linear input change in the INEX mutation ratio has a linear affect on the mean fitness of population. This experiment employed parametric control of the INEX mutation ratio while measuring the resulting computational effort to evolve a solution. We test three mutation selection modes including *none*. Each experiment increased the problem depth and corresponding problem complexity. Each data point presented in the following graphs is the average of 100 runs. The INEX ratio varies from 0.0 to 1.0 in steps of 0.01 Thus, each plot line is composed of 10,100 runs. For each problem depth, the CEAIS value computes the effort expended to evolve a solution. The following terms and calculations determine the minimum computational effort average for each acceptable solution. Mean, mean generation for an acceptable solution across 100 runs. Average Computational Effort (ACE) based on mean generation for all 100 runs. Number of Acceptable Solutions (NS). ACE counts generation 0 by adding 1 to the final generation value.

$$ACE = (MeanGeneration + 1) * PopulationSize \qquad (1)$$

$$ComputationalEffortAverageIndividualSolution(CEAIS) = \frac{ACE}{NS} \quad (2)$$

Low values of CEAIS indicate the minimum average computational effort per solution that corresponds to an optimal configuration of parameter settings. A prior experiment determined optimal crossover and mutation rates for SAMT-3.

Table 1. Experiment Parameters

Parameter	SAMT-1	SAMT-2	SAMT-3
Objective	2^{Depth}	2^{Depth}	2^{Depth}
Terminal Set	$0, 1$	$0, 1$	$0, 1$
Functional Set	$+, -$	$+, -$	$+, -$
Fitness Cases	1	1	1
Fitness	Value of tree	Value of tree	Value of tree
Population Size	10,000	1,000	4,000
Max Generations	1,000	1,000	1,000
Initialization Method	Half & Half	Full	Full
Initialization Depth Ramp	2-6	7	$maxdepth - 1$
Maximum Tree Depth	8	8	3,4,5,6,7,8
Crossover selection mode	fitness	fitness	fitness
Crossover rate	$0.9 - MR$	0.89	optimal
Reproduction selection	fitness	fitness	fitness
Reproduction rate	0.1	0.1	0.1
Mutation selection mode	(n),(r),(b)	(n),(r),(b)	(n),(r),(b)
Mutation rate	(n)=0	(r,b)=0.01	optimal
(MR)	(r,b)=0.01	0.11, 0.21,	for each
	0.11,0.21,0.31	0.31	problem size
Internal mutation ratio	0.9	0.9	variable
External mutation ratio	0.1	0.1	variable

7 Results

The following subsections give a brief description of experiment results. The legend in Figure 4 and 5 represent the four population selection mutation rates.

SAMT-1, we examine four mutation rates applied to a population size of 10,000. The mutation rates of 0.01, 0.11, 0.21 and 0.31, affect only the *best* individuals for mutation. The four rates select the *best* 100, 1100, 2100 and 3,100 individuals of the population. Results for the *none* selection mode which is equivalent to 0 nodes mutated, and for the *random* mutation selection mode produced no acceptable solutions and these data are subsequently omitted from SAMT-1 results. The numbers of nodes mutated at each mutation rate are 1, 2, 3, 4 and 5 are selected independently. The maximum value that can be obtained with a depth of 8 is 256, this represents all 255 internal nodes containing a + and all 256 external nodes contain a 1. Figure 4 and 5 show the results of the number nodes mutated for 0 to 5, the 0 indicates no mutation. Figure 4 plots the mean population fitness against the number of nodes mutated. The deleterious impact to fitness beyond 2 nodes is significant and severely limits mean fitness the population obtains.

Figure 5 plots the change in mean population fitness between each interval of the number of nodes mutated. The curves between the points reflect a cubic spline interpolation. The spline curves provide a projection that shows the trend of the data points. $y - axis$ values above 0 indicate improvement over the previous number of nodes mutated, values below 0 indicate a reduction in mean fitness.

Table 3 shows the results of each experiment run, the column labels indicate: NM - Nodes Mutated, MF column shows the mean fitness for all generations and the second value is the the number of acceptable solutions. The none trial had no mutation phase.

Table 2 shows the the rates used for each of the four trials for the random and best mutation selection methods.

Table 2. SAMT-1 Trial Rates

Trial	Crossover	Mutation	Reproduction
MF-1	0.89	0.01	0.10
MF-2	0.79	0.11	0.10
MF-3	0.69	0.21	0.10
MF-4	0.59	0.31	0.10

Table 3. SAMT-1 Result Summary

Mode	NM	MF-1	MF-2	MF-3	MF-4
none	0	0.004, 00	0.004, 00	0.004, 00	0.004, 00
random	1	0.004, 00	0.004, 00	0.004, 00	0.004, 00
random	2	0.004, 00	0.004, 00	0.004, 00	0.004, 00
random	3	0.004, 00	0.004, 00	0.004, 00	0.004, 00
random	4	0.004, 00	0.004, 00	0.004, 00	0.004, 00
random	5	0.004, 00	0.004, 00	0.004, 00	0.004, 00
best	1	0.588, 48	0.739, 66	0.842, 81	0.730, 66
best	2	0.527, 47	0.564, 52	0.405, 37	0.027, 01
best	3	0.414, 38	0.271, 23	0.024, 01	0.009, 00
best	4	0.241, 21	0.057, 04	0.008, 00	0.007, 00
best	5	0.004, 00	0.005, 00	0.004, 00	0.004, 00

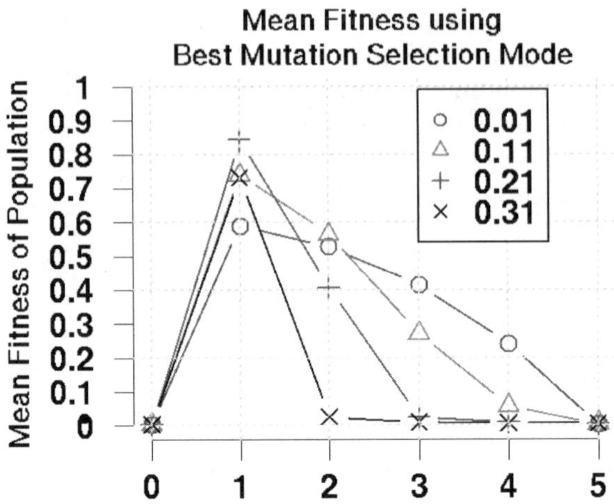

Fig. 4. SAMT-1: Absolute Scale of Mean Fitness versus Number of Nodes Mutated

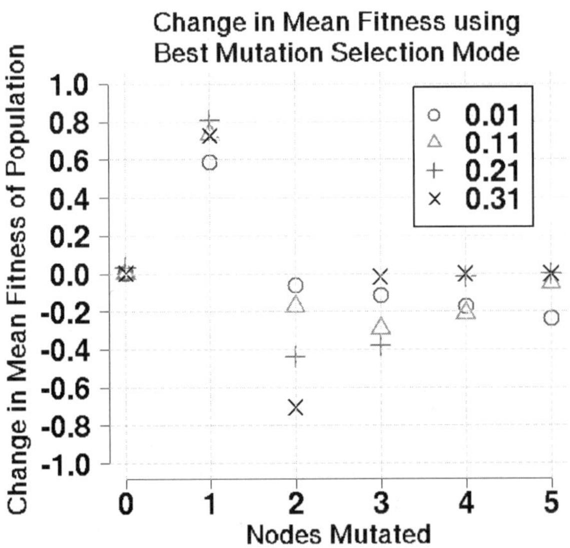

Fig. 5. SAMT-1: Relative Change in Mean Fitness versus Number of Nodes Mutated

For SAMT-2, we examine four mutation rates applied to a population size of 1000. The four rates randomly select 10, 110, 210 and 310 individuals for mutation. The mutation rates of 0.01, 0.11, 0.21 and 0.31, affect only the *best* individuals for mutation. Results for the *random* mutation selection mode produced no acceptable solutions, and have been omitted from SAMT-2 results. SAMT-2 results are given in Table 4. Figure 6 and 7 indicate no mutation mean population fitness by the $x - axis$ (n) label. Figure 6 and 7 show the impact of controlled replacement tree depth mutation. In Figure 6 the mean fitness improves from a depth 0 or size 1 replacement tree through a depth 3 replacement tree. Fitness decreases from depth 2 through depth 8. Figure 7 shows the relative change in fitness for each depth replacement tree. All positive contributions occur at depths of 0 and 1. Further increases in depth cause a negative fitness influence on the population relative to the previous value. Mutations beyond a depth of 4 evolved no acceptable solutions.

Table 4. SAMT-2 Result Summary

Mode	D	MF-1	MF-2	MF-3	MF-4
none	0	0.0040, 00	0.0040, 00	0.0040, 00	0.0040, 00
random	0	0.0040, 00	0.0039, 00	0.0039, 00	0.0039, 00
random	1	0.0040, 00	0.0040, 00	0.0040, 00	0.0040, 00
random	2	0.0040, 00	0.0040, 00	0.0040, 00	0.0040, 00
random	3	0.0040, 00	0.0040, 00	0.0041, 00	0.0041, 00
random	4	0.0040, 00	0.0041, 00	0.0041, 00	0.0041, 00
random	5	0.0040, 00	0.0042, 00	0.0042, 00	0.0042, 00
random	6	0.0041, 00	0.0043, 00	0.0042, 00	0.0043, 00
random	7	0.0042, 00	0.0045, 00	0.0043, 00	0.0043, 00
random	8	0.0044, 00	0.0045, 00	0.0045, 00	0.0045, 00
best	0	0.1692, 12	0.2353, 20	0.0482, 02	0.0112, 00
best	1	0.1889, 15	0.4405, 37	0.2503, 20	0.0661, 03
best	2	0.0700, 04	0.1692, 13	0.0376, 01	0.0146, 00
best	3	0.0000, 00	0.0246, 01	0.0118, 00	0.0093, 00
best	4	0.0000, 00	0.0189, 01	0.0085, 00	0.0072, 00
best	5	0.0000, 00	0.0065, 00	0.0066, 00	0.0062, 00
best	6	0.0000, 00	0.0056, 00	0.0057, 00	0.0063, 00
best	7	0.0000, 00	0.0053, 00	0.0056, 00	0.0056, 00
best	8	0.0000, 00	0.0046, 00	0.0045, 00	0.0046, 00

SAMT-2b has identical parameters to that of SAMT-2 with one exception, *keep_trying* is set to false or the default for some genetic programming environments. SAMT-2b results are given in Table 5. Figure 8 and 9 present the results of the mean fitness at each depth tested and the relative change in fitness between the current and prior depth. Both plots indicate a decreasing change in fitness. Due to the behavior of the genetic programming system treatment of mutation replacement subtrees we observe increasing fitness instead of decreasing. The reason postulated is that as the mutation replacement

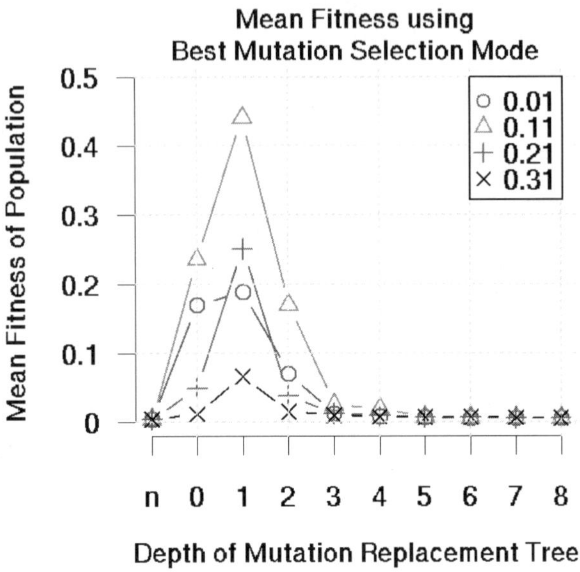

Fig. 6. SAMT-2: Absolute Scale of Mean Fitness versus Depth of Mutation Replacement Tree

subtree increases in depth it becomes less likely to meet the maximum depth constraints. The result is that less subtree mutations occur than are specified due to geometry constraints for the individual. The impact is fitness is reduced since fewer replacement subtrees fit the individual size constraints as the replacement subtree grows. So the absolute number of trees mutated decreases, the result is an increase in the mean fitness of the population as the depth increases.

In the SAMT-3 experiment, five increasingly complex versions of the MAX binary tree problem had replacement tree depths ranging from 3 to 8. Due to space limitations, only the depth 6 results appear. Figure 7 plots the CEAIS values for the MAX BTP of depth 6 against the internal/external selection ratio. The 5 point spline interpolation of Figure 7 shows the *best* CEAIS decreasing from 2402 to 1084 over the internal mutation rate range. *None* has a CEAIS value of 1830 and outperforms *random* across the entire internal mutation rate range. The *random* selection mode has CEAIS higher then *none* and has increasing CEAIS as internal mutation rate increases. Note the observation for this even depth problem is that the slope of the *best* CEAIS decreases with increasing internal mutation rate.

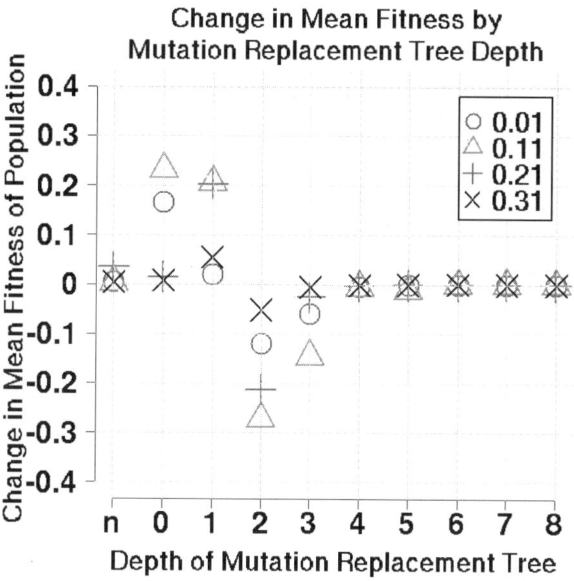

Fig. 7. SAMT-2: Relative Change in Mean Fitness versus Depth of Mutation Replacement Tree

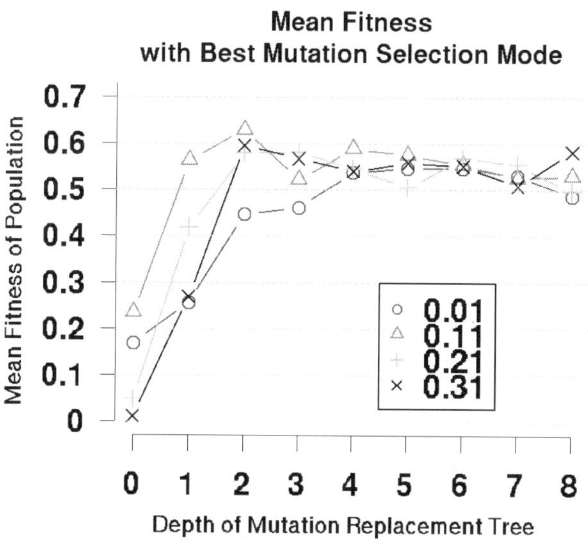

Fig. 8. SAMT-2b: Absolute Scale of Mean Fitness versus Depth of Mutation Replacement Tree

Table 5. Experiment SAMT-2b Result Summary

Mode	D	MF-1	MF-2	MF-3	MF-4
none	0	0.0040, 00	0.0040, 00	0.0040, 00	0.0040, 00
random	0	0.0039, 00	0.0039, 00	0.0039, 00	0.0039, 00
random	1	0.0040, 00	0.0040, 00	0.0040, 00	0.0040, 00
random	2	0.0040, 00	0.0040, 00	0.0040, 00	0.0040, 00
random	3	0.0040, 00	0.0040, 00	0.0041, 00	0.0041, 00
random	4	0.0040, 00	0.0041, 00	0.0041, 00	0.0041, 00
random	5	0.0040, 00	0.0041, 00	0.0042, 00	0.0042, 00
random	6	0.0041, 00	0.0042, 00	0.0042, 00	0.0043, 00
random	7	0.0042, 00	0.0043, 00	0.0043, 00	0.0043, 00
random	8	0.0043, 00	0.0044, 00	0.0045, 00	0.0045, 00
best	0	0.1692, 12	0.2353, 20	0.0482, 02	0.0113, 00
best	1	0.2566, 20	0.5673, 49	0.4177, 35	0.2689, 23
best	2	0.4464, 33	0.6296, 50	0.5787, 42	0.5960, 46
best	3	0.4610, 32	0.5223, 36	0.5817, 45	0.5677, 41
best	4	0.5376, 40	0.5889, 45	0.5458, 41	0.5406, 39
best	5	0.5469, 41	0.5757, 45	0.5059, 37	0.5592, 43
best	6	0.5474, 42	0.5547, 43	0.5686, 43	0.5521, 44
best	7	0.5309, 40	0.5261, 39	0.5550, 43	0.5092, 37
best	8	0.4862, 34	0.5317, 40	0.4968, 37	0.5829, 47

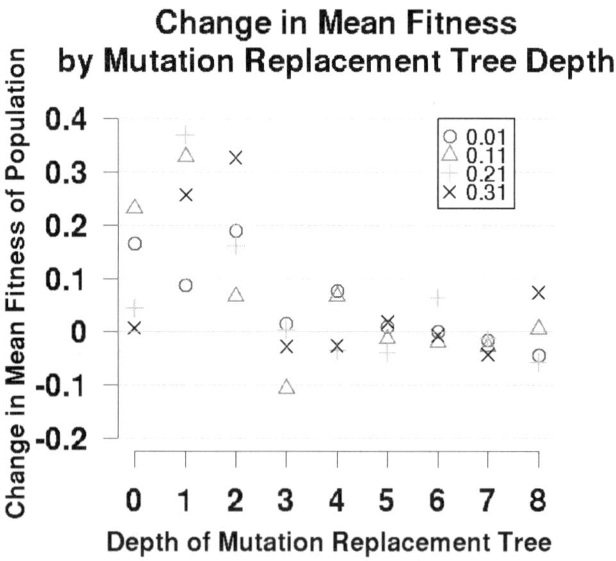

Fig. 9. SAMT-2b: Relative Change in Mean Fitness versus Depth of Mutation Replacement Tree

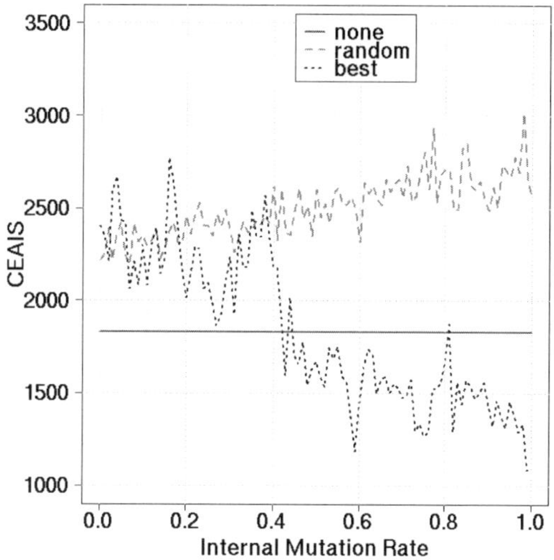

Fig. 10. SAMT-3: Depth 6, CEAIS versus ratio of mutation at non-terminal nodes, Linear-Log Plot

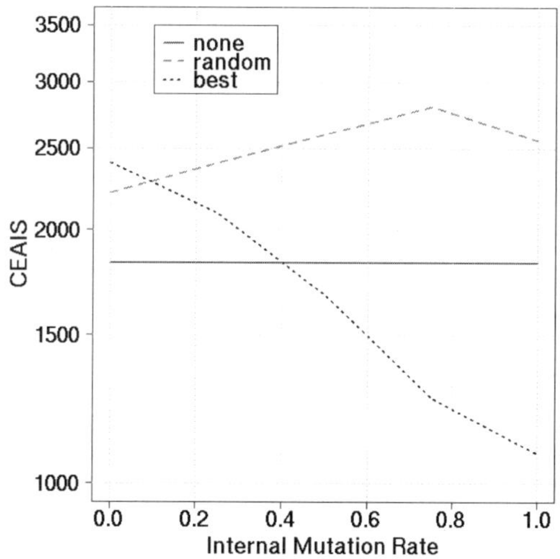

Fig. 11. SAMT-3: Depth 6, CEAIS versus ratio of mutations at non-terminal nodes, Linear-Log Plot, 5 point spline interpolation

Table 6 summarizes results from each experiment. The following abbreviations appear in the table header. Mean of *best* (MB), CEAIS values, and Standard deviation (SD) of *best* CEAIS values.

Table 6. SAMT-3 Experiment Summary

Depth	MB	SD
3	7	0.8
4	40	7.3
5	161	24.6
6	1827	415.4
7	12024	2023.2
8	8638296	10639.0

The following observations concern the *best* mutation selection mode in the cases examined above. For the odd depth MAX BTP tested, we note the internal mutation ratio range from 0.0 to 1.0 produces CEAIS values that decrease by 100%. The payoff for identification of the optimal INEX value needs careful balance with the effort of discovery. For the even depth problems we note a trend that identifies higher internal mutation rates providing significantly lower CEAIS values than low internal mutation ratio rates. Even depth problems show a consistent reduction of CEAIS values as the internal mutation ratio increases from 0.0 to 1.0. The even depth problems also have a higher CEAIS variance across the entire mutation rate range as a fraction of the mean than the odd depth problems. In this situation, it indicates the improvement with decreasing CEAIS values as the mutation rate changes from 0.0 to 1.0.

7.1 Observations

In SAMT-1 we found that the *best* mutation mode performed significantly better than *random* and *none* showing that mutation can improve performance. Mutating the best individuals results in the better performance, possibly because those individuals are mostly likely to have an affect on the future path of evolution. The critical points for mutation in four separate rate trials show the beneficial affects of mutation occur only for depth 1. Higher mutation rates accelerate the deleterious impact of multi-node mutation, although mutating two nodes is still better than none. *Mutation boost* is describes the significant improvement in evolved *best* solutions over the other two selection modes, *none* and *random*.

SAMT-2, 22.2% is the optimal tree replacement depth for the binary tree problem with a depth of 8. Depths 0 to 2 perform better than no mutation.

SAMT-3, Selection of the best individuals and performing mutation on INEX ratio of 50/50 should provide a consistent and neutral affect on the resulting computational effort. Significant variance is noted for the *best* mutation selection mode for even problems between internal mutation rates below and above 50% or 0.5. This suggests deleterious affects of mean population fitness for the MAX BTP using high external mutation rates. Optimal parameter tuning for the INEX rate leads to improved performance with even depth MAX BTP problems. For odd depth problems we observe improvements as the INEX rate approaches $\approx$0.60. When complexity reaches depth 7, a decrease of CEAIS shows higher INEX rates outperform lower INEX rates.

8 Summary

Parameter settings significantly influence the effectiveness of the structure altering mutation techniques evaluated and generally show a nonlinear response to population fitness and computational effort. Each experiment discovered unique genetic programming behavior using parametric analysis of values in isolation. The key results from the experiments:

- SAMT-1: The behavior of mutating 1 to 5 nodes is near linear when 1% to 11% of the population is mutated. When more of the population is subjected to mutation (21% and 31%) the relationship between the number of nodes mutated and the performance becomes increasingly nonlinear. The rate of decrease of mean population fitness increases with mutation rate as more nodes are mutated.
- SAMT-2: Varying the replacement tree depth produced a nonlinear response as measured by the mean population fitness. Performance with SAMT-2 peaks at approximately 22% of the maximum tree depth across all four mutation rates.
- SAMT-3: Increasing the rate at which non-terminal nodes are mutated results in a linear response of computational effort for the *random* and *best* mutation selection modes. The computational effort correlation with increasing the INEX rate was negative for *random* and positive for *best* mutation selection modes.
- SAMT-3: As the internal mutation selection ratio increases the computational effort decreases. This indicates that higher internal mutation selection rates are optimal for the *best* mutation selection mode on the MAX BTP. The average response of the population to *random*, *best* and *none* are near linear with respect to the internal mutation parameter value.

References

1. Edmund K. Burke, Steven Gustafson, and Kendall Kendall. Diversity in genetic programming: An analysis of measures and correlation with fitness. *IEEE Transactions on Evolutionary Computation*, 8(1):47–62, 2004.

2. Bill Punch Douglas Zongker. lil-gp 1.0 user's manual. Technical report, Michigan State University, 1995.
3. Agoston Endre Eiben, Robert Hinterding, and Zbigniew Michalewicz. Parameter control in evolutionary algorithms. *IEEE Transations on Evolutionary Computation*, 3(2):124–141, 1999.
4. Robert Feldt and Peter Nordin. Using factorial experiments to evaluate the effect of genetic programming parameters. In Riccardo Poli, Wolfgang Banzhaf, William B. Langdon, Julian F. Miller, Peter Nordin, and Terence C. Fogarty, editors, *Genetic Programming, Proceedings of EuroGP'2000*, volume 1802 of *LNCS*, pages 271–282, Edinburgh, 15-16 April 2000. Springer-Verlag.
5. Chris Gathercole and Peter Ross. An adverse interaction between crossover and restricted tree depth in genetic programming. In John R. Koza, David E. Goldberg, David B. Fogel, and Rick L. Riolo, editors, *Genetic Programming 1996: Proceedings of the First Annual Conference*, pages 291–296, Stanford University, CA, USA, July 1996. MIT Press.
6. John R. Koza. *Genetic Programming: On the Programming of Computers by Means of Natural Selection*, pages 106–107. MIT Press, Cambridge, MA, USA, 1992.
7. John R. Koza. *Genetic Programming: On the Programming of Computers by Means of Natural Selection*. MIT Press, Cambridge, MA, USA, 1992.
8. John R. Koza, Forrest H. Bennett III, David Andre, and Martin A. Keane. *Genetic Programming III: Darwinian Invention and Problem Solving*. Morgan Kaufmann Press, May 1999.
9. John R. Koza, Martin A. Keane, Matthew J. Streeter, William Mydlowec, Jessen Yu, and Guido Lanza. *Genetic Programming IV: Routine Human-Competitive Machine Intelligence*. Kluwer Academic Publishers, July 2003.
10. W. B. Langdon and R. Poli. An analysis of the MAX problem in genetic programming. In John R. Koza, Kalyanmoy Deb, Marco Dorigo, David B. Fogel, Max Garzon, Hitoshi Iba, and Rick L. Riolo, editors, *Genetic Programming 1997: Proceedings of the Second Annual Conference*, pages 222–230, Stanford University, CA, USA, 13-16 July 1997. Morgan Kaufmann.
11. Sean Luke. Modification point depth and genome growth in genetic programming. *Evolutionary Computation*, 11(1):67–106, 2003.
12. Sean Luke and Lee Spector. A comparison of crossover and mutation in genetic programming. In John R. Koza, Kalyanmoy Deb, Marco Dorigo, David B. Fogel, Max Garzon, Hitoshi Iba, and Rick L. Riolo, editors, *Genetic Programming 1997: Proc. of the Second Annual Conf.*, pages 240–248, San Francisco, CA, 1997. Morgan Kaufmann.
13. Brian J. Ross. Searching for search algorithms: Experiments in meta-search. Technical report, Brock University, December 2002. Technical Report CS-02-23.

Parameter Sweeps for Exploring Parameter Spaces of Genetic and Evolutionary Algorithms

Michael E. Samples, Matthew J. Byom, and Jason M. Daida

Center for the Study of Complex Systems and the Department of Atmospheric, Oceanic, and Space Sciences, University of Michigan. Ann Arbor, Michigan USA 48109-2143 (msamples,mjbyom,daida)@umich.edu

Summary. The quality of experimental research in Genetic and Evolutionary Computation can suffer from experiments that are too small to generalize results. Experiments can be too small by not being statistically significant with a given set of algorithm parameters, or by not identifying the effect of modifying an algorithm parameter, or by not understanding the causal relationships between different algorithm parameters. Parameter Sweep Experiments solve these problems by sampling many different parameter configurations enough times to achieve statistical significance and identify trends in parameter modifications. Unfortunately, Parameter Sweep Experiments have a high cost of labor to setup and run experiments and analyze data. As a result, these experiments are rarely conducted in Genetic and Evolutionary Computation research. This chapter argues that although Parameter Sweep Experiments are difficult to conduct, they are required to understand complicated nonlinear interactions between an algorithm's parameters. We present a software solution, *Commander*, that works in conjunction with a genetic and evolutionary algorithm to assist users in conducting Parameter Sweep Experiments in a distributed computing environment.

1 Introduction

In recent years, a number of authors have provided critiques of the existing experimental methodology in Genetic and Evolutionary Computation (GEC). This chapter continues that trend through an analysis of the design, execution, and data reduction stages of GEC experiments with respect to how the parameter values for GEC algorithms are chosen. Although genetic programming (GP) is referenced in specific examples, the methodology and tools presented in this chapter are applicable to both the wider GEC community and the even larger field of parameterized computer simulations. (For examples from GP, see [13, 14, 23, 26, 29].)

The execution of a GEC algorithm to solve a particular problem can be considered as a search in the space of all solutions expressible by a given representation (e.g., a tree or a vector). Observations that are made during the execution of a search are the result of many low-level, nonlinear interactions. The rules, which govern these low-level interactions and thus influence experimental results, are highly dependent

Michael E. Samples et al.: *Parameter Sweeps for Exploring Parameter Spaces of Genetic and Evolutionary Algorithms*, Studies in Computational Intelligence (SCI) **54**, 161–184 (2007)
www.springerlink.com

on algorithm configurations and parameter values. Slight changes to these rules can produce dramatically different outcomes of the search process.

To effectively use GEC algorithms as a search process, users should understand how potentially different outcomes are influenced by choices of algorithm parameters. Towards this end, one common goal is to characterize a mapping between arbitrary algorithm configurations and their corresponding outcomes on specific classes of problems. Since an exhaustive sampling of an infinitely large parameter space is computationally infeasible, users often rely instead on a partial sampling to generate a set of results from which predictions about untested configurations can be based.

Not surprisingly, a theory is often relied upon to provide a framework under which interpolations and extrapolations of algorithm outcomes can be made. Unfortunately, it is still quite possible that interpolation and extrapolation of a partial sampling—even when guided by theory—can remain ineffective. Too small of a sampling can miss something, if only because an underlying assumption was incorrectly applied.

In this chapter, we take the position that the effectiveness of interpolation and extrapolation can be increased by large, comprehensive, multivariate experiments. We further take the position that these experiments are challenging to conduct, particularly in GEC, because even small experiments can have demanding requirements for data storage and computational power. We examine similar experimental frameworks in other fields and propose a software solution that assists an investigator in administering these types of experiments in a distributed-computing environment.

In Section 2, we discuss the background of large computational experiments (particularly experimental frameworks), and discuss how a particular type of experiment is helpful in characterizing GEC dynamics. We also outline criteria for a distributed-computing software tool to assist in experiment administration. Section 3 presents our software solution, *Commander*, and discusses architectural concerns for robustness and efficient execution. Section 4 presents results from applying our tool to common GP test problems that show the utility of a methodology fostered by *Commander*. Section 5 discusses the results of this demonstration. Section 6 concludes, while Section 7 suggests future work.

2 Background

2.1 Previous Work

The experimental methodology that we propose is not limited to GEC. In fact, such experiments have often been practiced in other fields that involve the characterization of parameterized behavior. A diverse body of work—from fields such as high-energy physics [5], biomedical molecular modeling [10], analog circuit analysis [30], and agent-based modeling [3]—have all benefited from large parameter sweep experiments. While this methodology is not new to science, it is not common in GEC. (Some exceptions include Daida, Samples *et al.* [15, 16], Daida *et al.*, [17], Luke & Panait [23], and Luke & Spector [24].) This type of experimental methodology frequently involves large numbers of trials and can be computationally prohibitive. Such prohibitions are particularly relevant regarding empirical GEC research, where even small experiments can be computationally intensive. (For example, in [15] alone, empirical results required two CPU-years to generate and used several terabytes of data to store experimental results).

Our field is not the first to undertake computationally challenging problems and several existing solutions for distributed computing are worth mentioning. One of the earliest large-scale distributed computing projects, SETI@home [21], distributes radio data to hundreds of thousands of volunteers who donate their spare CPU hours to perform signal analysis. Another project, the Great Internet Mersenne Prime Search (GIMPS) [33], searches for large prime numbers of the form $2^n - 1$ for an integer n. Like SETI@home, GIMPS is distributed among many volunteers who run small portions of a larger experiment on their own desktop computers.

A number of software packages exist to perform such distributed computation, notably: BOINC [2], APST [10, 11], APSE [8], Drone [6], Condor [5, 32], and Nimrod/G [1]. In general, these packages are designed to distribute a list of computational tasks across a network to remote computers. A downside of these packages is that only some have support for the type of experiment we propose. For example, Drone, Nimrod/G, and APSE support this experiment type. However, APST, Condor, and BOINC all require the presence of custom external scripts to control an experiment.

Distributed computing frameworks should be robust to most common failures and operate transparently with respect to computer and network characteristics. APST, Condor, and Nimrod/G contain support for only a limited number of grid protocols. Drone does not support common grid protocols, but is designed for cluster environments where a user has a local login account. It has also been out of development for several years and is not robust to common network failures such as disconnections and local user interference. Although APSE shares very similar design goals to our project, robust behavior was not incorporated into its initial design and the closed-source nature of APSE has precluded its extension. Currently, BOINC is the most widespread utility for distributed public computing and its open-source structure allows many projects to be built upon it (see, e.g., Stardust@home [25] and Folding@home [28]).

2.2 Transitions

Modifying the parameter values of an algorithm frequently induces an observable change in that algorithm's behavior. Usually, minor parameter value modifications result in only small differences to algorithmic behavior. However, there can exist ranges of parameter values such that small changes in those parameter values can result in large, nonlinear changes in the observable behavior of an algorithm. We consider such parameter values to be within a *transition region*. Although such large-scale, nonlinear, behavioral transitions may be analogous to phase transitions in physical systems (e.g., water transitioning from liquid to solid phases), this remains to be rigorously studied. It is sufficient for the purposes of this chapter to understand that the mapping between parameters and algorithmic behavior may exhibit such transitions. This is discussed in detail in Section 5.

An example of transitions in GEC is shown in Figure 1 from the *Highlander* experiment initially discussed in [15]. In that paper, we developed a GP test problem to probe for limits of building block mixing and assembly. In the figure, β represents

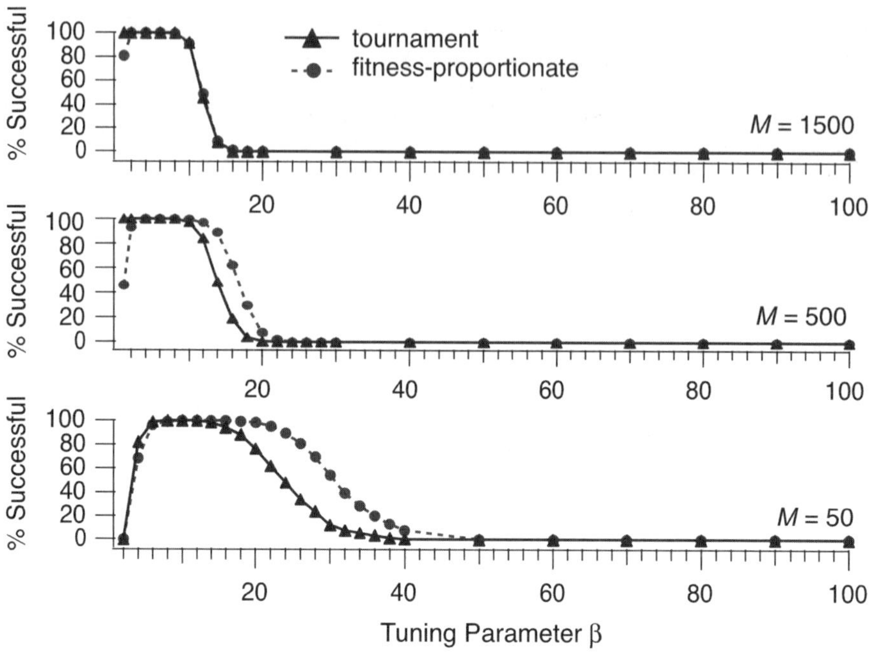

Fig. 1. Difficulty curves for the *Highlander* experiment. Each plot corresponds to a population size (M) and depicts the percentage of successful trials (out of 1000 trials) as a function of tuning parameter β. Parameter β values that correspond to easier settings have higher percentages of successful trials.

a tuning parameter in the fitness function that rewards GP trees with particular characteristics (chosen by β) of building block mixing and assembly.

Experimental results on the *Highlander* problem demonstrated at least two transitions: one based on the tuning parameter β, and the other based on the population size M. The β-transition yields highly successful trials for small values of β (e.g., when $M = 1500$ and $\beta < 10$). As β increases through a transition region, the problem rapidly becomes impossible to solve. Each point in each graph represents a number of successes from 1000 trials, yet GP was unable to solve the test problem for $\beta \geq 18$ when $M = 1500$. The second transition—the M-transition—changes the location and shape of the β-transition. Small values of M yield a wide and gradual transition while large values yield a narrow and steep transition. Note that the β-transition covers three orders-of-magnitude (from 1000 successful trials out of a 1000 total trials, to 0 successful trials out of 1000 trials). Further information on the *Highlander* test problem is available in this chapter's appendix.

As stated previously, experimentalists often rely upon a theory to assist in the interpolation and extrapolation of results to untested configurations. However, if this theory does not suggest the existence of transition regions, such transition regions might not be discovered except through direct experimental sampling of parameter configurations. Consequently, experiments that involve too few unique parameter configurations might not find any evidence of a transition region. As a result, theory that is developed from small experiments might only be valid within boundaries of unobserved (but nonetheless existent) transitions.

To determine the location of such transitions, general knowledge of an algorithm's behavior should be constructed by testing multiple classes of parameter configurations. If experiments were run under multiple configurations, one would be able to develop laws and theories that encapsulate knowledge of the entire parameter space and not merely the space on one side of a particular transition. One should observe an algorithm's behavior by conducting multiple tests and changing input parameters.

2.3 Notation

In this section, we define the terms that we use to describe parameter sweep experiments.

Definition 1. ***Parameter*** - *A parameter is broadly defined to be a modifiable algorithm-specific value that is capable of influencing some aspect of that algorithm's runtime behavior.*

In other words, a parameter is a controllable variable that influences observable data that is collected during the execution of an algorithm. Parameters can be numeric (e.g., population size $p = 1000$) or algorithmic (e.g., tournament selection, steady-state replacement, elitism). Frequently, the choice of a particular parameter value leads to more parameter choices (e.g., we also need to choose a tournament size under tournament selection).

Definition 2. ***Parameter Configuration*** - *The map of parameters to their corresponding values is called a parameter configuration and is denoted by the symbol* $\mathscr{C}$.

A parameter configuration is considered valid if it completely specifies the input to an algorithm without containing conflicting parameter values (i.e., a GEC algorithm cannot *simultaneously* employ fitness-proportionate and tournament selection as the selection method without some way of explicitly resolving what is meant by specifying both). We denote the set of all possible valid parameter configurations as $\mathscr{C}^*$. To refer to the value of a parameter, we use index notation (e.g., the population size of a particular configuration could be $\mathscr{C}[\text{population size}] = 1000$).

Definition 3. ***Trial*** - *A trial t in GEC research is defined to be the execution of a GEC algorithm to solve a particular problem $\mathscr{P}$ (e.g., regression or ant). A trial has a specified parameter configuration $\mathscr{C}_t$ and an initial random number generator (RNG) seed r_t.*

In common experimental practice, a trial is executed and observable behaviors are recorded for later analysis. In GEC research, this typically takes the form of simple values (e.g., each generation's average fitness), but more complicated audit procedures can be used to record more complicated observable states (e.g., a complete ancestry tree for each individual). Since a trial t is uniquely defined in terms of its problem $\mathscr{P}$, its parameter configuration $\mathscr{C}_t$ and its RNG seed r_t, we can denote the observations from a trial as $\Omega(\{\mathscr{P}, \mathscr{C}_t, r_t\})$, or shorter, as $\Omega(t)$ where $t = \{\mathscr{P}, \mathscr{C}_t, r_t\}$.

Definition 4. ***Datapoint*** - *A datapoint $\mathscr{D}$ is a set of trials with identical parameter configurations but unique RNG seeds.*

Since a nondeterministic algorithm's behavior for all possible RNG seeds is unobservable, an experimentalist needs to make do with taking observations from a finite—and thus observable—number of trials. A finite-size datapoint $\mathscr{D}$ consists of m trials $\{t_1, \ldots t_m\}$, where unique trials t_i and t_j use the same problem $\mathscr{P}$ and the same parameter configuration (i.e., $\mathscr{C}_{t_i} = \mathscr{C}_{t_j}$), but different RNG seeds. We can discuss the observations $\Omega_{\mathscr{D}}$ available from a datapoint $\mathscr{D}$, and note that the observations made from a datapoint are also capable of expressing statistics (e.g., the average fitness of the m best-of-trial individuals). In our notation, a datapoint is the smallest unit of experiment capable of conveying meaningful results regarding a particular parameter configuration. We denote the set of all datapoints in an experiment as $\mathscr{D}^*$.

Definition 5. ***Parameter Sweep Experiment*** - *A parameter sweep experiment (PSE) is conducted by examining multiple datapoints, i.e., by executing an algorithm numerous times with varying parameter configurations.*

No two trials in a PSE should share the same random number seed, and two trials would only share the same parameter configuration if they belong to the same datapoint. We denote the *size* of an experiment as the number of datapoints tested and the *dimension* of an experiment as the number of parameters altered across different parameter configurations. Parameter sweeps can be one-dimensional (e.g., each datapoint could have a different population size parameter value) or multidimensional (e.g., studying a set of datapoints in which each datapoint has a different 2-tuple of population size and maximum allowable generations). Multidimensional PSEs are unique in their ability to empirically reveal the effects of interactions between parameters.

2.4 Parameter Sweep Challenges

Large n-dimensional PSEs are more difficult to perform than small single-datapoint experiments for several key reasons:

- *Experimental Setup and Administration* - For small experiments, there is often only one parameter configuration being tested. Since that configuration does not generally change, it is common just to write a shell script that is capable of simply executing the associated GEC algorithm multiple times. For those experiments, parameters can be specified inside this script or precompiled into the GEC algorithm's runtime code. An investigator can distribute this experiment on multiple computers easily by merely running the script on each computer and manually collecting data from each computer after an experiment concludes.

  In a PSE, a shell script should be capable of generating a list of all parameter configurations under study. To do this, complex scripts allow for iteration over possible parameter configurations (a subset of the Cartesian product of the parameters). The simplest instantiation of PSE scripts is constructed with nested loops to iterate over desired parameter combinations, but the serial nature of such an instantiation is not robust to normal computer failures and does not utilize distributed computing resources. Satisfactory scripts are generally both lengthy and non-portable. Construction of a new experiment would generally involve large, time-consuming changes to the scripts.

- *Required Computation Time* - A single trial in GEC can be computationally expensive. However, because of the nondeterministic behavior of GEC algorithms, a datapoint should consist of many trials to collect enough data to understand the behavior of a particular parameter configuration. PSEs amplify this effect since, by design, they evaluate a larger number of datapoints. This effect can be somewhat ameliorated through the use of parallel processing, though scripting becomes more complicated because of provisions for network programming and robust network behavior.
- *Data Reduction* - Single datapoint experiments generally produce the same type of data since they are generated by a common parameter configuration. As a result, the extraction of relevant data is a straightforward process. In PSEs, the extraction of relevant data often necessitates the use of scripts to arrange the data in meaningful ways. Due to a possibly large number of configurations, a PSE's data sets should be labeled with the parameter configuration that generated it. Analysis scripts have another degree of complexity, because they interact with the tagging scheme used to identify data sets with their parameter configurations. In a similar fashion to a PSE's setup scripts, these data reduction scripts can be difficult to maintain between large experiments as configurations change.

For several years, our research group maintained a large collection of Perl scripts to partially automate the process of running PSEs. To reduce turnaround time between experiments, we would partition an experiment among several machines. Each machine's script would be manually executed on idle network machines in a high-use multi-user environment. From these experiences, we learned that distributing a computational load has its own unique challenges:

- *Distribution Algorithms* - When running distributed processes on multiple machines, it is essential to have some way of determining which tasks get assigned to which machines. Inefficient procedures waste valuable resources, but efficient procedures can be either time-consuming to set up manually or difficult to set up automatically. A recent increase in grid computing availability has suggested that distributed algorithms should take advantage of common grid protocols. This introduces an additional difficulty, since different grids can operate under different job-control protocols.
- *Robustness in Job Management* - While remotely executing a trial, there are a number of normal ways in which a machine or algorithmic process can fail without implying a flaw in experiment design. This means that a distributed experiment may not complete unless the experimental method gracefully handles such failures. In our group, this often involved the manual re-execution of trials until successful completion of an experiment. Such a procedure demands complex scripts with conditional behavior and methods for determining the validity of collected experimental data. Since the distribution and validation processes were written for a specific experiment, significant retooling time was spent on these scripts between experiments.
- *Data Collection* - Any data that is generated during a trial is typically stored locally on a machine. While this method poses no problems for small experiments, large distributed experiments should have some way of consolidating data from

all of the machines used in an experiment. The data transmission process is susceptible to errors and thus this process should be robustly designed.

Our research group has experienced all of these difficulties associated with the remote distribution of PSEs. Our scripts for conducting PSEs grew from collections of scripts to manage smaller experiments. When we needed to make modifications to an experiment, we would change the scripts. When trials inevitably failed (for reasons completely divorced from GEC, such as running out of disk space), we manually solved the problem and restarted portions of our experiment. We stress that the time-consuming aspect of experiment administration should not be underestimated, particularly when the addition of robust behavior to experiment scripts increases their complexity. These difficulties, which are inherent to parameter sweeps and distributed computation, led us to explore automated experiment administration.

2.5 Major Design Criteria

We have developed the following three criteria for a generic distributed parameter sweep engine. Namely, a sweep engine should:

- Facilitate the *easy setup* of new PSEs. In defining experiments, a user should not need to write a script to generate a list of parameter configuration combinations. Instead, a sweep engine should derive parameter configurations from a simple grammar provided by a user. Existing algorithms that a user may already possess should not require modification to be used in a PSE. There should be no reason that the program we are executing—such as a GEC algorithm—needs to be rewritten or extensively modified before it is used with the parameter sweep engine. Of the previous work, only Drone and Nimrod/G met these standards for easy experimental setup.
- Operate *transparently and seamlessly* as middleware across many types of computing networks. It should be able to take advantage of large-scale grid networks but should also be able to take advantage of local clusters and available desktop machines. In taking advantage of these remote resources, a user should not need to specify or submit experiments in different manners—all remote distribution should remain transparent to a user. This system should be robust; data integrity should be verified at multiple steps.
- Support *automated data reduction and analysis*. When an experiment concludes, a parameter sweep engine should perform initial steps of data reduction. A sweep engine should use the knowledge of the parameters under study to create graphs and charts as requested by a user in advance. Previously, only Nimrod/G attempted this feature. A simple, integrated analysis tool to create graphs and charts would allow a user to quickly see results from a PSE. This feature would be particularly valuable for GEC investigators in both problem feedback and final analysis. These standard graphical tools could encapsulate overwhelming amounts of information immediately available after a GEC experiment, thereby giving a user invaluable feedback about the correctness of the results.

In fulfilling each of the above design criteria, a parameter sweep engine should act as a simple-to-use generic tool for scientific exploration of parameterized computational problems. It is presumed that users of this tool have little interest in writing setup algorithms, distribution algorithms for multiple grid protocols, and large analysis scripts. As such, a parameter sweep engine should hide the tedious operational decisions from a user. This allows a user to spend time and effort on design and analysis instead of tedious experiment administration.

Although our software solution is not novel in either its ability to perform large PSEs or its ability to run processes on distributed systems, its blend of automated parameter sweep experimentation with transparent, robust, distributed computation is unique. In the next section, we discuss the implementation-level details of our software solution.

3 *Commander*: Parameter Sweep Engine

Commander is our software solution that meets the criteria developed in Section 2.5 for a distributed parameter sweep engine.

The *Commander* system consists of one computer running a host program and an arbitrary number of computers independently running a client program. The host is manually launched by a user and has complete knowledge of the current experiment. Clients may be started manually, through scripts on cluster machines, or through common grid protocols. Clients have no initial knowledge of any existing PSEs; instead they rely upon the host to provide them with instructions to perform trials. Unlike several previous solutions, *Commander* relies on no specific grid technology to distribute its workload. Instead, *Commander*'s distributed system is modeled with a master-worker farm architecture, popularized by the SETI@home project.

How the *Commander* system works can be briefly outlined as follows. A user creates an experiment configuration file that the *Commander* host parses to obtain necessary information about how to conduct a specific PSE. A user then launches the client software on an arbitrary number of computers. Each client is provided with the IP address of the host machine. Through a process described in Section 3.2, the clients individually begin to run trials to complete the PSE. As the clients complete each trial, results data is stored on the host machine for data reduction and subsequent analysis. Clients continue to perform trials until the PSE is finished.

To allow clients to run on as many platforms as possible, *Commander* was written entirely in Python, a cross-platform interpreted language. Consequently to run a client, a computer only needs to have a local Python interpreter and network access to the host computer. Our research group regularly runs clients on various Unix, Linux, and OSX platforms, spread across different grid architectures, cluster networks, and desktop computers. In Section 2.5, we described our three criteria for a distributed parameter sweep engine: experiment construction, support for distributed computing, and data validation and analysis. Figure 2 shows our design of *Commander*'s architecture in response to those criteria. The following sections subsequently detail how *Commander*'s design meets those criteria.

3.1 Experiment Construction with *Commander*

The first design criterion is that a parameter sweep engine should facilitate the easy setup of new PSEs. We consequently designed a simple procedure for a PSE to be run on *Commander*. A user creates a configuration file that *Commander* parses to create a new PSE. In this file, a user can define what parameter values to sample during a PSE, and the number of trials desired for each datapoint. Also in the configuration file, a user can define the location of a project repository that contains all project-specific files needed to run the experiment.

The project repository is commonly made available through traditional file repository software (such as CVS [20] or subversion [12]), and may be run on a different computer to conserve host bandwidth. The project repository contains four types of items:

- *Project Source Code* (Required) - This would include all of the files needed to interpret, compile, link, and/or execute an experiment's trial. This could include C++ source code, Java class files, data input files, and any required libraries. At runtime, a project's main executable—a GEC algorithm—should be able to accept as input either the name of a data file (containing runtime parameters) or a list of command-line runtime parameters. Since clients may run on any architecture, project algorithms should also be capable of running on any architecture. The use of source code in this process is particularly recommended because fresh compilations can sometimes take advantage of specialty hardware. Cross-platform compilation frequently requires more effort in algorithm construction, but a user can usually be aided by an array of compilation and configuration utilities.
- *Configuration Scripts* (Optional) - Before executing a PSE trial on a client computer, it may be necessary to perform platform-specific configuration and compilation routines. A user should include such custom configuration scripts, so that an algorithm would be ready to run after the successful conclusion of a configuration process. This stage is frequently accomplished through common build utilities, such as Make or Apache ant. As the complexity of the underlying source code increases, the configuration utilities also become more complicated, particularly when cross-platform compatibility is desired. Tools such as *GNU autoconf* and *automake* can greatly simplify this process and help programmers write portable configuration utilities. These configuration scripts could then perform additional work before attempting code compilation.
- *Validation Script* (Optional) - After the successful conclusion of a trial, a client can optionally perform verification that the data output appears correct. When operating on a heterogeneous (and sometimes hostile) multi-user network, trials may not correctly finish. Local users can kill processes and machines may halt or run out of memory. It is possible to have the centralized host check for data errors, but it is more efficient to have the numerous clients each perform their own validation procedure on any locally generated data before that data is returned to the host.
- *Analysis Measures* (Optional) - *Commander* can automatically perform basic types of data reduction and generate graphs to display results for analysis. Project-specific analysis measures would be included in the project repository to analyze the data produced by a user's specific project. Analysis measures would usually be used to compute a project-specific statistic from results collected for a datapoint. For example, in a GEC experiment, it may be useful to have an analysis measure which computes the average best-of-trial fitness for a datapoint. *Commander* could then employ such an analysis measure to display the average best-of-trial information for each datapoint in the PSE. These analysis measures are discussed in greater detail in Section 3.3. On a side

note, all statistics on experimental results presented in Section 4 were originally generated by the automatic execution of *Commander* analysis measures.

To finish writing a configuration file, a user would append information about the file structure of the project repository. Specifically, a user could include the names of any optional configuration scripts, the name of an algorithm executable (the resultant binary name after the configuration script is executed), and the names of the optional validation scripts.

For many projects, the project repository would not drastically change between experiments. A key differentiator between experiments lies in which datapoints are examined. When creating an experiment configuration file, a user defines the set of datapoints used in a PSE by defining the sets of values to which parameter configurations can be set. Experiment parameters can be of three types:

- *Option* - This type of parameter has no associated value. Every datapoint in a PSE would have this option and thus every trial that is executed would run with this parameter. An example of a command line option from MGP (a genetic programming engine) could be "`mgp -useLowMem`", which tells the engine to make tradeoffs favoring a small memory footprint over execution speed. Regardless of any other parameters subsequently issued, each trial that is conducted by *Commander* would subsequently be run with the "`-useLowMem`" flag.
- *Constant* - This type of parameter has a value that is constant for every datapoint in the PSE. An example from MGP could be "`mgp -maxGenerations 200`".
- *List* - The list parameters define the parameter configurations over which *Commander* will sweep. These are like constants in that they have a value. Unlike the constant parameters, however, different datapoints can have different parameter values. For example, sweeping over population size $\{1000, 5000\}$ in MGP would generate two unique datapoints with the command lines "`mgp -popSize 1000`" and "`mgp -popSize 5000`".

When creating experiments, *Commander* creates one datapoint for each element in the Cartesian product of the List-type values. Given that list of datapoints, *Commander* than adds to each datapoint the set of constant and option parameters.

After creating a set of datapoints for a PSE, *Commander* creates trial packages for each datapoint. A trial package consists of all information needed to configure, start, and analyze a single trial—e.g., a URL of a project repository, a parameter configuration, an RNG seed, and the names of the configuration, execution, and validation scripts. When the host and clients are operational, clients receive compressed trial packages and use the contained information to execute a trial.

Table 1. Sample parameter types in an experiment configuration file. This experiment would contain four datapoints, representing the Cartesian product of parameter sets "`-popSize`" and "`-selectionMethod`".

Parameter Type	Name	Value
Option	`useLowMem`	*None*
Constant	`maxGenerations`	200
List	`popSize`	{1000, 5000}
List	`selectionMethod`	{Tournament, Fitness-Proportionate}

While it is possible to have only one trial package per datapoint, it is necessary in most GEC experiments to create multiple trial packages to achieve accuracy in later statistical analysis. After a user specifies the number of trials per datapoint, *Commander* adds trial packages with unique RNG seeds to each datapoint. For example, Table 1 contains a parameter sweep encoding for four datapoints. If the number of trials per datapoint is ten, then *Commander* would create 40 trial packages, each with unique RNG seeds. The assignment of RNG seeds is particularly important in a distributed computing environment: if two clients simultaneously run the same trial package, the results will not be identical unless both applications use the same RNG seed. Consequently, if the host receives different results for the same trial package, it might not be able to determine which trial package's data to accept without introducing sampling bias. Since an RNG seed is permanently associated with each trial package, every client that would complete the same trial would return the same data in correspondence with the trial package's RNG seed.

Commander manages RNG seeds automatically. If at a later date this experiment is rerun, there is no danger that *Commander* would accidentally reassign the same RNG seeds. However, experiment files are saved for reuse and *Commander* does allow the reuse of old RNG seeds at a user's option.

After generating trial packages for each datapoint, *Commander* saves the experiment to a local directory on the host machine for later recovery. At this point, the host is able to start serving the experiment and only waits for clients to connect and request trial packages.

3.2 *Commander*'s Distributed System Architecture

The second design criterion is that a parameter sweep engine should operate transparently and seamlessly as middleware across many types of computer networks. Consequently, *Commander* uses a master-worker architecture that has been designed for cross-platform compatibility and robust behavior to network failures.

In general, master-worker architectures have been shown to be particularly robust and efficient (e.g., in [19]) when the master relies on no single worker and each worker can complete trials without employing complicated scheduling algorithms on the host. These conditions are generally satisfied in a PSE because individual trials are decoherent—i.e., the order of trial execution does not influence the PSE results—and workers may be efficient by starting a new trial immediately after the conclusion of a current trial without requiring the host to rerun a scheduling algorithm.

The master-worker architecture of *Commander* functions as follows. The host maintains both a list of uncompleted trial packages and a filesystem directory of data packages that are generated by the results from completed trial packages. *Commander* client programs are launched on an arbitrary number of computers. Each client first conducts an initialization procedure, which includes creating a temporary directory somewhere on that client machine's filesystem. The location of this temporary directory is machine-specific, and can depend on how a client was locally installed. Clients use this directory to store project-specific files and any data files produced during a trial.

When ready to begin executing trials, each client connects to the host using an XML-RPC protocol and downloads a randomly-selected trial package. Upon receipt of a trial package, each client subsequently processes the package's contents to obtain

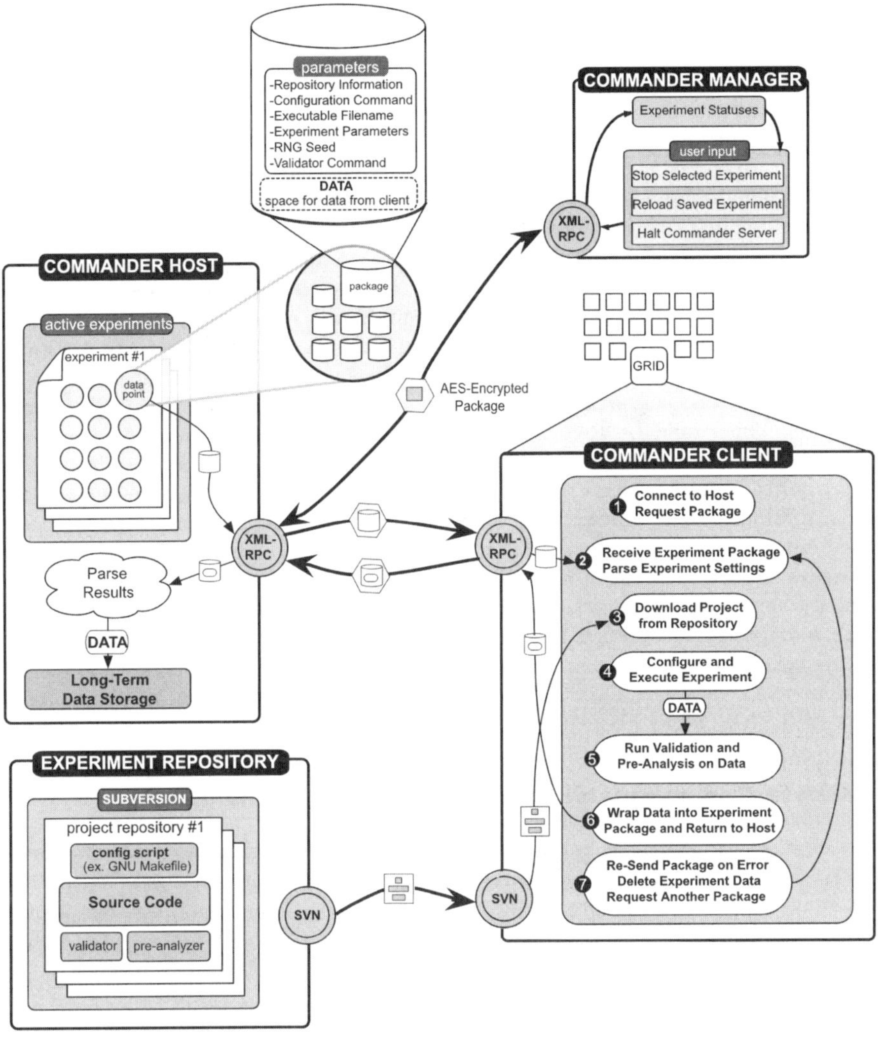

Fig. 2. *Commander*'s Architectural Diagram. *Commander* clients connect to the host, receive trial packages, configure packages, run trials, and return data packages. The master-worker architecture allows an arbitrary number of clients to be used in running experiments.

algorithm source code from the project repository, configure the source code, and execute an algorithm with the specified parameter configuration and RNG seed. In the event of an error, a client returns any error messages to the host and requests a new trial package.

During the execution of a trial, an algorithm needs to write all output files to that client's temporary directory (which was created by *Commander*). After a trial concludes and any data is analyzed for correctness, optional data reduction scripts are executed. A client then places all available results in a data package that is subsequently transferred back to the host using the XML-RPC protocol. After the host receives this data package, the data is stored in the host's local data directory for later analysis. The host then removes the corresponding trial package from the list of uncompleted trial packages.

After successful transfer of the data package, a client deletes all local copies of its data, but temporarily retains its copy of the project repository since it would likely need that information again for another trial. Each client repeats the process of "*fetch trial package, decrypt, fetch experiment files, configure, execute, validate, analyze, encrypt, and return data package*" until no more uncompleted trial packages remain on the host. Each client then shuts down, removing all project-specific material and any working directories for the purpose of returning a client's local filesystem's directory structure back to an original state.

In some cases, a user may desire secure communications between clients and the host. This would be particularly important on a multi-user network, such as those found on many university campuses. Host-client message encryption would be important for three reasons. First, encryption can help thwart man-in-the-middle attacks, which could induce clients to execute arbitrary code. Second, it guarantees that only trusted clients (i.e., those with the appropriate AES key) would be able to return results to the host machine. Third, the security of confidential algorithm source code would be restricted to trusted clients. *Commander* optionally supports AES encryption, but requires that AES keys be manually distributed to client computers before the start of an experiment. Our research group uses AES keys to secure communications on our university's computer network. We normally distribute AES keys through a shared filesystem.

3.3 Data Reduction and Analysis with *Commander*

The third design criterion is that a parameter sweep engine should support automated data reduction and analysis. Consequently, *Commander* has provisions that automatically collate and process data. Furthermore, *Commander* also has provisions for arbitrary visualizations to be generated. To understand how this is done, we describe a formal approach that allows for an otherwise intuitive process to be automated.

Although visualizations between parameters and resulting output data are helpful, the high-dimensional nature of GEC parameters and observations precludes the existence of a simple visualization strategy. *Commander* attempts to assist a researcher by preparing certain visualizations that are available for browsing immediately after the conclusion of an experiment. The types of graphs that are produced are based on mathematical constructs that are defined by a user and specified within an experiment configuration file.

In mathematics, a graph (or more precisely, a map) is defined as a set of points,

$$G = \left\{ (d,r) \,\middle|\, (d,r) \in D \times R, \text{ and } P(a,b) \text{ is True} \right\}, \tag{1}$$

for a domain D, a range R, and a property P that determines whether the point (d,r) exists in the graph. Every datapoint $\mathscr{D} \in \mathscr{D}^*$ is represented by a (d,r)-tuple in each graph type. The domain and range of each datapoint's representation is a user-defined function of the form:

$$F : \mathscr{D}^* \rightarrow D \times R. \tag{2}$$

Usually, a user's domain and range mapping takes the form:

$$F(\mathscr{D}) = \Big(F_D\big(\mathscr{C}_\mathscr{D}, \Omega_\mathscr{D}\big)\,,\; F_R\big(\mathscr{C}_\mathscr{D}, \Omega_\mathscr{D}\big)\Big), \tag{3}$$

where user-written functions F_D and F_R operate on both a datapoint $\mathscr{D}$'s parameter configuration $\mathscr{C}_\mathscr{D}$ and the collection of observations $\Omega(\mathscr{D})$, which is available from all of the datapoint's trials.

After mapping each datapoint to domain and range values, *Commander* could create a single graph defined as:

$$G = \big\{F(\mathscr{D}) \,\big|\, \mathscr{D} \in \mathscr{D}^*\big\}.$$

While such a conglomeration would display characteristics from every data-point, it is only useful for graphs where the domain is a low-dimensional space. A common technique to visualize domain-range relationships embedded in higher dimensions is to use multiple graphs. Each datapoint is represented in only one graph, and the choice of graph membership is made based on its domain classification. For example, to view 4D data, one could construct a set of 3D graphs where each graph expresses a particular value for some fourth dimensional parameter in the domain.

Commander supports such higher-dimensional graph visualizations using mathematical concepts of partitions and equivalency relations. A set S of nonempty sets of datapoints is called a partition of $\mathscr{D}^*$ if for some sets of datapoints $s_1, s_2 \in S$, $s_1 \cap s_2 = \emptyset$ and $\cup_{s \in S} s = \mathscr{D}^*$. A user defines an equivalence relation EQ such that whenever datapoints $\mathscr{D}_1, \mathscr{D}_2 \in \mathscr{D}^*$ are equivalent (i.e., $\mathscr{D}_1 \equiv_{EQ} \mathscr{D}_2$), they belong to the same partition set. Since the partition set is fully specified by EQ, we can denote the partition set as S_{EQ}. Then, for each set of datapoints $s \in S_{EQ}$, *Commander* defines a graph:

$$G_s = \Big\{F(\mathscr{D}) \,\Big|\, \mathscr{D} \in s \subseteq \mathscr{D}^*\Big\}. \tag{4}$$

For each created graph G_s, *Commander* saves a list of the graph's points to disk in the host's local directory where such data can later be viewed and interpreted by other programs.

These formalisms do not by themselves present a new way of examining experimental results. However, the development of a formal treatment for interpreting datapoints allows *Commander* to use a generic process for each PSE. This helps to simplify analysis methods for investigators who would otherwise need to develop scripts to handle data set partitioning and domain and range classification for each datapoint.

To revisit the earlier example from Table 1, we could choose to partition $\mathscr{D}^*$ using an equivalency relation where two datapoints are considered equivalent if and only if they have the same value for the "`selectionMethod`" parameter. This would produce two partition sets: one partition set would contain datapoints with "`selectionMethod`" equal to *Tournament* while the other partition set would contain datapoints with "`selectionMethod`" equal to *Fitness-Proportionate*. We could then define the domain function F_D as returning the value for the parameter configuration "`popSize`". Such a configuration would create two 2D graphs, each of which would have values for "`popSize`" as their domain.

4 Parameter Sweep Examples

In this section, we demonstrate the use of *Commander* on a sample PSE. For this experiment, we examine the effect of selection and replacement methods on two common GP test problems.

A number of authors [4, 7, 9, 18, 27, 31] have studied the behavior of selection and replacement methods from both mathematical and empirical perspectives. Although their discussions of factors such as fitness distribution, takeover time, selective pressure, and loss of diversity have fostered an understanding of these methods' basic behaviors, the effects of combining various selection and replacement strategies turns out to be not well understood. For example, although an empirical study could reveal population convergence traits of an algorithm configured to use tournament selection with steady-state replacement, that empirical study would reveal little information about the behavior of a similar algorithm configured to use fitness-proportionate selection with a generational replacement strategy. In general, as a larger number of parameter values are modified between experiments, the results become increasingly difficult to predict.

To demonstrate the utility of PSEs in understanding the nonlinear interactions between various GEC parameters, we studied two well-known GP problems while sweeping across various parameter values of different selection and replacement methods. We used 6-input multiplexer and 4-bit parity as test problems because they have been shown to be tunably-difficult under varying population size and maximum number of generations (e.g., [24]). We configured 6-input multiplexer to use logical NAND and NOR as functions while 4-bit parity used AND, OR, NAND, and NOR. Our selection methods included tournament ($q = 7$) and fitness-proportionate selection and our replacement methods included both a generational and a steady-state strategy. Trials were conducted with the MGP engine and input parameters were chosen from the available selection methods, the available replacement methods, and different combinations of population size and maximum number of generations. The PSE's parameters are described in Table 2.

In this experiment, *Commander* swept over 2 problems, 2 selection methods, 2 replacement methods, 10 population sizes and 10 maximum allowable generations for a total of 800 datapoints. Each datapoint used 100 trials for statistical validity. *Commander* distributed the 80,000 trial packages over 50 clients. The experiment took 2 CPU-months, but was completed, with analysis, within 3 days.

Table 2. Parameter configurations for parameter sweep experiments with the common GP test problems 6-input multiplexer and 4-bit parity.

Type	Field Name	Value
Constant	Max tree depth	512
List	Population Size	$\{2^x \mid 2 \leq x \leq 11, x \in \mathbb{Z}\}$
List	Max Generations	$\{2^x \mid 0 \leq x \leq 9, x \in \mathbb{Z}\}$
List	Selection Method	{tournament, fitness-proportionate}
List	Replacement Method	{generational, steady-state}
Commander Setting	Trials per datapoint	100

Table 3 contains the total number of successful trials for each configuration class, grouped by selection and replacement methods. The results clearly show that some configurations are more difficult than others by a few orders of magnitude.

Figures 3 and 4 represent the total number of successes with a more fine-grained analysis in which the number of successes is measured as a function of population size, maximum allowed generations, and choice of selection and replacement methods. The data presented in Figures 3 and 4 was produced using the procedure for automated data reduction as described in Section 3.3. For these figures, the range value of each datapoint is merely the number of successes summed over all trials associated with that datapoint. The domain value of each datapoint is represented by the parameter configuration combination, but partitioned into separate graphs based on the selection and replacement methods. Datapoint-based statistics are naturally algorithm-specific due to the data formatting policies of the algorithm under study. Although new experimental algorithms used in a PSE usually incorporate new domain and range functions, those domain and range functions can be reused in later experiments with the same algorithm.

5 Discussion

Observation: Knowledge of selection or replacement methods by themselves is not sufficient in predicting which combinations of the two will yield the better performance. The relative ordering of parameter configurations based on success rate in Table 3 cannot be explained by simple linear effects acting independently on each parameter. For both the multiplexer and parity test problems, tournament selection outperformed fitness-proportionate selection in every configuration class. However, a similar statement cannot be made for replacement methods. When paired with tournament selection, generational replacement achieved more successes than steady-state replacement. However, when paired with fitness-proportionate selection, steady-state replacement outperformed generational replacement. Researchers who assume that parameter choices exert linear and independent influences on the results would be surprised to learn that generational replacement has a beneficial effect on success when tournament selection is used but an adverse effect when fitness-proportionate selection is used. Although it is beyond the scope of this chapter to analyze the cause of this, this example does illustrate the need to conduct PSEs to characterize the mapping between parameter configurations and results. Otherwise, a GEC experiment runs the risk of being too small.

Table 3. Total number of successes out of 10,000 possible trial successes per configuration summed for each configuration class and presented for both 4-bit parity and 6-input multiplexer. The configurations exhibit the same ordering of success counts for both test problems.

Configuration Class	Parity Success	Multiplexer Success
Tournament, Generational	1066	1739
Tournament, Steady State	780	1609
Fitness-Proportionate, Steady State	43	76
Fitness-Proportionate, Generational	4	21

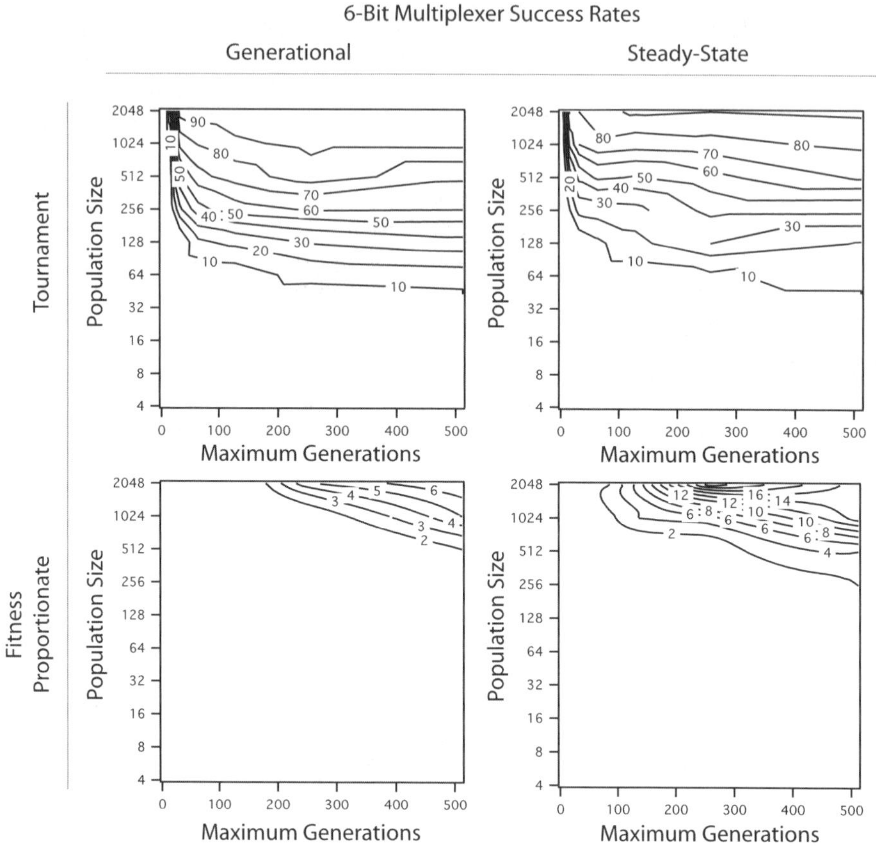

Fig. 3. Contour plots of success counts for the GP test problem 6-input multiplexer. Success counts are represented as a function of membership in a parameter configuration class. Data was collected from a PSE described in Table 2 using 40,000 trials.

In general, there are three ways in which GEC experiments can be too small:

- The stochastic nature of GEC algorithms implies that observations collected from merely a few trials may not have a confidence level high enough to be statistically significant. The use of larger numbers of trials allows a researcher to generate statistics that approximate the real unobservable behavior of a particular parameter configuration.
- It is difficult to characterize the unknown (and likely nonlinear) behavior that is associated with a set of parameters with only a small number of datapoints. To characterize and predict nonlinear behavior, one needs to sample enough values to understand the degree of nonlinearity involved. This is particularly true if modification to the value of a parameter can induce an abrupt transition on the observed behavior. If the mapping between an algorithm's parameter configurations and search behavior contains abrupt, nonlinear transitions, there is the possibility that theory derived from experimental results may be conditional

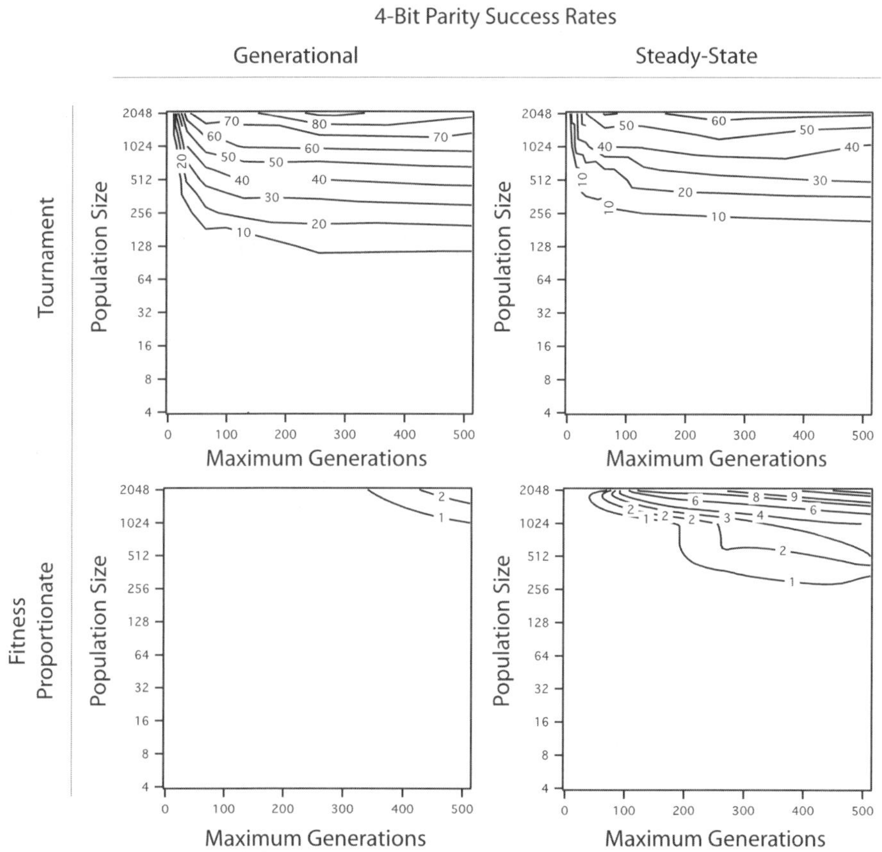

Fig. 4. Contour plots of success counts for the GP test problem 4-bit parity. Success counts are represented as a function of membership in a parameter configuration class. Data was collected from a PSE described in Table 2 using 40,000 trials.

only on a certain set of parameter values. Consequently, the failure to identify nonlinear behaviors because of an inadequate sampling of a parameter space may lead to flawed assumptions regarding corresponding theories. Extrapolation of theory across multiple transitions first requires that the location of the transitions be identified—a process that can only happen while sampling multiple datapoints for each parameter. Conversely, conducting experiments with many datapoints increases an investigator's ability to locate transitions and account for them when developing theory.

- It is difficult to understand correlations between parameters without conducting PSEs with multiple parameters. Consequently, experiments that involve too few algorithmic parameters might not be capable of generating results from which inferences about causality and correlations can be made. As demonstrated by this chapter's case study, parameters exhibit correlated and dependent behavior in influencing experimental results. Such results are based not solely on the behavior of one parameter, but on the coupled behavior of a collection of parameters that together influenced outcomes. A researcher who develops a theory of behavioral mappings between parameter configurations and results should account for such correlations between the parameters. Such correlations may also exhibit nonlinear effects resulting in transitions in higher-dimensional spaces. Without a theory in place, these multidimensional transitions can only be located by performing PSEs.

In highly nonlinear spaces, even minor modifications to an algorithm's parameter configurations can result in large qualitative changes in an experiment's results. Table 3 exhibits orders-of-magnitude improvements in trial success rate merely by switching the selection and replacement methods used in each experiment. This suggests that researchers may need to reconsider their notions of significant improvements in GEC research. Small performance improvements might not be significant when compared to the order-of-magnitude changes that can be discovered through the use of a PSE to locate transitions. Furthermore, it is difficult to judge the efficacy of GEC algorithm modifications without using PSEs because any benefits gained from such modifications might be dependent on a specific class of configuration settings. Consequently, such modifications could lose their advantage when an algorithm is configured with different parameter values.

We are aware that these suggestions for improved GEC methodology involve significant work with standard tools. Consequently, to provide a "big-picture view," we have presented a parameter sweep engine, *Commander*, to automatically perform many of the tedious tasks involved with administrating, executing, and analyzing PSEs. The process of constructing and analyzing experiments should not be tedious. Likewise, there is no fundamental reason to suggest that network trial distribution should be an inherently difficult task. We hope that tools like *Commander* enable and encourage the community to perform large PSEs, especially since *Commander* was intentionally designed to easily integrate with existing approaches to conducting GEC research (e.g., it has already been used with the popular GEC algorithms ECJ [22] and lilgp [34]).

6 Conclusions

Parameter sweep experiments are useful. The presence of nonlinearities inherent to the lower-level interactions of GEC algorithms implies that small experiments can yield an incomplete view of GEC dynamics. Large PSEs can examine multiple configurations to determine influential parameter settings. This process can lead to a more complete mapping of parameter configurations to results and a deeper understanding of GEC theory.

Parameter sweep experiments are also challenging. This methodology requires that a GEC algorithm can accept externally specified parameter values and that a researcher can define a script that can iterate over collections of parameter configurations. Conducting these experiments is expensive in terms of both human labor and computational costs because these experiments have a high administration overhead and can require many CPU hours to conduct an inherently large number of trials. An increased number of trials allows for a greater chance that errors can occur, which in turn could invalidate results. Further detection and resolution of these errors requires more scripts and more human labor. Distributed computation can alleviate the pains of lengthy trials, but introduces time-consuming code development for network distribution and robust data collection.

Commander is our solution to a generic parameter sweep engine. Commander is designed to automate as much of the experimentation process as possible by alleviating tedious human tasks, providing robust remote trial distribution, and ensuring validity in data collection. Through simple interfaces, *Commander* allows researchers to quickly define new experiments, execute them, and use analysis measures to perform immediate data reduction and quickly see results. This software addresses many of the challenges commonly associated with large PSEs. This should encourage researchers to perform more experiments that examine nonlinear correlations between multiple parameters in the configurations of GEC algorithms.

Commander has been publicly released under an open-source GPL license. Versioned downloads and current development status can be found on our website at `http://sitemaker.umich.edu/umacers/commander`

7 Future Work

There are numerous ways to increase the effectiveness of automated experimentation. Currently, *Commander* conducts a factorial design experiment that is based on a user's parameter inputs. Future versions could support other optional design-of-experiment techniques, such as fractional factorial design or Latin Squares to reduce the requisite number of datapoints for a particular experiment. Further automation is possible if data analysis stages are allowed to provide feedback in *Commander*'s subsequent choice of datapoint parameter configurations. Future versions of *Commander* could perform analysis of variance tests and statistical correlation tests to determine how many times to sample a stochastic datapoint and which new parameter configurations to sample for new datapoints. Such variance and correlation statistics could also be reported to a user without necessitating direct requests for such analysis.

Finally, observational data frequently exists in high-dimensional results spaces, which makes analysis difficult. Standardized techniques exist for data dimension-reduction (e.g., principal component analysis or manifold learning), which could take advantage of correlations in a results space to identify the structure of the mapping between parameter configurations and results. Such a structural mapping could then provide a deeper theoretical insight into the behavior of the algorithm under study.

8 Acknowledgments

The authors would like to thank the other members in our research group, UM-ACERS, for computational support and helpful comments on an earlier draft of this chapter. The first author would like to thank his colleagues at the Toyota Technical Center for their discussions. The first and third authors gratefully acknowledge I. Kristo.

Appendix: The *Highlander* problem

The *Highlander* GP problem was designed to test the limits of building block mixing and assembly by coercing a GP system to attempt to construct a solution with certain characteristics, as determined by a tunable parameter $\beta \in [0, 1]$.

The *Highlander* problem proceeded as follows. An initial population of trees was constructed from the arity-2 function set $\{Join\}$ and the terminal set $\{x\}$. During the course of evolution, these trees were not evaluated for content. Instead, each tree was assigned fitness based on the number of ancestral building blocks that participated in the composition of the current tree.

To audit this procedure, every node of every tree in the initial population was assigned a unique integer ID number. Subsequent generations were constructed using only the crossover operation—no mutation was applied. The crossover operation was modified to preserve the integer ID of each node. Consequently, it was possible to examine a tree in a later generation and determine the initial population ancestor of each node in that tree.

Fitness was based on the fraction of initial population nodes found within each tree. In particular, a percentage p was calculated, representing the percentage of nodes from the initial population found within the current tree. Raw fitness was established as

$$f = |\beta - p|. \tag{5}$$

In this fashion, a tree attains maximum fitness when $p = \beta$, i.e., when the fraction of initial population genetic material within a tree reaches a user-specified value. A value of $\beta \approx 1$ would cause GP to attempt to evolve a solution using all of the nodes available from the initial population. Likewise, a value of $\beta \approx 0$ would cause GP to evolve a solution using very little initial population material.

We found that the *Highlander* problem exhibits at least two transition regions. As shown in Figure 1, there are values of β for which GP can easily find a solution. This region of the settings for β is, however, small—there is a much larger region of β-values for which GP is unable to evolve a successful solution. A second transition—based on the population size—was identified that determined the behavior of the β-transition. The findings presented shed light on the ease at which a GP system may incorporate initial population material into any single tree of a later generation. Specifically, the *Highlander* problem suggests a critical point at the β-transition as an upper bound for the fraction of initial population material available to a later generation solution.

We refer the interested reader to our full discussions in [15].

References

1. D. Abramson, R. Sosic, J. Giddy, and M. Cope. The laboratory bench: Distributed computing for parametised simulations. In *Proceedings from the 1994 Conference on Parallel Computing and Transputers*, pages 12–27, 1994.
2. David Anderson. Boinc documentation. `http://boinc.berkeley.edu`, 2005.
3. Robert Axelrod. The dissemination of culture: A model with local convergence and global polarization. *Journal of Conflict Resolutions*, 41(2):203–226, 1997.

4. Thomas Bäck. Selective pressure in evolutionary algorithms: A characterization of selection mechanisms. In *Proceedings of the First IEEE Conference on Evolutionary Computation*, pages 57–62, Piscataway, NJ, 1994. IEEE Press.
5. J. Basney, M. Livny, and P. Mazzanti. Harnessing the capacity of computational grids for high energy physics. In *Proceedings of the 2000 Conference on Computing in High Energy and Nuclear Physics*, 2000.
6. Ted Belding. Drone manual. `http://drone.sourceforge.net`, 2005.
7. Tobias Blickle and Lothar Thiele. A comparison of selection schemes used in genetic algorithms. TIK-Report 11, Technische Informatik und Kommunikationsnetze Institut, Computer Engineering and Networks Laboratory, ETH, Swiss Federal Institute of Technology, Gloriastrasse 35, 8092 Zurich, Switzerland, December 1995.
8. Sven A. Brueckner and H. Van Dyke Parunak. Resource-aware exploration of the emergent dynamics of simulated systems. In *Proceedings of the Second International Joint Conference on Autonomous Agents and Multiagent Systems*, pages 781–788. ACM Press, 2003.
9. Erick Cantú-Paz. Order statistics and selection methods of evolutionary algorithms. *Information Processing Letters*, 82(1):15–22, 2002.
10. Henri Casanova, T. Bartol, J. Stiles, and Francine Berman. Distributing mcell simulations on the grid. *International Journal of High Performance Computing Applications*, 15(3):243–257, 2001.
11. Henri Casanova, Graziano Obertelli, Francine Bermand, and Richard Wolski. The apples parameter sweep template: User-level middleware for the grid. In *Proceedings of Super Computing 2000*, pages 75–76, 2000.
12. Ben Collins-Sussman, Brian Fitzpatrick, and C. Pilato. Version control with subversion. `http://svnbook.red-bean.com`, 2005.
13. Jason M. Daida, Derrick S. Ampy, Michael Ratanasavetavadhana, Hsiaolei Li, and Omar A. Chaudhri. Challenges with verification, repeatability, and meaningful comparison in genetic programming: Gibson's magic. In *Proceedings of the Genetic and Evolutionary Computation Conference*, volume 2, pages 1851–1858, Orlando, Florida, USA, 13-17 July 1999. Morgan Kaufmann.
14. Jason M. Daida, Steven Ross, Jeffrey McClain, Derrick Ampy, and Michael Holczer. Challenges with verification, repeatability, and meaningful comparisons in genetic programming. In *Genetic Programming 1997: Proc. of the Second Annual Conference*, pages 64–69, Stanford University, CA, USA, 13-16 July 1997. Morgan Kaufmann.
15. Jason M. Daida, Michael E. Samples, and Matthew J. Byom. Probing for limits to building block mixing with a tunably-difficult problem for genetic programming. In *GECCO 2005: Proceedings of the 2005 conference on Genetic and Evolutionary Computation*, volume 2, pages 1713–1720, Washington DC, USA, 25-29 June 2005. ACM Press.
16. Jason M. Daida, Michael E. Samples, Bryan Hart, Jeffry Halim, and Aditya Kumar. Demonstrating constraints to diversity with a tunably difficulty problem for genetic programming. In *Proceedings of the 2004 IEEE Congress on Evolutionary Computation*, pages 1217–1224, Portland, Oregon, 20-23 June 2004. IEEE Press.
17. Jason M. Daida, David Ward, Adam Hilss, Stephen Long, and Mark Hodges. Visualizing the loss of diversity in genetic programming. In *Proceedings of the 2004 IEEE Congress on Evolutionary Computation*, pages 1225–1232, Portland, Oregon, 20-23 June 2004. IEEE Press.

18. David Goldberg and Kalyanmoy Deb. A comparative analysis of selection schemes used in genetic algorithms. In *Proceedings of the 1991 Foundations of Genetic Algorithms*, pages 69–93, San Mateo, California, 1991. Morgan Kauffmann.
19. Jean-Pierre Goux, Sanjeev Kulkarni, Jeff Linderoth, and Michael Yoder. An enabling framework for master-worker applications on the computational grid. In *Proceedings from the 9th IEEE Symposium on High Performance Distributed Computing*, pages 43–50. IEEE Press, 2000.
20. Dick Grune, Brian Berliner, and Jeff Polk. Concurrent versions system. `http://nongnu.org/cvs`, 2006.
21. Eric Korpela, Dan Werthimer, David Anderson, Jeff Cobb, and Matt Lebofsky. SETI@Home - Massively Distributed Computing for SETI. *Computing in Science and Engineering*, 3(1):78–83, 2001.
22. Sean Luke. ECJ Manual. `http://cs.gmu.edu/~eclab/projects/ecj/`, 2005.
23. Sean Luke and Liviu Panait. Is the perfect the enemy of the good? In *GECCO 2002: Proceedings of the Genetic and Evolutionary Computation Conference*, pages 820–828, New York, 9-13 July 2002. Morgan Kaufmann Publishers.
24. Sean Luke and Lee Spector. A revised comparison of crossover and mutation in genetic programming. In *Genetic Programming 1998: Proceedings of the Third Annual Conference*, pages 208–213, University of Wisconsin, Madison, Wisconsin, USA, 1998. Morgan Kaufmann.
25. Bryan Mendez. Stardust@home. `http://stardustathome.ssl.berkeley.edu`, 2006.
26. Mark M. Meysenburg and James A. Foster. Random generator quality and GP performance. In *Proceedings of the Genetic and Evolutionary Computation Conference*, volume 2, pages 1121–1126, Orlando, Florida, USA, 13-17 July 1999. Morgan Kaufmann.
27. Tatsuyo Motoki. Calculating the expected loss of diversity of selection schemes. *Evolutionary Computation*, 10(4):397–422, 2002.
28. Vijay Pande. Folding@home. `http://folding.stanford.edu`, 2006.
29. Norman Paterson and Michael Livesey. Performance comparison in genetic programming. In *Late Breaking Papers at the 2000 Genetic and Evolutionary Computation Conference*, pages 253–260, Las Vegas, Nevada, USA, 8 July 2000.
30. T. Quarles, D. Pederson, R. Newton, A. Sangiovanni-Vincentelli, and Christopher Wayne. Spice documentation. `http://bwrc.eecs.berkeley.edu/Classes/IcBook/SPICE/`.
31. Gilbert Syswerda. A study of reproduction in generational and steady state genetic algorithms. In *Foundations of Genetic Algorithms 1990*, pages 94–101, San Mateo, USA, 15 July 1991. Morgan Kauffman.
32. D. Thain and M. Livny. *Building Reliable Clients and Servers*, chapter 19, pages 285–318. Morgan Kauffman, 2003.
33. George Woltman. Great internet mersenne prime search. `www.mersenne.org`, 2006.
34. Doug Zongker and Dr. Bill Punch. lilgp. `http://garage.cse.msu.edu/software/lil-gp/`, 1998.

Adaptive Population Sizing Schemes in Genetic Algorithms

Fernando G. Lobo[1,2] and Cláudio F. Lima[1]

[1] UAlg Informatics Lab,DEEI-FCT, University of Algarve,
Campus de Gambelas, 8000-117 Faro, Portugal
[2] Also a member of IMAR - Centro de Modelação Ecológica.
flobo@ualg.pt
clima@ualg.pt

Summary. This chapter presents a review of adaptive population sizing schemes used in genetic algorithms. We start by briefly revisiting theoretical models which rely on a facetwise design decomposition, and then move on to various self-adjusting population sizing schemes that have been proposed in the literature. For each method, the major advantages and disadvantages are discussed. The chapter ends with recommendations for those who design and compare self-adjusting population sizing mechanisms for genetic and evolutionary algorithms.

1 Introduction

In a traditional genetic algorithm (GA), the population size is set by the user to a fixed value at the beginning of the search and remains constant through the entire run. Having to specify this initial parameter value is problematic in many ways. If it is too small the GA may not be able to reach high quality solutions. If it is too large the GA spends unnecessary computational resources. Finding an adequate population size is a difficult task. It has been shown, both theoretically and empirically, that the optimal size is something that differs from problem to problem. Moreover, some researchers have observed that at different stages of a single run, different population sizes might be optimal.

A somewhat widely accepted intuition behind population sizing is that it should be set proportionally to the problem's size and difficulty. However, problem difficulty is very hard to estimate for real-world problems, which brings us back to the difficulty of setting the appropriate population size. Faced with such difficulties, many users end up either using a so-called "standard setting" (50-100 individuals), guessing a number, or doing some experimentation with a number of different sizes to see which one works best. Guessing right is pure luck, and most likely a user guesses wrong by choosing a population size that is either too small or too large for the problem at hand. The scenario is illustrated in Figure 1.

F.G. Lobo and C.F. Lima: *Adaptive Population Sizing Schemes in Genetic Algorithms*, Studies in Computational Intelligence (SCI) **54**, 185–204 (2007)
www.springerlink.com

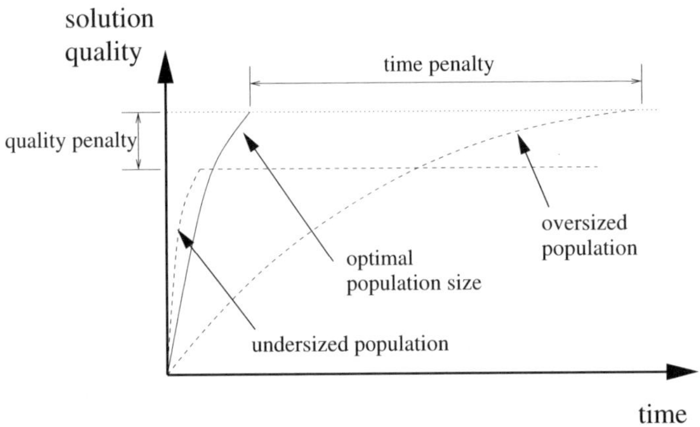

Fig. 1. Population sizing in genetic algorithms. Too small and the user pays a quality penalty. Too large and the user pays a time penalty. (Copied with permission from Lobo and Goldberg [17])

Based on these observations, researchers have suggested various schemes that try to learn a good population size during the GA run itself. The various techniques basically differ in the motivation that supports the reasoning behind each one. Typically, methods that adapt the population size are based either on biological grounds, population sizing theory, or simply empirical evidence.

The rest of this chapter is organized as follows. We start by briefly reviewing theoretical work on population sizing which is largely based on Goldberg's facetwise decomposition for designing competent GAs and relies on the notion of a building block (BB) [7]. After that, we review a number of techniques that have been proposed to adjust the population size during the GA run itself. The advantages and disadvantages of each method are highlighted along the way. The chapter ends with recommendations for those who propose and compare adaptive population sizing schemes for genetic and evolutionary algorithms.

2 A Brief Review of Population Sizing Theory

In this section we review relevant theoretical studies that help our understanding on the role that the population size has in terms of GA performance and solution quality. Most of these studies use a facetwise approach to get a better insight in population sizing dynamics, and for that some assumptions are made: (1) work with selectorecombinative GAs (no use of mutation), (2) use fixed-length and binary-coded strings, (3) use fixed-size and non-overlapping populations, and (4) solve stationary objective functions. Although these assumptions are made for computational and analytical tractability, most of them can be relaxed with small or no changes.

Despite these assumptions and the operational simplicity of GAs, they are still complex systems to analyze. To have a better understanding on how to design better GAs a decomposition approach [7] has been taken by many. In this way, it is assumed that the problem to be solved is additively decomposable in a number of m subfunctions. Each subfunction maps to k decision variables and corresponds to a building block partition. Under this definition, it is possible to have 2^k BBs in a partition, one of which is superior to the others and belongs to the global optimal solution.

We start by reviewing the work on initial supply of BBs. Next, the issue of deciding well between competing BBs is revisited and major results are presented under the form of *dimensionless models* that relate the population size (N) with certain problem features, such as problem size (ℓ), BB size (k), number of subfunctions (or BB partitions) (m), and others. Finally, the population size requirements for correct identification and mixing of BBs is addressed in the context of *Probabilistic Model Building Genetic Algorithms* (PMBGAs), also known as *Estimation of Distribution Algorithms* (EDAs) [16, 31].

2.1 Initial Supply of Building Blocks

When using selectorecombinative GAs (no mutation), the only source of diversity is the supply of raw BBs in the initial generation. Once the randomly initialized population is filled up with enough BBs, it is more likely that the GA will be able to propagate and mix them, converging to good solutions. A simple supply model considers the number of BBs present in the initial random population (generated under a uniform distribution), where the probability to generate a single BB of size k is $1/2^k$, for binary encodings. Therefore, the initial supply of BBs can be estimated as $x_0 = N/2^k$. From this simple relation it is possible to notice that the population required to supply all possible BBs grows exponentially with the BB size k. This suggests that problems with short BBs require smaller populations than the ones with longer BBs.

2.2 Decision Making Between Competing Building Blocks

The second aspect where population size plays an important role is in the decision-making between competing BBs. Holland [13] early recognized this issue as a statistical decision problem. He recasted this problem as a cluster of parallel and interconnected two-armed bandit problems. Although this is an idealization of the GA behavior, his results gave an optimistic bound on the allocation of trials in a GA. De Jong [4] also used this analogy and presented equations for trial allocation that removed some of the assumptions made by Holland [13]. He also recognized the role of signal-to-noise ratio on the population sizing question. Years later, Goldberg and Rudnick [9] presented a method to calculate the variance of schema fitness using Walsh transforms, and also proposed an estimate for the population size based on the variance of fitness. Following this work, Goldberg, Deb, and Clark [8] proposed a decision-based model that gives a conservative bound on the convergence quality of selectorecombinative GAs for a given population size. The model focuses on the correct decision between the best BB and its closer competitor in a given partition. This decision process takes place in the presence of noise that comes from the other partitions.

Let us consider a competition between two individuals that contain the best BB and the second best BB (in the partition of interest). The GA will probably choose the individual with the best BB given its higher fitness contribution to the overall fitness. However, in a single trial, the wrong decision can also be made because the remaining $m - 1$ partitions also contribute to the fitness of the individuals in competition, and act as noise in the decision making process in the particular partition. By increasing the population size, it is possible to make this decision error as small as possible.

Based on this observation, Goldberg et al. [8] derived a population sizing equation which gave, for additive decomposable problems with uniformly scaled BBs , the required population size (in order to correctly solve each BB with a specified error probability α) in terms of the BB size k, the total number of BBs m, and the fitness difference d between the best and second best BBs.

Goldberg et al. [8] confirmed that the model conservatively estimates the convergence quality of the GA. This somewhat pessimistic estimation is due to the fact that the model approximates the behavior of the GA by the outcome of the first generation. If the wrong BBs are selected in the initial generation the model assumes that the GA will be unable to recover from the error.

A few years later, Harik, Cantu-Paz, Goldberg, and Miller [10] refined this model by incorporating the initial BB supply as well as cumulative effects of decision making over an entire GA run. Their model was based on the well-known gambler's ruin problem and resulted in the following population sizing equation

$$N = -2^{k-1} \ln(\alpha) \sqrt{\pi (m-1)} \frac{\sigma_{BB}}{d}, \quad (1)$$

where σ_{BB} is the fitness variance within each BB partition, and α is the failure rate. Experimental results from the original paper show that this equation is a reliable estimate for population sizing on decomposable problems with uniformly-scaled and near-uniformly-scaled building blocks. Additionally, the authors also proposed an expression to population sizing in the presence of exogenous noise as well as for different selection pressures.

However, if the fitness contribution to the overall fitness function is exponentially-scaled, the model can not estimate the population size correctly. For that case, Thierens, Goldberg, and Pereira [39], for BBs with unit size, and later, Lobo, Goldberg, and Pelikan [18], for a general BB size, developed a model based on the concepts of domino convergence and random genetic drift.

In this kind of problems, BBs converge in a sequential manner from the most salient BBs to the least salient ones. The least salient BBs only get the attention of the selection operator later in the run, and may be lost from the population due to chance variation alone, a process known as *random genetic drift*. It has been shown that the drift process is the crucial aspect regarding population sizing for this type of problems. That is, the population size has to be large enough to prevent the low salient BBs from getting extinct from the population during the GA run. The theoretical models for exponentially-scaled BBs reveal that the population size should grow as $\mathcal{O}(m)$, instead of $\mathcal{O}(\sqrt{m})$ for the uniformly-scaled case.

2.3 Proper Identification and Mixing of Building Blocks

The population sizing models discussed above assume perfect mixing between BBs, which in practice might not occur when using fixed, problem-independent recombination operators. The issue of correct identification and mixing of BBs can be related to the population size if we consider a recent class of GAs known as estimation of distribution algorithms (EDAs) [16, 31]. In EDAs the underlying problem structure is learned on-the-fly for proper identification and mixing of BBs. EDAs have very strong connections with Data Mining algorithms. Both interpret the population as data and use techniques to infer patterns in that data that are associated with high fitness. Without sufficient data it is not possible to do accurate inference. Fortunately, within a GA framework, having more data translates into working with larger populations.

EDAs use probabilistic models to induce non-linearities between variables in order to detect the proper decomposition of the problem. In order to be able to do that the EDA requires a minimal population size to detect the correlations. Pelikan, Goldberg, and Cantu-Paz [20] were the first to approach this issue, reporting that for the Bayesian Optimization Algorithm (BOA) the population size in order for the model to be of required quality grows at least linearly with the number of BBs (m) in a problem.

A more detailed analysis was made later by Pelikan, Sastry, and Goldberg [24] to investigate the critical population size to discovers good dependencies (between variables that are correlated) and bad dependencies (between variables that are independent) with BOA. They concluded that the population size should be greater than $\mathcal{O}(\ell^{1.05})$ and smaller than $\mathcal{O}(\ell^{2.1})$, where ℓ is the problem size. The upper bound has to be respected in order to avoid that the probabilistic model discover dependencies between independent variables.

Typically, the population size requirements for initial supply and decision-making are bounded by the requirements for correct model building.

Before finishing this part of the chapter, we would like to stress that although not trivial to put in practice, the theoretical models on population sizing are very important and crucial for understanding the role of the population in a GA. Among other things, an important lesson of those models is that setting the population size to a fixed value regardless of the problem's size and difficulty, is a mistake that is likely to result in poor performance.

3 Adaptive Population Sizing Schemes

This section reviews a number of techniques that have been developed to adjust the population size during the GA run itself. The motivation for developing these schemes are essentially threefold: (1) the recognition that an adequate sizing of the population is problem dependent and is hard to estimate, (2) the observation that a GA with an adaptive population sizing scheme may yield a better solution quality and better performance than a GA with a fixed (and poorly set) population size, and (3) to make life easier for users by eliminating the population sizing parameter.

The exposition of the various methods follows in chronological order, starting with Smith and Smuda's [28] 1993 work and ending with Yu et al. [33] 2005 work. Incidentally, both works are attempts to incorporate existing population sizing theory in an algorithmic form.

3.1 Population Sizing Through Estimates of Schema Variances (1993)

Smith and Smuda [28, 29] tried to transfer the population sizing theory of Goldberg et al. [8] to a more practical context by incorporating their population sizing equation in the GA itself. The authors wanted to remove the parameters of the GA, and starting with the population size seemed like a logical first step because experience had shown that the GA was quite robust regarding the other parameters.

Based on Goldberg et al.'s theoretical work, Smith and Smuda suggested an algorithm for autonomously adjusting the population size as the search progresses. They argued that the parameters of the GA are hard to relate to the user's need. That is, the user typically does not know how the population size affects the solution quality of the problem. The authors suggested that the parameters should have more meaning from the user's point of view, and proposed an algorithm that only required the user to specify a desired accuracy for the overall search, something equivalent to the building block's signal from Goldberg et al.'s equation. In order to do that, Smith and Smuda defined the *expected selection loss* between two competing schemata as the probability that the GA makes a selection error (selection chooses the lower fit schema) weighted by their fitness difference. Then, they suggested an algorithm that does online estimates of schema fitness variances, and sizes the population so that the expected selection loss approximates the user's specified target accuracy.

Although they suggested and implemented such a scheme, the resulting algorithm had several limitations. First, it is hard to relate the user's specified selection target loss with the actual accuracy of the GA in solving the problem. Second, the estimation of schema fitness variances is very noisy, because the GA samples many schemata simultaneously. Third, the population sizing model [8] assumes that the building block mixing is going to be nearly perfect, which may not occur in practice. Nonetheless, the work was very important because up to that point, adaptive population sizing schemes were practically inexistent.

3.2 Population Sizing in GAVaPS (1994)

Arabas, Michalewicz, and Mulawka [1] proposed the *Genetic Algorithm with Varying Population Size* (GAVaPS). As opposed to Smith and Smuda's work, this method does not rely on the population sizing theory [8]. Instead, it relies on the concept of *age* and *lifetime* of an individual. When an individual is created, either during the initial generation or through a variation operator, it has age zero. This step corresponds to the birth of the individual. Then, for each generation that the individual stays alive its age is incremented by 1.

At birth, every individual is assigned a lifetime which corresponds to the number of generations that the individual stays alive in the population. When the age exceeds the individual's lifetime, the individual dies and is eliminated from the population. At every generation, a fraction ρ (called the *reproduction ratio*) of the current population is allowed to reproduce. Every individual of the population has an equal probability of being chosen for reproduction. Thus, GAVaPS does not have an explicit selection operator as traditional GAs do. Instead, selection is achieved indirectly through the lifetime that is assigned to individuals. Those with above-average fitness have higher lifetimes than those with below-average fitness. The idea is that the better an individual is, the more time it should be allowed to stay in the population, and therefore increase the chance to propagate its traits to future individuals.

Arabas et al. mentioned that different lifetime strategies could be implemented based on the idea of reinforcing individuals with high fitness and restricting individuals with low fitness, and suggested three different strategies: proportional, linear, and bi-linear allocation. All those strategies rely on two parameters, *MinLT* and *MaxLT*, which correspond to the minimum and maximum lifetime value allowable for an individual. The proportional strategy is inspired in the roulette wheel selection scheme, the linear strategy relates the fitness of an individual with the best fitness found so far, the bi-linear strategy is a compromise between the first two strategies. For completeness, and because it is also used in the APGA algorithm (see later in the chapter), the details of the bi-linear strategy used to compute the lifetime of an individual i is shown below.

$$LT(i) = \begin{cases} MinLT + \eta \frac{fit[i] - MinFit}{AvgFit - MinFit} & \text{if } AvgFit \geq fit[i] \\ \frac{MinLT + MaxLT}{2} + \eta \frac{fit[i] - AvgFit}{MaxFit - AvgFit} & \text{if } AvgFit < fit[i] \end{cases}$$

where $\eta = (MaxLT - MinLT)/2$, $MinFit$, $MaxFit$, and $AvgFit$ are the minimum, maximum, and average fitness of the population, and $fit[i]$ is the fitness of individual i. Figure 2 shows the pseudo-code of GAVaPS.

Arabas et al. argued that this approach seems natural because the aging process is well-known in all natural environments. In their original work, the authors tested GAVaPS on four test functions, compared its performance with that of a simple GA using a fixed population size, and observed that GAVaPS seemed to incorporate a self-tuning process of the population size.

GAVaPS requires the user to specify the initial population size, but the authors referred that GAVaPS is robust with respect to that, i.e., the initial population size seemed to have no influence on the performance for the test functions chosen. The same thing did not hold with the reproduction ratio ρ which was found to have a large influence in the performance. Work done more recently [31, 5] also reveals that GAVaPS is very sensitive to the reproduction ratio parameter.

For the *MinLT* and *MaxLT* parameters, Arabas et al. used 1 and 7 respectively. However, they did not give recommendations on how those parameters should be set.

The ideas presented in GAVaPS were later explored by Fernandes and Rosa [6] and by Bäck, Eiben, and Van der Vaart [3] (more about that later).

3.3 Strategy Adaptation by Competing Subpopulations (1994, 1996)

Schlierkamp-Voosen and Mühlenbein [26] proposed an adaptive scheme which is inspired by the biological observation that well adapted species increase and poorly adapted species decrease when competing for the same resource (e.g. food).

In their scheme there are S competing populations, each running a different search strategy. The search strategies run independently in each population, and at regular intervals the populations compete with each other. The idea is to increase the size of the best performing population, and decrease the size of the others.

```
procedure GAVaPS
begin
  t = 0;
  initialize pop(t);
  evaluate pop(t);
  while not termination-condition do
  begin
     t = t+1;
     pop(t) = pop(t-1);
     increase age of pop(t) members by 1;
     cross and mutate pop(t);
     evaluate pop(t);
     remove from pop(t) all individuals
        with age greater than their lifetime;
  end
end
```

Fig. 2. Pseudocode of the genetic algorithm with varying population size (GAVaPS).

In order to do so, the authors define a *quality criterion* which scores the performance of a population, as well as a *gain criterion* which gives the amount of increase or decrease in the size of the competing populations. In their scheme, the sum of the sizes of all populations remains constant through time. After each competition, in addition to adapting the population sizes, the best performing individual migrates to the other populations, giving them a chance to perform better for future competitions.

Later on, the same authors [27] extended the model by allowing a *consumption factor* γ to be assigned to each population. The idea was motivated by the recognition that different search strategies seem to require different population sizes to perform efficiently. For example, the authors recognized that mutation is more efficient with small populations and recombination works best with large populations. As opposed to the basic competition model, the extended one allowed the total population size to change during the whole simulation. According to the authors, the extended competition scheme is useful for location good regions of attraction with a breadth search algorithm, and also useful to fine-tune solutions using a more direct search method.

The scheme suggested by Schlierkamp-Voosen and Mühlenbein is, in some sense, an hybrid algorithm consisting of multiple search strategies, each utilizing different resources (different population sizes and potentially different variation operators). The overall scheme reinforces those strategies that appear to work well, and penalizes those that appear to work poorly.

The basic competition model has a severe limitation that the sum of the population sizes of all strategies remained constant during the whole simulation. The extended model allowed the total population size to change during the whole simulation but it is not clear how sensitive the algorithm is with respect to the consumption factor γ, the loss factor used in the gain criterion, and the evaluation interval between successive competitions. On the other hand, the idea of competing populations is interesting and has subsequently been explored by others [12, 11].

3.4 Population Sizing in SAGA (1996)

Hinterding, Michalewicz, and Peachey [12] proposed the *Self-Adaptive Genetic Algorithm* (SAGA) which included an adaptive population sizing scheme. The basic idea is to run three independent GAs, each with its own population (call them $P1$, $P2$, $P3$ with $size(P1) < size(P2) < size(P3)$) which are initially set to 50, 100, and 200, and have rules that adjust the population sizes in order to "optimize" the performance of the mid-sized population $P2$.

Hinterding et al. allowed the populations to range between 10 and 1000 individuals, and made sure that there should always be a difference of at least 20 individuals in their sizes in order to avoid population sizes which are too similar.

At regular intervals (an *epoch* in the authors' terminology), the best fitness found in each GA is used as a criterion to adjust the population sizes. The adjusting mechanisms are based on a number of rules. Specifically, there is a "move-apart" rule that states that two populations should be moved apart when the fitness of their best individuals are within a threshold $\epsilon = 10^{-9}$. In such cases, moving the populations apart means halving the size of the smallest population, and doubling the size of the largest population. This situation also holds for the case that all three populations are within the ϵ fitness threshold. In such case, the mid-size population is left untouched, while the others grow and shrink.

In case the fitness values of the best individuals of the three populations are not within ϵ distance, the following rules apply (below, `f` stands for the fitness of the best individual):

```
if f(P1) < f(P2) < f(P3): move right

if f(P3) < f(P2) < f(P1): move left

if f(P1) < f(P3) < f(P2)
or f(P2) < f(P1) < f(P3): compress left

if f(P2) < f(P3) < f(P1)
or f(P3) < f(P1) < f(P2): compress right
```

where `move right`, `move left`, `compress left`, and `compress right` are defined as follows:

```
move right: size(P1) = size(P2);
            size(P2) = size(P3);
            size(P3) = size(P3)*2;
```

```
move left: size(P1) = size(P1)/2;
           size(P2) = size(P1);
           size(P3) = size(P2);

compress left: size(P1) = (size(P1)+size(P2))/2;

compress right: size(P3) = (size(P2)+size(P3))/2;
```

The reasoning behind these rules is to increase (or decrease) all the population sizes if the "ideal" population size appears to be out of the range of the current population sizes. On the other hand, if the "ideal" population size appears to be within the current population sizes, then the rules should adjust the size of the worst performing population so that its size becomes closer to the size of the other two populations.

The time span of an epoch was chosen by Hinterding et al. as 1000 fitness function evaluations for each GA. The authors also described different strategies that either allowed the populations to mix or to continue running separately after each epoch.

The major drawback of SAGA is, in our opinion, the time span parameter as well as the existence of a maximum population size bound. It is not clear why 1000 fitness evaluations is utilized for the time span. Also, setting an upper bound for the population size goes against all existing theoretical work on population sizing [8, 10, 39, 18, 20, 24]. On the other hand, the population resizing heuristics introduced by SAGA are quite interesting and have some similarities with those implemented later by the parameter-less GA technique that we are about to explore next.

3.5 Population Sizing in the Parameter-less GA (1999)

The parameter-less GA introduced by Harik and Lobo [11] was developed with the assumption that solution quality grows monotonously with the population size. That is, if after some number of generations t, a GA is able to reach a solution quality q with some population size N_1, then (in a statistical sense) it would also be able to reach at least that same solution quality had it started with a population size $N_2 > N_1$. Based on that observation, Harik and Lobo suggested a scheme that simulates an unbounded (potentially infinite) number of populations running in "parallel" with exponentially increasing sizes. Their scheme gives preference to smaller sized populations by allowing them do do more function evaluations than the larger ones. The rationale for taking such a decision relies on the observation that all other things being equal, a smaller sized population should be preferred. After all, if a GA can reach a good solution quality (goodness depends on the stopping criteria, of course) with a population size N_1, then why bother spending more efforts with a population size $N_2 > N_1$.

Initially the algorithm only has one population whose size is a very small number N_0 (say 4). As time goes by, new populations are spawned and some can be deleted (more about that later). Thus, at any given point in time, the algorithm maintains a collection of populations. The size of each new population is twice as large as the previous last size. The parameter-less GA does have one parameter (although Harik and Lobo fixed its value to 4). That parameter (call it m) tells how much

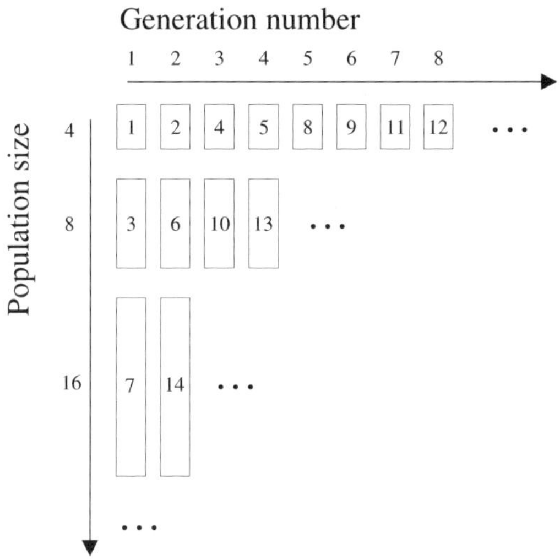

Fig. 3. The parameter-less GA simulates an unbounded number of populations. The numbers inside the rectangles (populations) denote the sequence in which the generations are executed by the algorithm. The next step of the algorithm (step 15, not shown in the figure) would spawn a new population with size 32.

preference the algorithm gives to smaller populations. Specifically, it tells how many generations are done by a population before the next immediate larger population has a chance to do a single generation.

An example is helpful to illustrate the mechanism. Figure 3 depicts an example with $m = 2$. Notice that the number of populations is unbounded. The figure shows the state of the parameter-less GA after 14 iterations. At that point, the algorithm maintains three populations with sizes 4, 8, and 16. Notice how a population of a given size does $m = 2$ more generations than the next larger population.

This special sequence can be implemented with a m-ary counter as suggested by the original authors [11], and also with a somewhat simpler implementation as suggested by Pelikan and Lin [22].

In addition to maintaining an unbounded collection of populations, the algorithm uses a heuristic to eliminate populations when certain events occur. In particular, when the average fitness of a population is greater than the average fitness of a smaller sized population, the algorithm eliminates the smaller sized one. The rationale for taking this decision is based on the observation that the larger population appears to be performing better than the smaller one, and it is doing so with less computational resources (recall that the algorithm gives preference to smaller populations). Thus, whenever such an event occurs, Harik and Lobo reasoned that as a strong evidence that the size of the smaller population was not large enough, and the algorithm should not waste any more time with it. By doing so, the algorithm maintains an invariant that the average fitness of the populations are in decreasing order, with smaller sized populations having higher average fitness than larger populations.

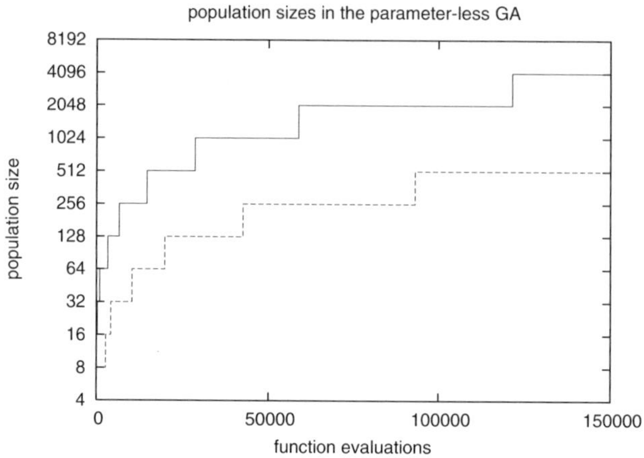

Fig. 4. Range of population sizes maintained by the parameter-less GA on a typical run. The dashed line is the minimum population size, and the solid line is the maximum population size maintained by the algorithm at any given point in time. (Copied with permission from [19])

In the absence of mutation, the parameter-less GA also eliminates populations that converge (convergence meaning that the whole population consists of copies of identical individuals) since its not possible to generate new individuals thereafter.

Figure 4 shows the population sizes maintained by the parameter-less GA as time goes by on a typical run. In the figure there are two lines. The bottom one is for the lowest sized population and the line above it is for the largest sized population maintained by the algorithm at any given point in time. Notice how early in the run, the algorithm is using small sized populations, but fairly quickly detects that those small sizes are not enough and eliminates them. Each vertical step in the lower line corresponds to an event where a population is deleted. Regarding the top line, each vertical step corresponds to an event where a new population is being created for the first time. For example, on the particular run shown in Figure 4, the population of size 1024 was first created when the algorithm had already spent around 30 thousand evaluations. The lower line represents in some sense the current guess of the parameter-less GA as to what might be a good population size for the problem at hand.

Pelikan and Lobo [34] did a theoretical analysis for the worst-case performance of the parameter-less GA as compared to a GA that starts with an optimal population size. The worst-case scenario occurs when populations are never eliminated (either never converge, or are never overtaken by a larger population). That work revealed that a value of $m = 2$ is the better one in terms of the time complexity of the worst-case performance, yielding an increase in the number of fitness function evaluations of only a logarithmic factor when compared with a GA that uses an optimal fixed population size.

This population sizing technique has been used not only with traditional GAs but also with EDAs [16, 22, 15].

Recently, a related scheme where the population size is increased by a factor of 2 at each restart has been suggested in the context of Evolution Strategies [2]. The major difference regarding the parameter-less GA technique occurs in the criteria to decide when a restart should be conducted.

The technique used in the parameter-less GA for removing the fixed population size parameter is very effective, can be used with most type of GAs, and has only a small overhead when compared with a GA that starts with an optimal fixed population size. The mechanism is able to adapt the population size to different orders of magnitude, solving problems that need small populations as well as problems that need very large populations.

A criticism that can be made to the technique is the fact that the algorithm keeps increasing the population size forever, something that goes against the common intuition that GAs need large populations at the beginning and small populations at the end of the run. The parameter-less technique, however, assumes that there is no way to know if a given population size is large enough to do a sufficiently wide search at the beginning. Thus, is keeps increasing the population size as time goes by. Another criticism that can be made to the technique is the existence of the m parameter. In the absence of mutation, Harik and Lobo suggested a value of 4 as a compromise for problems where the population takes a long time to converge relative to those where the population converges quickly. Pelikan and Lobo [34], however, have shown that a value of $m = 2$ is the best in terms of the worst-case time complexity of the algorithm, even tough the worst-case scenario is very unlikely to occur in practice. In any case, the algorithm is relatively robust regarding the m parameter, and there are sound recommendations on how to specify it.

3.6 Population Sizing in APGA (2000)

The *Genetic Algorithm with Adaptive Population Size* (APGA) proposed by Bäck, Eiben, and Van der Vaart [3] is a slight variation of the GAVaPS algorithm. The difference between the two is that APGA is a steady-state GA, the best individual in the population does not get older, and in addition to the selection pressure obtained indirectly by the lifetime mechanism, APGA also uses an explicit selection operator for choosing individuals to reproduce. Thus, APGA uses a stronger selection pressure than GAVaPS. An algorithmic description of APGA is presented in Figure 5.

Rather than using 1 and 7 as in the original GAVaPS, Bäck et al. set the values of *MinLT* and *MaxLT* to 1 and 11, because according to them, initial runs with different values indicated that *MaxLT=11* delivers good performance. APGA also needs the initial population size to be specified (Bäck et al. used 60 individuals in their experiments).

Similarly to GAVaPS, APGA can also use different lifetime strategies. Bäck et al. [3] used a bi-linear strategy similar to the one proposed for GAVaPS.

Although some researchers [3, 5] have claimed this method to be very effective at adapting the population size, a recent study [19] shows that APGA is not able to adapt the population size. Specifically, it is given a formal proof that after $MaxLT$ generations (which corresponds to 2 $MaxLT$ fitness evaluations), the population size in APGA is upper bounded by 2 $MaxLT+1$, and from that point on until the end of the search, the population size does not raise beyond that bound [19]. Independently of the problem being solved, the newly introduced parameters, $MinLT$ and $MaxLT$, simply act as the actual population size parameter of a traditional GA.

```
procedure APGA
begin
  t = 0;
  initialize pop(t);
  evaluate pop(t);
  compute RLT for all members of pop(t);
  while not termination-condition do
  begin
     t = t+1;
     pop(t) = pop(t-1);
     decrement RLT by 1 for all but
         the best member of pop(t);
     select 2 individuals from pop(t);
     cross and mutate the 2 individuals;
     evaluate the 2 individuals;
     insert the 2 offspring into pop(t);
     remove from pop(t) those members with RLT=0;
     compute RLT for the 2 new members of pop(t);
  end
end
```

Fig. 5. Pseudocode of the genetic algorithm with adaptive population size (APGA). RLT stands for *remaining lifetime*.

3.7 Population Sizing in PRoFIGA (2004)

Eiben, Marchiori, and Valkó [5] and Valkó [31] proposed the *Population Resizing on Fitness Improvement GA* (PRoFIGA).

PRoFIGA is similar to a traditional GA but at the end of the typical selection, reproduction, and mutation steps, the population size can grow or shrink based on an improvement of the best fitness contained in the population. The population grows either when (1) there is an improvement in best fitness, or (2) there is no improvement in the best fitness for a "long time". If neither of the above occurs, the population size shrinks by a small percentage (1-5%). According to the authors, the motivation behind PRoFIGA is to use large population sizes for exploration and small population sizes for exploitation. In their work, the growth rate X for a population is given by,

$$X = increaseFactor \times (maxEvalNum - currEvalNum) \times \frac{maxFitness_{new} - maxFitness_{old}}{initMaxFitness}$$

where $increaseFactor$ is a parameter in the interval (0,1), $maxEvalNum$ is the maximum number of fitness evaluations allowed for the whole run (or a very large number if one does not know it ahead of time), $currEvalNum$ is the current evaluation number, and $maxFitness_{new}$, $maxFitness_{old}$, and $initMaxFitness$, are the best fitness values in the current, previous, and initial generation.

A user of PRoFIGA also has to specify an initial population size, as well as a minimum and maximum population sizes in which the algorithm must operate. Thus, although PRoFIGA eliminates the traditional fixed population sizing parameter, it introduces 6 new parameters: (1) initial population size, (2) $increaseFactor$, (3) the so-called "long time without improvement" (V), (4) $decreaseFactor$, (5) $minPopSize$, and (6) $maxPopSize$.

In their comparative experiments, the authors used $initialPopSize = 100$, $increaseFactor = 0.1$, $V = 500$, $decreaseFactor = 0.4$, $minPopSize = 15$, $maxPop\ Size = 1000$. Unfortunately, it is not clear from a user's perspective, how should the newly introduced parameters should be set in order to solve a particular problem.

In their study, Eiben et al. did an empirical comparison of PRoFIGA with other adaptive population sizing schemes, including APGA and the parameter-less GA. The winner of the competition was APGA, followed by PRoFIGA. Those claims, however, have been shown to be unjustified [19].

3.8 Population Sizing Based on Noise and Substructural Measurements (2005)

Yu, Sastry, and Goldberg [33] proposed a method to adjust the population using noise and substructural measurements. Their approach focus on estimating the population size according with the population size models described in Section 2. The parameters of the population size model are estimated through the substructural information learned by linkage model-building techniques used in EDAs [16, 31]. This methodology was demonstrated using the dependency structure matrix genetic algorithm (DSMGA) [32].

The population sizing model used, which incorporates decision-making and model-building requirements [8, 20, 24, 25], predicts that the population size required to solve a decomposable problem with m subfunctions (or BB partitions), each of size k, with an error probability of $\alpha = 1/m$, is given by

$$N = c_n \, 2^k \, \frac{\sigma_{BB}^2}{d^2} \, m \, \ln(m), \tag{2}$$

where c_n is a problem-dependent constant.

Given the linkage model learned by DSMGA (or other EDA) that identifies which variables form a BB partition, the estimation of the above parameters becomes a straightforward process. The number or BB partitions m and their respective size k are directly extracted from the linkage model, where if different sizes of BBs are detected the average size is considered.

The fitness variance of BBs (σ_{BB}^2) and the fitness signal between the two best BBs (d) are calculated in a simple way according with the structure of the model learned by DSMGA. First, Yu et al. [33] defined the fitness of a BB as the average fitness of the correspondent schema relatively to the average fitness of the population. Thus, the variance of fitness in a BB partition is given by the variance of the fitnesses of all possible BBs in that partition. Similarly, the fitness signal d is given by the difference between the two BBs with highest fitness in the partition of interest. Again, in the presence of different values for σ_{BB}^2 and d, the average over all BB partitions is the value considered for the population sizing model.

After the linkage model is built the size of the offspring population is estimated using the above methodology. Yu et al. verified that this method can correctly adapt the population size on-the-fly to the ideal size, where the number of evaluations spent by the GA with adaptive population size remains practically constant when compared with a GA with a fixed population size, as the initial population size is increased. However, for very small initial population sizes (much smaller than the ideal size) both algorithms failed to converge. To deal with this situation, Yu et al. proposed a more robust scheme where new individuals (that are randomly generated) are inserted into the offspring population to ensure diversity when the parent population is much smaller than the offspring population.

Although the accuracy of the present method relies on the same assumptions of the population sizing model—problems additively decomposable in uniformly-scaled subfunctions—it is an important step towards black-box optimization with GAs, where both operators and parameters are automatically updated on-the-fly according to the underlying problem structure. For further details about this method please refer to Chapter 11.

4 Summary and Conclusions

This chapter has made a review of population sizing in genetic algorithms. We have addressed both theoretical models which rely on a facetwise decomposition of GAs, as well as automated population sizing schemes that have been proposed by various researchers in the field.

Some adaptive population sizing schemes that have been proposed were developed having in mind the population sizing theory for genetic algorithms as a foundation to engineer the adaptive algorithms, other methods were primarily inspired by what happens in Nature, and included concepts such as *age* and *lifetime* of an individual, as well as competition among species for limited resources.

Several authors have argued that it is beneficial to have a varying population size through a GA run because it seems that the population sizing requirements differ depending on the stage of the search. Typically, it is argued that the GA needs a larger population size early on in order to do a breadth search to find good regions of the search space, and then the GA needs less population size at the end of the run in order to fine-tune the solutions.

We agree with the argument above. However, the critical aspect is to discover a population size which is large enough to find those good regions of attraction. Without sufficient knowledge about the problem it is very difficult to know what should be a large enough value. At this point, we would like to stand back and give some recommendations for authors that propose and compare adaptive population sizing schemes.

- **Do not enforce a bound for the maximum population size.** Several adaptive population sizing algorithms have substituted the fixed population size parameter by an interval *MinPopsize* and *MaxPopsize* where the adaptive algorithm is supposed to work. We strongly believe that the upper bound for population sizing should not exist.
  Both theoretical and empirical work have shown that the population sizing requirement for GAs grow with respect to the problem's size and difficulty. Thus, it is not rational to impose an upper bound on the size of the population. In other words, a *MaxPopsize* parameter should be eliminated from adaptive

population sizing schemes. If that does not happen, then the user ends up having to guess a value for *MaxPopsize* more or less in the same way that he has to guess a value for the fixed population size parameter of a traditional GA. In our opinion, there are only two reasons for setting an upper bound on the size of the population:

1. when the maximum number of fitness function evaluations allowed for the whole simulation, T, is known in advance. In such case, the upper bound for the size of the population could be set to a value which is a function of T (and obviously less than T). As an extreme case, one generation of a population of size T would terminate the GA run, and that would correspond to random search.
2. when there are memory constraints that do not allow our computers to run populations larger than a certain amount.

- **Test on problems with known population sizing requirements.** The different adaptive algorithms that we have reviewed were tested by their authors on different test problems. It is our strong belief that the test suit should include problem instances of varying size and difficulty and whose optimal population sizing requirements are well known. That is the only reasonable way to assess if a given algorithm is capable or not of properly adapting the population size to solve the problem at hand. Moreover, the problems should have different population sizing requirements, some requiring small and others requiring large population sizes. A good self-adjusting population sizing algorithm should be able to detect both cases. We suggest that, among other problems, the class of additive decomposable problems, both with uniform and exponentially scaled building blocks, as well as problems with external sources of noise, should be included in the test suite because their population sizing requirements are well understood from a theoretical point of view.
- **Do fair comparisons.** There are several ways to make comparative studies on the performance of algorithms. Two of the most used methods are (1) given a fixed computational time T measure the solution quality Q, and (2) given a solution quality Q, measure the time T that the algorithm needs to reach that quality. Both components should be addressed on comparative studies to conclude about the efficiency and robustness of different adaptive population sizing methods. Otherwise, comparisons may be unfair and totally insignificant.
- **Do scalability analysis.** It is important to perform time and space complexity analysis for the proposed algorithms, even if they can only be done on a particular class of problems. The analysis should be complemented by computer experiments that verify how the algorithm scales up with problems of different size.
- **Make life easier for users.** All the adaptive population sizing algorithms that we have reviewed in this paper eliminate the need to specify a fixed population size for the GA run. Unfortunately, eliminating one parameter and introducing several other parameters is certainly not making life any easier from a user's perspective. If new parameters are introduced, either there should be recommendations on how to set them, or the adaptive algorithm has to be robust enough so that the performance is not much affected by these new parameters.

Acknowledgments

The authors thank the support of the Portuguese Foundation for Science and Technology (FCT/MCES) under grants POSI/SRI/42065/2001, POSC/EEA-ESE/61218/2004, and SFRH/BD/16980/2004.

We would also like to thank Kumara Sastry for his helpful discussions about population sizing theory.

References

1. J. Arabas, Z. Michalewicz, and J. Mulawka. GAVaPS – a genetic algorithm with varying population size. In *Proc. of the First IEEE Conf. on Evolutionary Computation*, pages 73–78, Piscataway, NJ, 1994. IEEE Press.
2. A. Auger and H. Nikolaus. A restart CMA evolution strategy with increasing population size. In D. Corne et al., editors, *Proceedings of the 2005 IEEE Congress on Evolutionary Computation*, volume 2, pages 1769–1776. IEEE Press, 2005.
3. Thomas Bäck, A. E. Eiben, and N. A. L. van der Vaart. An empirical study on GAs "without parameters". In *Parallel Problem Solving from Nature, PPSN VI*, pages 315–324, 2000.
4. K. A. De Jong. *An analysis of the behavior of a class of genetic adaptive systems.* PhD thesis, University of Michigan, Ann Arbor, 1975.
5. A. E. Eiben, E. Marchiori, and V. A. Valko. Evolutionary algorithms with on-the-fly population size adjustment. In X. Yao et al., editors, *Parallel Problem Solving from Nature PPSN VIII, LNCS 3242*, pages 41–50. Springer, 2004.
6. Carlos Fernandes and Agostinho Rosa. A study on non-random mating and varying population size in genetic algorithms using a royal road function. In *Proceedings of the 2001 Congress on Evolutionary Computation CEC2001*, pages 60–66. IEEE Press, 2001.
7. D. E. Goldberg. *The Design of Innovation - Lessons from and for Competent Genetic Algorithms.* Kluwer Academic Publishers, Norwell, MA, 2002.
8. D. E. Goldberg, K. Deb, and J. H. Clark. Genetic algorithms, noise, and the sizing of populations. *Complex Systems*, 6:333–362, 1992. Also IlliGAL Report No. 91010.
9. D. E. Goldberg and M. Rudnick. Genetic algorithms and the variance of fitness. *Complex Systems*, 5(3):265–278, 1991.
10. G. Harik, E. Cantú-Paz, D. E. Goldberg, and B. L. Miller. The gambler's ruin problem, genetic algorithms, and the sizing of populations. *Evolutionary Computation*, 7(3):231–253, 1999.
11. G. R. Harik and F. G. Lobo. A parameter-less genetic algorithm. In W. Banzhaf et al., editors, *Proceedings of the Genetic and Evolutionary Computation Conference GECCO-99*, pages 258–265, San Francisco, CA, 1999. Morgan Kaufmann.
12. Robert Hinterding, Zbigniew Michalewicz, and Thomas C. Peachey. Self-adaptive genetic algorithm for numeric functions. In *Parallel Problem Solving from Nature, PPSN IV*, pages 420–429, 1996.
13. John H. Holland. Genetic algorithms and the optimal allocation of trials. *SIAM Journal of Computing*, 2(2):88–105, June 1973.

14. Pedro Larrañaga and Jose A. Lozano, editors. *Estimation of distribution algorithms: a new tool for Evolutionary Computation.* Kluwer Academic Publishers, Boston, MA, 2002.
15. C.F. Lima and F.G. Lobo. Parameter-less optimization with the extended compact genetic algorithm and iterated local search. In K. Deb et al., editors, *Proceedings of the Genetic and Evolutionary Computation Conference (GECCO-2004), Part I, LNCS 3102*, pages 1328–1339. Springer, 2004.
16. F.G. Lobo. *The parameter-less genetic algorithm: Rational and automated parameter selection for simplified genetic algorithm operation.* PhD thesis, University of Lisbon, Portugal, 2000. Also IlliGAL Report No. 2000030.
17. F.G. Lobo and D.E. Goldberg. The parameter-less genetic algorithm in practice. *Information Sciences*, 167:217–232, 2004.
18. F.G. Lobo, D.E. Goldberg, and M.Pelikan. Time complexity of genetic algorithms on exponentially scaled problems. In Darrell Whitley, David Goldberg, Erick Cantu-Paz, Lee Spector, Ian Parmee, and Hans-Georg Beyer, editors, *Proceedings of the Genetic and Evolutionary Computation Conference (GECCO-2000)*, pages 151–158, Las Vegas, Nevada, 2000. Morgan Kaufmann.
19. F.G. Lobo and C.F. Lima. Revisiting evolutionary algorithms with on-the-fly population size adjustment. In *Proceedings of the ACM Genetic and Evolutionary Computation Conference (GECCO-2006)*, pages 1241–1248. ACM Press, 2006.
20. Martin Pelikan, David E. Goldberg, and Erick Cantú-Paz. Bayesian optimization algorithm, population sizing, and time to convergence. In Darrell Whitley, David Goldberg, Erick Cantu-Paz, Lee Spector, Ian Parmee, and Hans-Georg Beyer, editors, *Proceedings of the Genetic and Evolutionary Computation Conference (GECCO-2000)*, pages 275–282, Las Vegas, Nevada, 10-12 July 2000. Morgan Kaufmann. Also IlliGAL Report No. 2000001.
21. Martin Pelikan, David E. Goldberg, and Fernando Lobo. A survey of optimization by building and using probabilistic models. *Computational Optimization and Applications*, 21(1):5–20, 2002. Also IlliGAL Report No. 99018.
22. Martin Pelikan and Tz-Kai Lin. Parameter-less hierarchical BOA. In K. Deb et al., editors, *Proceedings of the Genetic and Evolutionary Computation Conference (GECCO-2004), Part II, LNCS 3103*, pages 24–35. Springer, 2004.
23. Martin Pelikan and Fernando G. Lobo. Parameter-less genetic algorithm: A worst-case time and space complexity analysis. IlliGAL Report No. 99014, University of Illinois at Urbana-Champaign, Illinois Genetic Algorithms Laboratory, Urbana, IL, 1999.
24. Martin Pelikan, Kumara Sastry, and David E. Goldberg. Scalability of the bayesian optimization algorithm. *International Journal of Approximate Reasoning*, 31(3):221–258, 2003. Also IlliGAL Report No. 2002024.
25. K. Sastry and D.E. Goldberg. Designing competent mutation operators via probabilistic model building of neighborhoods. In K. Deb and et al., editors, *Proceedings of the Genetic and Evolutionary Computation Conference (GECCO-2004), Part II, LNCS 3103*, pages 114–125. Springer, 2004. Also IlliGAL Report No. 2004006.
26. Dirk Schlierkamp-Voosen and Heinz Mühlenbein. Strategy adaption by competing subpopulations. In *Parallel Problem Solving from Nature, PPSN III*, pages 199–208, 1994.

27. Dirk Schlierkamp-Voosen and Heinz Mühlenbein. Adaptation of population sizes by competing subpopulations. In *Proceedings of 1996 IEEE International Conference on Evolutionary Computation*, pages 330–335, 1996.
28. R. E. Smith. Adaptively resizing populations: An algorithm and analysis. Technical Report No. 93001, The University of Alabama, Tuscaloosa, AL, 1993.
29. R. E. Smith and E. Smuda. Adaptively resizing populations: Algorithm, analysis, and first results. *Complex Systems*, 9:47–72, 1995.
30. Dirk Thierens, David E. Goldberg, and Angela G. Pereira. Domino convergence, drift, and the temporal-salience structure of problems. In *Proceedings of 1998 IEEE International Conference on Evolutionary Computation*, pages 535–540, Piscataway, NJ, 1998. IEEE Press.
31. V. A. Valkó. Self-calibrating Evolutionary Algorithms: Adaptive Population Size. Master's thesis, Free University Amsterdam, 2003.
32. T.-L. Yu, D. E. Goldberg, A. Yassine, and Y.-P. Chen. A genetic algorithm design inspired by organizational theory: Pilot study of a dependency structure matrix driven genetic algorithm. In *Proceedings of the Artificial Neural Networks in Engineering 2003 (ANNIE 2003)*, pages 327–332, 2003. Also IlliGAL Report No. 2003007.
33. T.-L. Yu, K. Sastry, and D. E. Goldberg. Online population size adjusting using noise and substructural measurements. In D. Corne et al., editors, *Proceedings of the 2005 IEEE Congress on Evolutionary Computation*, volume 3, pages 2491–2498. IEEE Press, 2005.

Population Sizing to Go: Online Adaptation Using Noise and Substructural Measurements

Tian-Li Yu[1,2], Kumara Sastry[1,3], and David E. Goldberg[1,3]

[1] Illinois Genetic Algorithm Laboratory
[2] Department of Computer Science
[3] Department of Industrial & Enterprise Systems Engineering
University of Illinois at Urbana-Champaign
Urbana, IL 61801
tianliyu@illigal.ge.uiuc.edu
kumara@illigal.ge.uiuc.edu
deg@illigal.ge.uiuc.edu

Summary. This chapter presents an online population size adjustment scheme for genetic algorithms. It utilizes substructural identification techniques to calculate the parameters used in facetwise population-sizing models. The methodology is demonstrated using the dependency structure matrix genetic algorithm on a set of boundedly-difficult problems. Empirical results indicate that the proposed method is both efficient and robust. If the initial population size is too large, the proposed method automatically decreases the population size, and thereby yields significant savings in the number of function evaluations required to obtain high-quality solutions; on the other hand, if the initial population size is too small, the proposed scheme increases the population size on-the-fly, thereby avoiding premature convergence.

1 Introduction

One of the challenges faced by genetic and evolutionary algorithm (GEA) users is making appropriate choices of codings, operators, and parameter values. Several linkage-learning methods have been developed to relieve GEA practitioners from having to develop problem-specific operators and encodings [5]. These linkage-learning methods automatically identify key sub-structures of the underlying search problem and solve them quickly, reliably, and accurately. Additionally, facetwise models have been developed to estimate the genetic-algorithm (GA) parameter values such as population size, run duration, crossover and mutation probabilities, and selection pressure [6]. Subsequently, the results of the facetwise models have been used in the design of so-called parameterless GEAs [10, 12, 16] to alleviate the guesswork out of parameter settings.

T.-L. Yu et al.: *Population Sizing to Go: Online Adaptation Using Noise and Substructural Measurements*, Studies in Computational Intelligence (SCI) **54**, 205–223 (2007)
www.springerlink.com

However, the parameterless GEAs do not completely use the population-sizing models and are restricted to having multiple populations with exponential increasing population sizes. While some of the factors in the population-sizing models are not known *a priori*, they can be estimated on-the-fly with linkage-learning GEAs in a straightforward manner. Therefore, the purpose of this chapter is to develop an on-line population-sizing estimate which automatically identifies values of key components of the population-sizing model. Specifically, we perform an online measurement of key factors of existing facetwise population-sizing models and alter the population size on-the-fly in accordance with the population-sizing models. Effective linkage identification via linkage-learning GEAs enables us to perform sub-structural measurements and to compute different factors of the facetwise population-sizing models.

We demonstrate the effectiveness of the proposed method using design structure matrix GA [28]—a linkage-learning GEA that is inspired by organization theory—on a class of boundedly difficult test problems. We consider population sizing dictated by building-block supply [7], accurate decision making [3, 9], and accurate linkage learning [14, 15, 19, 20] in this study. However, we note that the proposed method is not restricted to the above models, but other population-sizing models can be easily incorporated. We demonstrate that the proposed method is not only robust, but also efficient in terms of number of function evaluations required to obtain high quality solutions.

This chapter is organized as follows. Section 2 revisits some related work concerning adjusting population size on-the-fly. Then we give background for several facetwise population-sizing models in Section 3. A brief introduction of dependency structure matrix genetic algorithm (DSMGA) is given in Section 4. The population size needed for DSMGA is empirically verified with the population-sizing model for linkage learning. Section 5 describes the basic idea of adjusting population size on-the-fly by utilizing population-sizing models, including how to calculate parameters needed by population-sizing models, and how to generate individuals for the next generation to maintain both efficiency and robustness. Section 6 proposes a scheme to improve both the robustness and efficiency when the GA starts from a small population. It also demonstrates how to choose a reasonable initial population size. Finally, section 7 concludes this chapter.

2 Related Work

Since the inception of GAs, numerous studies have been conducted on on-line and off-line parameter settings. A detailed survey of which is beyond the scope of this study and and the interested reader is referred elsewhere [10, 12]. Since we propose an on-line population resizing mechanisms for linkage-learning GEAs in this chapter, we will briefly review related work on adaptive resizing of population sizes in the remainder of this section.

One of the early effort on resizing the population during the GA run can be backtracked to Smith and Smuda [23, 22]. They dynamically resized the population to match a target selection loss L_t. They calculated the estimated selection loss $\hat{L}$ for each mating pair. By adopting a sigmoid function, the fraction $\frac{\hat{L}}{L_t}$ is used to conduct a growth factor, which is then used to adjust the population size. One of the weakness of the methodology proposed by Smith and Smuda [23, 22] is that users need to determine three parameter values: the target selection loss, and two parameters (α and β) in the sigmoid function used to compute the growth factor.

Harik and Lobo [10] proposed a parameter-less GA, which is composed of two parts: eliminating selection pressure and crossover rate, and eliminating population size. The parameter-less GA eliminates population size by simulating a set of populations of different sizes; starting with a population of base size, the subsequent populations are twice the size of the previous one. Then, the GA operators are applied to each population in a special sequence. For every two GA generations for a populations of a particular size, a single GA generation is performed on a population of twice the size. Therefore, the number of function evaluations performed on each population are roughly the same. The parameter-less GA terminates a population when no improvement is expected from that population. The parameter-less GA stops when no improvement is expected from the larger populations as well. Suppose that a regular GA with an optimal population-size setting needs n^*_{fe} number of function evaluations for a particular problem. It is shown that the parameter-less GA uses no more than $O(n^*_{fe} \log(n^*_{fe}))$ [34] number of function evaluations.

The parameter-less GA technique has been applied not only to simple GAs [10], but also to linkage-learning GAs such as the extended compact GA (eCGA) [12], and the hierarchical Bayesian optimization algorithm (hBOA) [16].

The method used in this chapter to estimate the fitness variance of building blocks is similar to Smith and Smuda's work [23, 22], but most factors/components are estimated on-the-fly. Though our method requires an initial population size as we demonstrate in the result the performance—in terms of number of function evaluations and solution quality—is not very dependent on the choice of that value. Moreover, there is a significantly large sweet-spot for the on-line population adjustment scheme to be working with a wide range of the initial population size. The method will be detailed in Section 5.

3 Population-Sizing Models

Facetwise and dimensional models have been very effective not only in the design of genetic algorithms, but also in understanding GA dynamics and mechanisms. Since our methodology depends on the facetwise models of population sizing, we briefly outline the models dictated by building block supply, decision making, and accurate linkage learning in the remainder of this section.

3.1 Building-Block Supply Model

The first step towards understanding population sizing is to tackle the issue of building-block (BB) supply, where the minimum population size required to ensure the presence of at least on copy of all raw schemata is modeled. Holland [11] estimated the number of BBs that receive at least a specified number of trials using Poisson distribution. A later study [4] calculated the same quantity more exactly using binomial distribution and studied their effects on population sizing in serial and parallel computation. Reeves [18] proposed a population sizing model for supply of alphabets with fixed cardinality. Recently, Goldberg *et al.* [7] developed facetwise models for ensuring BB supply in the initial population for genetic algorithms. They considered a population of fixed-length strings consisting alphabets of arbitrary cardinality χ. Goldberg *et al* predicted that the population size required to ensure the presence of all competing building blocks with a tolerance of $\epsilon = 1/m$ is given by

$$n = \chi^k \left(k \log \chi + \log m\right), \tag{1}$$

where χ is the alphabet cardinality, k is BB size, and m is the number of BBs.

3.2 Gambler's Ruin Population-Sizing Model

Goldberg *et al.* [3] proposed population-sizing models for correctly deciding between competing BBs. They incorporated noise arising from other partitions into their model. However, they assumed that if wrong BBs were chosen in the first generation, the GAs would be unable to recover from the error. Harik *et al.* [9] refined the above model by incorporating cumulative effects of decision making over time rather than in first generation only. Harik *et al.* [9] modeled the decision making between the best and second best BBs in a partition as a gambler's ruin problem. Here we use an approximate form of the gambler's ruin population-sizing model:

$$n = \frac{\sqrt{\pi}}{2} \frac{\sigma_{BB}}{d} 2^k \sqrt{m} \log m, \tag{2}$$

where k is the BB size, m is the number of BBs, d is the signal between the competing BBs, and σ_{BB} is the fitness variance of a building block. The above equation assumes a failure probability, $\alpha = 1/m$.

3.3 Model-Building+Decision Making Population Sizing

Facetwise models for incorporating the effects of model building and BB-wise decision making on the population size have been analyzed for estimation of distribution algorithms (EDAs) in general, and Bayesian optimization algorithm and extended compact genetic algorithm in particular [14, 15, 19, 20]. The population-sizing model which incorporates the effect of model-building and its accuracy on the population sizing of the GA, and predicts the population size required to solve a problem with m building blocks of size k with a failure rate of $\alpha = 1/m$, is given by

$$n = c_n \cdot 2^k \left(\frac{\sigma_{BB}}{d}\right)^2 m \log m, \tag{3}$$

where n is the population size, c_n is a problem-dependent constant, k is the BB length, α is the probability of failure.

4 An Introduction to DSMGA

This section gives a brief introduction to the DSMGA. Readers who are interested in DSM clustering are referred to Yu *et al.* [27]. For more details about DSMGA, please refer to [28].

4.1 DSM and DSM Clustering

A DSM is a matrix where each entry d_{ij} represents the dependency between node i and node j [24, 26]. Entries d_{ij} can be real numbers or integers. The larger the d_{ij} is, the higher the dependency is between node i and node j. The diagonal entries (d_{ii}) have no significance and are usually set to zero or blacked-out. Figure 1 gives an example of DSM.

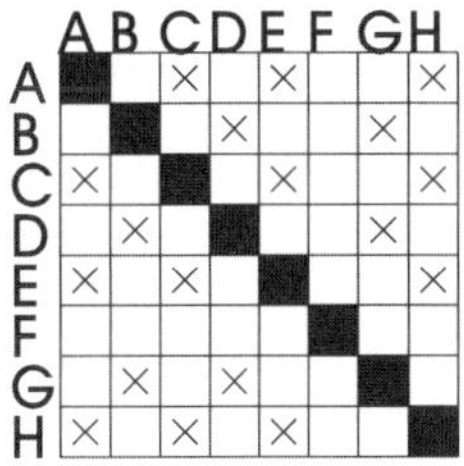

Fig. 1. A DSM. "x" means that dependency exists; the blank squares means no dependency. This figure illustrates, for example, that A and B are independent, and that A and C are dependent. Clustering is not so obviously at the first glance.

The goal of DSM clustering is to find subsets of DSM elements (i.e., clusters) so that nodes within a cluster are maximally dependent and clusters are minimally interacting. DSM clustering is a sophisticated task which requires expertise [21]. For example, it is not intuitive how to cluster the DSM in Figure 1. However, after we reorder the nodes (Figure 2), it is easily seen that the DSM can be cleanly clustered into three parts: {B, D, G}, {A, C, E, H}, and {F}.

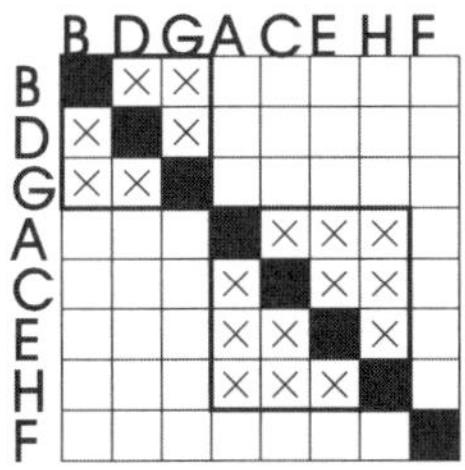

Fig. 2. The same DSM in Figure 1 but after reordered. The DSM can be cleanly clustered as ((B,D,G)(A,C,E,H),(F)).

Yu *et al.* [27] proposed the following objective function by using the minimal description length principle.

$$f_{DSM}(M) = (n_c \log(n_c) + \log(n_n)\Sigma_{i=1}^{n_c} cl_i) + (|S|(2\log(n_n) + 1)), \tag{4}$$

where f measures the description length that a model M needs to describe a given DSM; n_c is the number of clusters in M; n_n is the number of nodes of the DSM; cl_i is the size of the i-th cluster in M; S is the set of mis-described data of M. The above objective function has shown capable to cluster a given DSM, and the clustering results competes with human experts.

4.2 Utilizing DSM Clustering to Identify BBs: The DSMGA

The DSM clustering algorithm can be thought as an linkage-identification algorithm which turns pair-wise linkage information into high-order linkage information, and DSMGA [28] is motivated based on this thought.

DSMGA utilizes statistical analysis to estimate the fitness of order-2 schemata. Based on a nonlinearity-detection method similar to LINC [13], a DSM where each entry d_{ij} represents the linkage between gene i and gene j is created. By applying the DSM clustering algorithm, the pair-wise linkage information is turned into BB information, which is then utilized to achieve the BB-wise crossover [25]. The DSMGA has shown capable to correctly identify BBs and efficiently solve problems with uniformly-scaled BBs.

4.3 Population Sizing for DSMGA

Unlike EDAs, the linkage model in DSMGA is not exactly a probability density function. Nevertheless, empirical finding shows that the population-sizing model (Equation 3) for EDAs predicts well the population size needed by DSMGA (Figure 3). The test function is an (m,k)-trap [1], where $m = 10$ and $k = 3$. The k-bit trap function is defined as follows:

$$trap_k(u) = \begin{cases} 1.0 & u = k \\ (1-d)\frac{k-u-1}{k-1} & \text{otherwise} \end{cases}, \tag{5}$$

where u is unitary (the number of 1's among the k bits) and d is the minimal signal between the two most competing schemata. In this chapter, d is set to 0.2. By taking average, the constant c_n is empirically found to be roughly 0.24 for the (m, k)-trap where $m = 10$ and $k = 3$. This test function is also used for the experiments in later sections.

5 The Basic Idea of Population Size Adjustment

This section describes how to calculate parameters needed in population-sizing models and how to generate individuals for the next generation so that both efficiency and robustness are maintained.

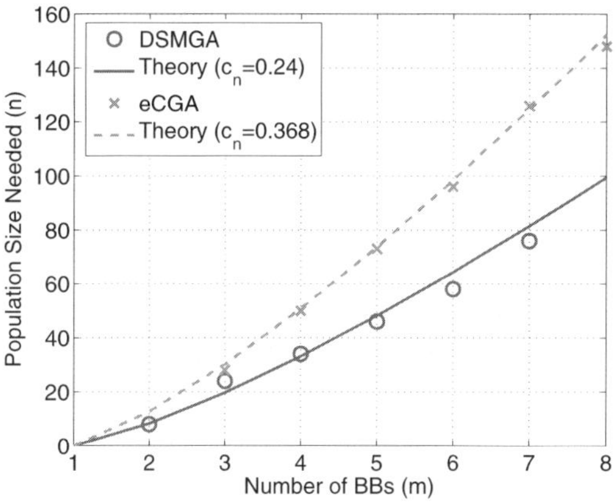

Fig. 3. The population-sizing model for EDAs predicts well the population size needed by DSMGA.

5.1 Population Size Estimation: Calculating Parameters in Population-Sizing Models

The BB information given by DSMGA tells us explicitly which genes form a BB. Therefore, calculating the number of BBs (m) and the size of BBs (k) is straightforward given the DSM clustering arrangement in DSMGA. For example, suppose that DSMGA tells us that $BB_1 = \{x_2, x_4\}$ and $BB_2 = \{x_1, x_3, x_5\}$ (where x's are genes), we know that $k_1 = 2$, $k_2 = 3$, and $m = 2$. For problems with the same size of BBs, we can simply take the average and then the estimated parameters would be $\hat{k} = 2.5$ and $\hat{m} = 2$.

The calculations of the fitness variance of BBs (σ_{BB}) and the signal size between competing BBs (d) are not any more difficult. First, define $H_{BB,i}$ as the schemata [11] which have allele values (0 or 1) for those genes in that BB and wild cards (*) for any other positions; i is just an enumerated number. In the previous example, $BB_1 = \{x_2, x_4\}$, we have $H_{BB_1,1} = *0*0*$, $H_{BB_1,2} = *0*1*$, $H_{BB_1,3} = *1*0*$, $H_{BB_1,4} = *1*1*$. Note that there are 2^k such schemata for a BB of size k. Define $f(H)$ as the fitness value for schema H. We can then express σ_{BB} and d as follows:

$$\sigma_{BB}^2 = \text{var}_i[f(H_{BB,i})], \quad (6)$$

$$d = \max_i[f(H_{BB,i})] - \text{second\_max}_i[f(H_{BB,i})], \quad (7)$$

where var is the variance function, and here we assume a maximization problem. The fitness of a schema can be estimated from the current population P by examining each individual x_i: $\hat{f}(H) = \text{mean}_{i,x_i \in H}[f(x_i)]$, where 'mean' is the arithmetic averaging function.

5.2 Population Size Adjustment: First Attempt and Empirical Results

After estimating the population size for the next generation, the most straightforward way to generation the next population is to simply apply GA operators to generate offspring of the desired size.

We test the proposed method on a (m,k) deceptive trap problem, where $m = 10$ and $k = 3$. We compare the performance of a DSMGA with on-line population-size adjustmentADJ and a DSMGA with fixed population sizes (FIX). In particular, we focus on the efficiency and robustness of each of the two GA methodologies. By efficiency we mean the number of function evaluations that the GAs need to converge to a solution of predefined quality. By robustness we mean that the rate that the GAs fail to converge. The termination condition is that on average $(m-1)/m =$ 90% percent of BBs converge correctly. The failure rate used in the population size estimation is set to $1/m = 10\%$. All results are averaged over 30 independent runs.

Figure 4 shows the number of function evaluations that the GAs need to converge solutions with 90% correct BBs by varying the initial population size (n_0). Given the empirical finding that $c_n \simeq 0.24$ for DSMGA on this test function and the failure rate 0.1, the population size needed by FIX is roughly 137. As shown in Figure 4, FIX consumes the least number of function evaluation (n_{fe}) when the initial population size n_0 is set close to $n^* = 137$ (we do not count the case for $n_0 = 100$, since FIX does not usually converge for that fixed population size). If the population size in FIX is less than the required size (even about 30% less), the GA converges to a local optimum and therefore does not yield high quality solutions. On the other hand, if the population size is greater than the required size, the number of function evaluations grows linearly with the initial population size n_0, thereby wasting significant amount of computation resources unnecessarily. In contrast to FIX, ADJ is able to properly shrink the population size if n_0 is large, and hence consumes roughly a constant number of function evaluations when n_0 is slight larger than n^*.

We can estimate the trend of the behavior for these two algorithms. Let n^*_{fe} be the number of function evaluations needed for FIX when $n_0 = n^*$. Empirically, $n^*_{fe} \simeq$ 2700. Following the facetwise modeling approach [6], we assume the running duration is nearly independent of the population size. For $n_0 = 600$, we can compute that the number of generations that FIX runs before convergence $g \simeq 14.8$. Therefore, we can estimate the number of function evaluations need for FIX for a fixed population size n_0 is roughly (the **FIX-trendline** in Figure 4):

$$n_{fe-FIX} = g \cdot n_0. \tag{8}$$

In ADJ, for a large initial population size n_0, we can assume that linkage model is quite accurate because n_0 is large. Hence, ADJ should estimate the population size accurate (close to n^*) for the next generation. Hereafter, we assume that ADJ performs in a similar behavior of FIX at an optimal population size n^*. Therefore, we model the total number of function evaluations needed for ADJ is the number of function evaluations in the first generation (n_0) plus the number of function evaluations needed for the optimal population size setting (n^*_{fe}) is roughly (the **ADJ-trendline** in Figure 4):

$$n_{fe-ADJ} = n^*_{fe} + n_0. \tag{9}$$

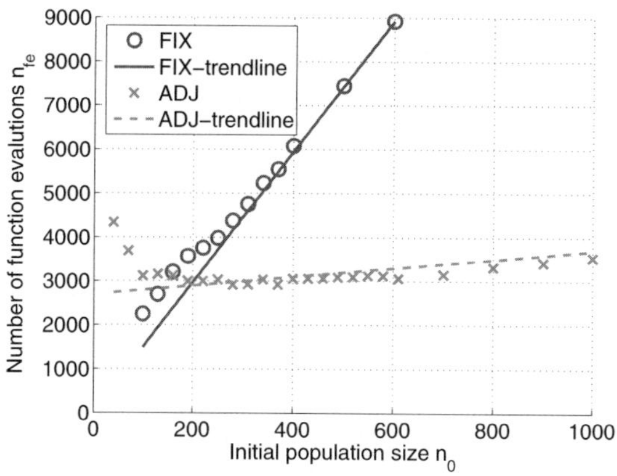

Fig. 4. Number of function evaluations consumed by DSMGAs with a fixed population (FIX) and with the population adjustment scheme (ADJ). ADJ needs slightly more number of function evaluations than FIX when a near-optimal population size (n^*) is used. For large initial population size, however, ADJ constantly outperforms FIX. For a too small initial population size, both FIX and ADJ do not converge.

Comparing Equations 8 and 9, we can see a dramatic reduction in the number of function evaluations when the on-line population size adjustment scheme is applied,

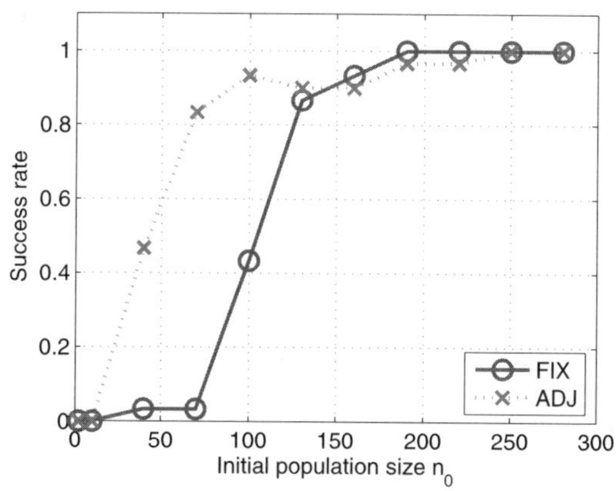

Fig. 5. The failure rates for FIX and ADJ for different n_0. FIX rarely converges for $n_0 < n^*$ as the population size predicts. ADJ is slightly more robust than FIX, but it still suffers from insufficient BB supply and the lack of diversity for small n_0.

especially for a large initial population size. Now we examine the algorithms from another direction: robustness for small initial population size. Figure 5 shows the rate that FIX and ADJ fail to converge for different small n_0. For small n_0, FIX rarely converges. According to the population-sizing theory, FIX does not have enough BB supply and cannot decide well when $n_0 < n^*$. When the initial population size is very small, $n_0 \ll n^*$, both FIX and ADJ fail to yield high-quality solutions. However, when the initial population is slightly increased (but still $n_0 < n^*$), the failure rate of ADJ decreases rapidly, while FIX consistently fails to yield high-quality solutions. The robustness of ADJ can be further enhanced by ensuring sufficient BB supply and maintaining diversity, especially in the early stage of the GA run (the method will be described in the next section). To understand the reason for ADJ to fail on a very small initial population size, imagine that $n^* = 1000$, and ADJ starts from $n_0 = 10$. Even if the linkage model is luckily accurate and estimates the the population size for the next generation should be 1000, ADJ generates the 1000 offspring chromosomes from only the 10 parent chromosomes. As can be anticipated, the newly generated population lacks of diversity and ADJ converges prematurely.

6 A More Robust Scheme for Small Initial Population Size

The online population size adjustment scheme described in the previous section allows users to start with a large population which guarantees good solution quality without spending too many extra function evaluations. While the on-line population adjustment scheme in the previous section (ADJ) is significantly more robust and efficient than a GA with a fixed population size, the performance—both in terms of efficiency and robustness—can be further enhanced when the GA practitioner sets the initial population size to be much smaller than the required size. This section proposes an new-individual-injection scheme to improve the robustness of the population size adjustment scheme.

6.1 New Individual Injection Scheme

As mentioned in the previous section, in ADJ the new population is generated via the variation operators—crossover and mutation—of DSMGA. While this is a straightforward method, it does not alleviate the problem of insufficient BB supply and the lack of diversity in the population. To circumvent these problems, new individuals need to be introduced into the population at the right time. One of the most intrinsic way is to introduce new individuals whenever the estimated population size is larger than the current population size.

In the previous example, suppose that $n^* = 1000$, $n_0 = 10$, and the GA estimates the population size for the next generation should be 1000. Then the GA generates 10 offspring chromosomes by recombining the 10 parent chromosomes and the generate the other 990 offspring chromosome by random initialization. However, the injection of randomly generated solutions into the population, especially during the later stages of the GA run has one sever drawback. Due to random initialization, the average solution quality goes down and the fitness variance goes up causing an elongation of the run duration [2]. As shown in Figure 6, whenever randomly initialized solutions are injected into the population, the proportion of correct BBs in the population is reduced causing a delay in the convergence.

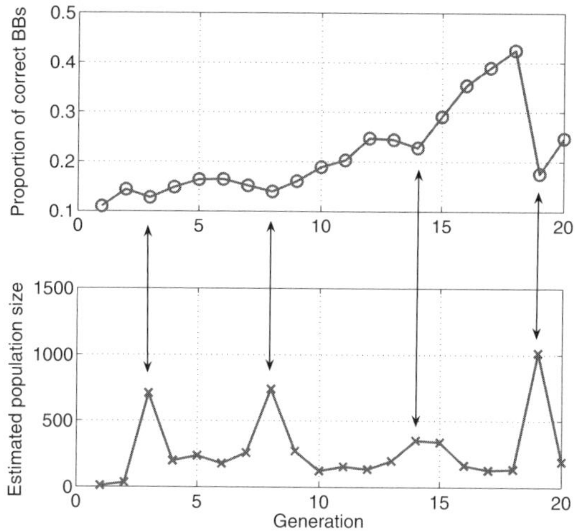

Fig. 6. One run of the GA with the introducing-new-individual scheme switched on all the time. The GA suffers from slow convergence because new individuals keep joining. The arrows indicates that whenever new individuals joins the population, the number of correct BBs goes down.

To alleviate the elongation in the run duration and also to ensure adequate supply of BBs, we should inject randomly initialized solutions into the population during the initial stages of the GA run and then create the new solutions only through recombination in the later stages. Using recombination during the later stages of the GA run not only circumvents the delay in convergence, but also significantly improves the exchange of building blocks and thereby allowing rapid creation of good quality solutions. One choice for the switching time can be the generation when the estimated population size for the next generation is smaller than the current population size. Intuitively this reduction in the population size indicates that the GA has sufficient supply of raw BBs and good decision making can be ensured. Therefore, after the first time that the estimated population size is smaller than the current one, we do not inject any new individual into the population.

The proposed scheme not only should improve the robustness, but also the efficiency, especially for small initial population size, n_0. As shown in Figure 4, the number of function evaluations needed for ADJ increases substantially for small n_0. It is because that for small initial population size, the linkage model is not accurate, and hence the estimation of the order of BBs, k, can be much larger than that required. Noting that the population size predicted by the population-sizing model (Equation 3) is exponential to k, if the value of k is overestimated, then the population size estimated on-the-fly is significantly larger than that required. However, one underlying assumption in the design of competent GAs is to solve boundedly difficult problems in polynomial time [6]. So, the bounded-difficulty assumption implies that GAs are efficient in solving those problems where the *necessary*[4] interactions are of lower order [6].

By the assumption of bounded difficulty, it is reasonable to limit the order of BBs during the model-building phase.

6.2 Empirical Results

Similar to the previous section, we use the (m,k)-trap functions with $m = 10$ and $k = 3$ for the experiments. Two different GAs are compared: (1)DSMGA + population-size-adjustment, and (2) DSMGA + population-size-adjustment + new-individual-injection. For simplicity, we call the first GA **ADJ** and the second **ADJ+INJECT**. The parameter settings are identical to those used in the previous section. Based on the assumption of bounded difficulty, we set the maximal order of BBs to 10 for ADJ+INJECT. All results are averaged out of 30 independent runs.

Figure 7 shows the number of function evaluations that the GAs need to converge by varying the initial population size (n_0). As shown in the figure, the number of function evaluations needed for ADJ and ADJ+INJECT are roughly the same for a wide range of initial population size n_0.

Figure 8 shows the failure rate of ADJ and ADJ+INJECT. As can be seen in the figure, ADJ+INJECT is significantly more robust than ADJ. Even for $n_0 = 2$ (which is unreasonably small), ADJ+INJECT still converges to the global optimum 83.3% of the time. The reason that a very small population size still works for ADJ+INJECT is as follows. For small population size, the pair-wise dependency detection is not accurate. The inaccuracy usually overestimate k than required. As a result, the estimated population size for the next generation is usually larger than the initial population size, which is small. Since the population size increases, the injection scheme introduces more newly-initialized individuals. If the population size estimated is still not large enough for the model building, the proposed population adjustment scheme would keep increasing the population size until the population size is large enough to give an accurate estimation. More number of function evaluations might be needed for small initial population size; nevertheless, the proposed method is extremely robust, and works with a large range of initial population sizes.

Figure 9 illustrates the population estimation behavior of ADJ+INJECT for different initial population size n_0. We can observe that for the first five generations, a smaller initial population yields larger estimated population sizes. However, after that, the estimated population sizes are roughly the same for different n_0. That is because once the population size is large enough to build an accurate model, the population estimation becomes accurate after that. An important and interesting

[4] Note that this does not mean that GAs can only solve problems with lower-order interactions. Problems such as hierarchical and spin-glasses have long-range interactions, but the necessary interactions required to solve the problem are still lower-order.

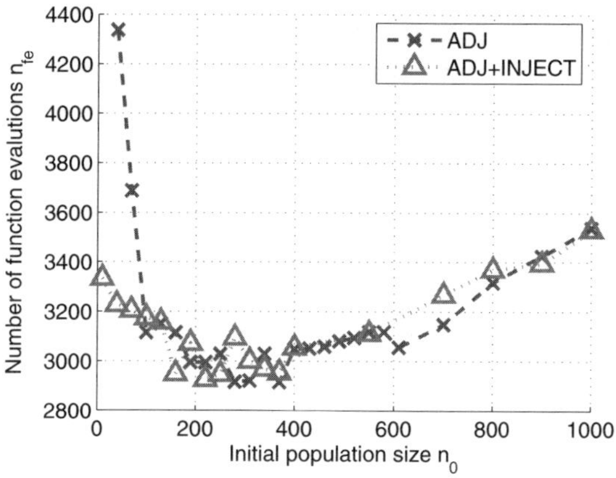

Fig. 7. Number of function evaluations consumed by DSMGAs with the population adjustment scheme (ADJ) and with population adjustment scheme + the injection scheme (ADJ+INJECT). The number of function evaluations needed for both algorithms are roughly the same.

observation is that at the beginning, the linkage model indicates that more individuals are needed. Towards the end of the GA run, the population loses the diversity

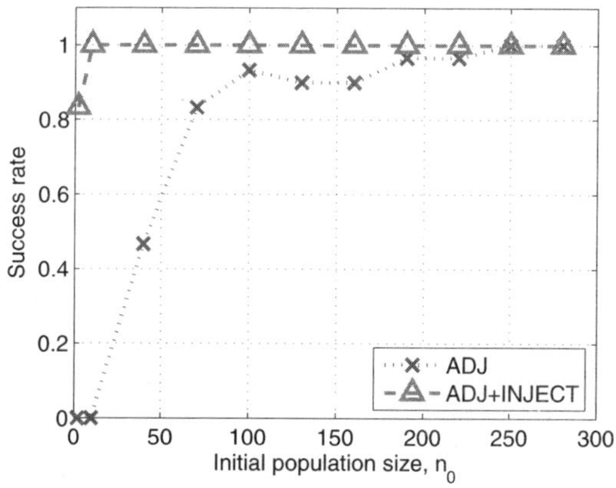

Fig. 8. The failure rates for ADJ and ADJ+INJECT for different n_0. ADJ+INJECT is significantly more robust than ADJ. ADJ+INJECT successfully converges for a wide range of initial population size n_0.

(fitness variance of BB in the current population is small), and hence not many individuals are needed. Since the population size is adjusted in accordance with the facetwise models, we obtain high-quality solutions requiring about the same number of function evaluations over a wide range of initial population sizes.

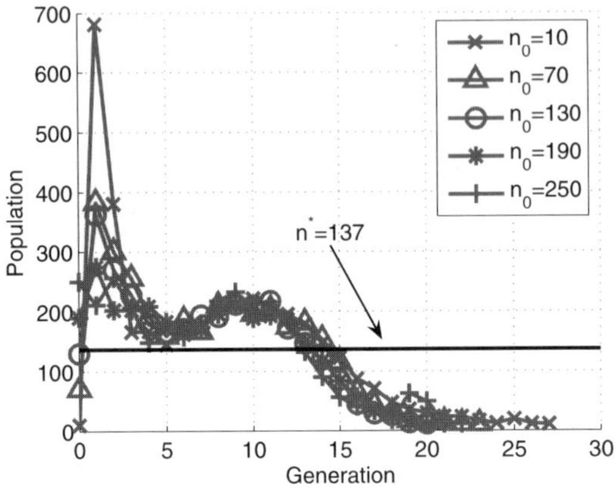

Fig. 9. The population estimation behavior of ADJ+INJECT for different initial population size n_0. At the beginning of the GA runs, a smaller n_0 yields a larger estimated population size. The estimated population sizes are less dependent on n_0 when the linkage model is more accurate after first few generations.

One other thing that we could try in the future is to apply this scheme to other linkage-learning GEAs (*e.g.* eCGA [8]) and test the scalability. We could also add an update rule to the population-sizing scheme. For example, if the estimated population size is $\hat{n}$ and the current population size is n_t, the next population size is decided by $n_{t+1} = \alpha n_t + (1-\alpha)\hat{n}$. Although the update rule introduce a new parameter α, it might add some values of stability.

One question remaining is how to set the initial population size. While we have shown that even when using a population size as small as 2, we obtain the global optimum with a high probability ($\sim$80%). Nevertheless, we need to obtain high-quality solutions with very high probability ($\sim$1) using reasonable number of function evaluations (close to theoretical minimum). We need some guideline to set the initial population size which we discuss in the following section.

6.3 Setting the Initial Population Size

Empirical results show that the online population adjustment scheme (ADJ+INJECT) works with a wide range of the initial population sizes. In real-world applications, we suggest to use the population size predicted by the gambler's ruin model [9] of the OneMax problem (call it n_{OneMax}) as an initial guess n_0. The reasons are that (1) n_0 provides as a lower bound of the population size needed since the OneMax problem is one of the simplest problems for GAs, and (2) yet n_0 is still large enough for the GAs with the online population adjustment scheme to converge.

Figure 10 shows the minimal population sizes n_{min} needed for ADJ+INJECT to converge. The test functions are (m, k)-traps with $k = 3$ and $k = 5$. The convergence requirement is that the GAs converge to the global optimum 30 times out of 30 independent runs. It is observable that n_{min} is slightly larger for a larger k. Nevertheless, for problems with bounded difficulty, the population size predicted by the gambler's ruin model [9] of the OneMax problem should be large enough to ensure the GA convergence as showed in the figure.

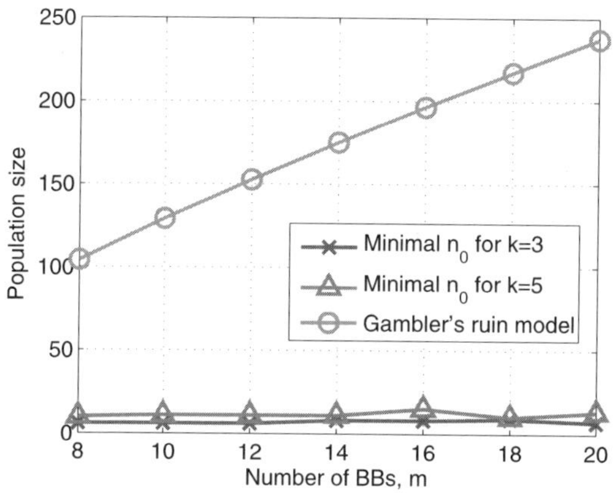

Fig. 10. The minimal population sizes needed for ADJ+INJECT to converge. The success rate is held constantly at 1.

However, recall that a smaller initial population size, we may require more number of function evaluations (Figure 7). A good initial guess should also yields roughly the same number of function evaluations compared to the optimum population size n^* predicted by the EDA population-sizing model (Equation 3). Figure 11 shows the number of function evaluations needed for the GA with initial population sizes set to n_{OneMax} and n^*. The order of BBs is 3. As Figure 11 indicates, n_{OneMax} is a good initial guess since it results in almost the same number of function evaluations as n^* does.

7 Conclusion

This chapter presents an online population size adjustment scheme based on main parameters estimated by facetwise population-sizing models. Given the linkage

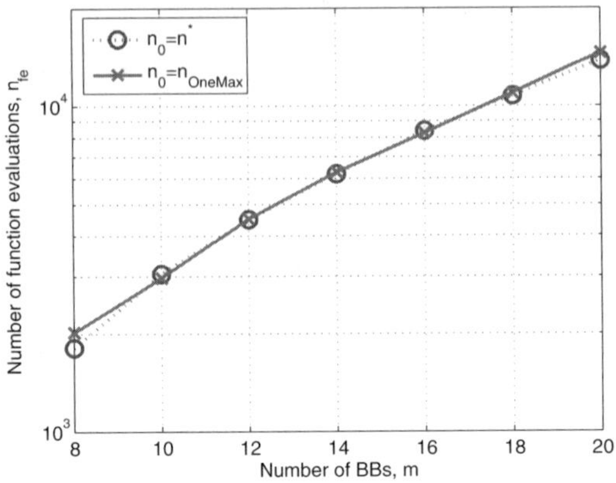

Fig. 11. Number of function evaluations needed for the initial population size set to (1) what the EDA population-sizing model predicted for the trap problem (n^*) and (2) what the gambler's ruin model predicted for the OneMax problem (n_{OneMax}). This figure suggests that n_{OneMax} provides a good initial guess. The success rate is held constantly at 1.

model, we calculate the parameters required by population-sizing models and estimate the population size in every generation according to the linkage-learning population-sizing model (Equation 3). Although we adopts the linkage-learning population-sizing model in this chapter, the proposed method is not limited to this particular population-sizing model.

When the estimated population size for the next generation is larger than the current population size, two methods are conducted to generate the individuals for the next generation. When the population size is more than required, the individuals are generated by ordinary GA recombination and variation operators. When the population size is less than required, the individuals are generated by random initialization to supply enough BBs. In this chapter, the first shrinkage of the population size estimated is adopted as the signal which indicates that the population size is more than required. In addition, we set a limit of maximal order of BBs for model building to prevent the estimated population size from being too large.

The proposed method provides sufficient BB supply at the beginning of the GA run even for a small initial population size, and concentrate on mixing to ensure good solution quality toward the end of the GA run. The proposed method is shown to have both efficiency and robustness. With the proposed online population size adjustment scheme, the number of function evaluations needed grows slowly with large initial population size, and the GA has a significantly higher probability to converges even with a very small initial population size. Even with an unreasonably small initial population size of 2, the GA converges to the global optimum of the (m,k)-trap, where $m = 10$ and $k = 3$, with a high probability of about 80%. With just a slightly larger initial population size, the GA converges to the high-quality solutions with a probability close to 1 and does not consume much more number of function evaluations. Empirical result also shows that the population size estimated by the gambler's ruin model for the OneMax problem, n_{OneMax}, provides a good initial population size for the proposed method. By setting $n_0 = n_{OneMax}$, the GA consumes almost the same number of function evaluations as predicted by the GA theory, and it always converges to the global optimum during the experiments.

Empirical results demonstrate that the GA requires a larger population size at the beginning to ensure good supply, decision making and exchange of important building blocks. On the other hand, at the later stages of the GA run, only a smaller population size is required since the population loses diversity and a number of the subproblems are solved. The proposed method enables competent GAs to use a small population size close to that required at every generation and yields high quality solutions using only a small number of function evaluations.

The work in this chapter suggests the possibility of taking theoretical facetwise models into practice. We believe that the online population sizing scheme is ready to use for real-world applications. Although in this chapter, we have only tested the scheme on problems with uniformly-scaled BBs with the same orders, there is no reason not to believe that the proposed method also works on problems with non-uniform salience with different orders by modifying the population-sizing model or plugging in a new one. In addition, once the problem structure is identified, we should be able to choose the most appropriate population-sizing model for the given problem. To conclude, this chapter presents a new direction of practical population-sizing and shows preliminary but promising results. We believe that more attention should be drawn in this direction to make GAs more accessible to real-world applications as well as to better understand population-sizing in GAs.

Acknowledgments

This work was sponsored by the Air Force Office of Scientific Research, Air Force Materiel Command, USAF under grant FA 9550-06-1-0096, and the National Science Foundation under ITR grant DMR-03-25939 at the Materials Computation Center. The US Government is authorized to reproduce and distribute reprints for Government purposes notwithstanding any copyright notation thereon.

The views and conclusions contained herein are those of the authors and should not be interpreted as necessarily representing the official policies or endorsements, either expressed or implied, of the Air Force Office of Scientific Research, the Technology Research, Education, and Commercialization Center, the Office of Naval Research, or the U.S. Government.

References

1. K. Deb and D. E. Goldberg. Analyzing deception in trap functions. *Foundations of Genetic Algorithms 2*, pages 93–108, 1993.
2. D. E. Goldberg. Using time efficiently: Genetic-evolutionary algorithms and the continuation problem. *Proceedings of the Genetic and Evolutionary Computation Conference*, pages 212–219, 1999.

3. D. E. Goldberg, K. Deb, and J. H. Clark. Genetic algorithms, noise, and the sizing of populations. *Complex Systems*, 6:333–362, 1992. (Also IlliGAL Report No. 91010).
4. David E. Goldberg. Sizing populations for serial and parallel genetic algorithms. *Proceedings of the Third International Conference on Genetic Algorithms*, pages 70–79, 1989.
5. David E. Goldberg. The race, the hurdle, and the sweet spot: Lessons from genetic algorithms for the automation of design innovation and creativity. In P. Bentley, editor, *Evolutionary Design by Computers*, chapter 4, pages 105–118. Morgan Kaufmann, San Mateo, CA, 1999.
6. David E. Goldberg. *The design of innovation: Lessons from and for competent genetic algorithms.* Kluwer Academic Publishers, Boston, MA, 2002.
7. David E. Goldberg, Kumara Sastry, and Thomas Latoza. On the supply of building blocks. *Proceedings of the Genetic and Evolutionary Computation Conference*, pages 336–342, 2001.
8. G. Harik. Linkage Learning via Probabilistic Modeling in the ECGA. IlliGAL Report No. 99010, University of Illinois at Urbana-Champaign, Urbana, IL, February 1999.
9. G. Harik, E. Cantú-Paz, D. E. Goldberg, and B. L. Miller. The gambler's ruin problem, genetic algorithms, and the sizing of populations. *Evolutionary Computation*, 7(3):231–253, 1999. (Also IlliGAL Report No. 96004).
10. G.R. Harik and F.G. Lobo. A parameter-less genetic algorithm. *Proceedings of the Genetic and Evolutionary Computation Conference*, pages 258–265, 1999.
11. J. H. Holland. *Adaptation in natural and artificial systems.* University of Michigan Press, Ann Arbor, MI, 1975.
12. Fernando Lobo. *The parameter-less genetic algorithm: Rational and automated parameter selection for simplified genetic algorithm operation.* phdthesis, University of Lisbon, Portugal, July 2000.
13. M. Munetomo and D. E. Goldberg. Identifying linkage groups by nonlinearity/non-monotonicity detection. *Proceedings of the Genetic and Evolutionary Computation Conference 1999: Volume 1*, pages 433–440, 1999.
14. M. Pelikan, D. E. Goldberg, and E. Cantú-Paz. Bayesian optimization algorithm, population sizing, and time to convergence. *Proceedings of the Genetic and Evolutionary Computation Conference*, pages 275–282, 2000. (Also IlliGAL Report No. 2000001).
15. M. Pelikan, K. Sastry, and D.E. Goldberg. Scalability of the Bayesian optimization algorithm. *International Journal of Approximate Reasoning*, 31(3):221–258, 2003. (Also IlliGAL Report No. 2002024).
16. Martin Pelikan and Tz-Kai Lin. Parameter-less hierarchical boa. *Proceedings of Genetic and Evolutionary Computation Conference 2004 (GECCO-2004)*, pages 24–35, 2004.
17. Martin Pelikan and Fernando Lobo. Parameter-less genetic algorithm: A worst-case time and space complexity analysis. IlliGAL Report No. 99014, Illinois Genetic Algorithms Laboratory, University of Illinois at Urbana-Champaign, Urbana, IL, 1999.
18. C.R. Reeves. Using genetic algorithms with small populations. *Proceedings of the Fifth International Conference on Genetic Algorithms*, pages 92–99, 1993.
19. K. Sastry and D. E. Goldberg. On extended compact genetic algorithm. *Late-Breaking Paper at the Genetic and Evolutionary Computation Conference*, pages 352–359, 2000. (Also IlliGAL Report No. 2000026).

20. K. Sastry and D. E. Goldberg. Designing competent mutation operators via probabilistic model building of neighborhoods. *Proceedings of the 2004 Genetic and Evolutionary Computation Conference*, 2:114–125, 2004. Also IlliGAL Report No. 2004006.
21. D. Sharman, A. Yassine, and P. Carlile. Characterizing modular architectures. *ASME 14th International Conference, DTM-34024*, Sept. 2002.
22. R. E. Smith and E. Smuda. Adaptively resizing populations: Algorithm, analysis, and first results. *Complex Systems*, 1(9):47–72, 1995.
23. Robert E. Smith. Adaptively resizing populations: An algorithm and analysis. *Proceedings of the Fifth International Conference on Genetic Algorithms*, page 653, 1993.
24. Donald V. Steward. The design structure system: A method for managing the design of complex systems. *IEEE Transactions on Engineering Managment*, 28:77–74, 1981.
25. D. Thierens and D.E. Goldberg. Convergence models of genetic algorithm selection schemes. In *Parallel Problem Solving from Nature, PPSN III*, pages 119–129, 1994.
26. A. Yassine, Donald R. Falkenburg, and Ken Chelst. Engineering design management: An informatoin structure approach. *International Journal of production research*, 37(13):2957–2975, 1999.
27. T.-L. Yu, A. Yassine, and D.E. Goldberg. A genetic algorithm for developing modular product architectures. *Proceedings of the ASME 2003 International Design Engineering Technical Conferences, 15th International Conference on Design Theory and Methodology (DETC 2003)*, pages DTM–48657, 2003. (Also IlliGAL Report No. 2003024).
28. Tian-Li Yu, David E. Goldberg, Ali Yassine, and Ying-ping Chen. Genetic algorithm design inspired by organizational theory: Pilot study of a dependency structure matrix driven genetic algorithm. *Proceedings of Artificial Neural Networks in Engineering 2003 (ANNIE 2003)*, pages 327–332, 2003. (Also IlliGAL Report No. 2003007).

Parameter-less Hierarchical Bayesian Optimization Algorithm

Martin Pelikan[1], Alexander Hartmann[2], and Tz-Kai Lin[1]

[1] Missouri Estimation of Distribution Algorithms Laboratory (MEDAL) and
Dept. of Mathematics and Computer Science, 320 CCB
University of Missouri–St. Louis
One University Blvd., St. Louis, MO 63121
pelikan@cs.umsl.edu
tlkq4@studentmail.umsl.edu

[2] Institut für Theoretische Physik
Universität Göttingen
Friedrich-Hund-Platz 1
37077 Göttingen, Germany
hartmann@physik.uni-goettingen.de

Summary. This chapter describes the parameter-less hierarchical Bayesian optimization algorithm (hBOA), which enables the use of hBOA without the need for tuning parameters for solving each problem instance. There are three crucial parameters in hBOA: (1) the selection pressure, (2) the window size for restricted tournaments, and (3) the population size. Although both the selection pressure and the window size influence hBOA performance, the asymptotic time complexity of hBOA should not change with typical choices of these two parameters. However, there is no population size that is suitable for all problems of interest because the population size strongly depends on problem size and problem difficulty. To eliminate the population size, the parameter-less hBOA adopts the population-sizing technique of the parameter-less genetic algorithm. Based on the existing theory, the parameter-less hBOA should be able to solve nearly decomposable and hierarchical problems in a quadratic or subquadratic number of function evaluations without the need for setting any parameters whatsoever. A number of experiments are presented to verify scalability of the parameter-less hBOA. The chapter also discusses how the parameter-less population-sizing scheme can be incorporated into other estimation of distribution algorithms (EDAs).

1 Introduction

When the hierarchical Bayesian optimization algorithm (hBOA) was designed [28, 30], it was argued that hBOA can solve difficult nearly decomposable and hierarchical problems without the need for setting any parameters. As a result, to solve

M. Pelikan et al.: *Parameter-less Hierarchical Bayesian Optimization Algorithm*, Studies in Computational Intelligence (SCI) **54**, 225–239 (2007)
www.springerlink.com

a new optimization problem, it should be sufficient to plug the new problem into hBOA, press the start button, and wait until hBOA figures out where the optimum is. It was argued that all hBOA parameters except for the population size can be set to their default values without affecting good scalability of hBOA [35, 26, 35]. However, all experiments assumed that the population size is set optimally to obtain best performance while retaining reliable convergence [35, 26]. That is why the dream of having a fully parameter-less optimizer for the entire class of nearly decomposable and hierarchical problems remained one parameter away from reality.

The purpose of this chapter is to propose a *fully parameter-less hBOA* by implementing the parameter-less population-sizing scheme of the parameter-less genetic algorithm (GA) [11] in hBOA. The parameter-less hBOA simulates a collection of populations of different sizes in parallel, starting with a small, base population size. Each next population is twice as large as the previous one. Each population is allowed to proceed for at least twice as many iterations as the next bigger population. This ensures that although there is no upper bound on the number of populations simulated in parallel and on their size, the overall computational overhead is still reasonable compared to the case with an optimal population size. In fact, theory exists that shows that the parameter-less population sizing scheme in the parameter-less GA does not increase the number of function evaluations until convergence by more than a logarithmic factor [34]. As a result, hBOA can be expected to perform within a logarithmic factor from the case with the optimal population size. This is verified with experiments on nearly decomposable and hierarchical problems. The parameter-less hBOA is also tested in a hybrid algorithm created by combining the parameter-less hBOA with cluster exact approximation (CEA) [15] on the problem of finding ground states of two-dimensional Ising spin glasses with $\pm J$ couplings and periodic boundary conditions. The results indicate that although the performance of the parameter-less hBOA+CEA is expected to be within a logarithmic factor of the performance of hBOA+CEA with optimal population size, the hybrid setting may increase the difference between these two cases by affecting the division of labor between the local and global searchers.

The chapter starts with a brief description of the hierarchical BOA (hBOA) in Section 2. Section 3 describes the parameter-less hBOA. Section 4 describes, presents and discusses experimental results. Section 5 outlines how the parameter-less population-sizing scheme can be used in many other EDAs. Finally, Section 6 summarizes and concludes the chapter.

2 Hierarchical Bayesian Optimization Algorithm

Estimation of distribution algorithms (EDAs) [24, 31, 16, 4] replace traditional variation operators of genetic and evolutionary algorithms by a two-step procedure:

1. Build a probabilistic model of promising solutions after selection.
2. Sample the built probabilistic model to generate new solutions.

By replacing standard crossover and mutation operators with machine learning techniques that allow automatic discovery of problem regularities from populations of promising solutions, EDAs can solve broad classes of difficult problems quickly, accurately, and reliably, including many problems intractable using other optimization methods [28, 30].

For an overview of EDAs, please see references [31] and [16]. EDAs are also referred to as the probabilistic model-building genetic algorithms (PMBGAs) [31] and iterated density-estimation algorithms (IDEAs) [3]. The remainder of this section describes the hierarchical Bayesian optimization algorithm (hBOA) [28, 30, 26], which is one of the most advanced and powerful EDAs for discrete representations.

2.1 Hierarchical BOA (hBOA)

The hierarchical Bayesian optimization algorithm (hBOA) [28, 30, 26] evolves a population of candidate solutions represented by fixed-length discrete strings over a finite alphabet. Usually, binary strings are considered but any other finite-base alphabet can be used analogically.

The first population of candidate solutions is usually generated at random according to a uniform distribution over all candidate solutions. The population is updated for a number of iterations using two basic operators: (1) selection and (2) variation. The selection operator selects better solutions at the expense of the worse ones from the current population, yielding a population of promising candidates. In this chapter, binary tournament selection is used. Binary tournament selection selects one solution at a time by first choosing two random candidate solutions from the current population and then selecting the best solution out of this subset. Random tournaments are repeated until the selected population has the same size as the original population and thus each candidate solution is expected to participate in two tournaments.

The variation operator starts by learning a probabilistic model of the selected solutions. hBOA uses Bayesian networks with local structures [5] to model promising solutions. The variation operator then proceeds by sampling the probabilistic model to generate new solutions.

The new solutions are incorporated into the original population using the restricted tournament replacement (RTR) [12], which ensures that useful diversity in the population is maintained for long periods of time. RTR has one parameter $w > 1$ called window size. RTR incorporates each new solution X into the original population (the one before selection) using the following three steps:

1. Select a random subset W of w solutions from the original population.
2. Denote Y the candidate solution in W that is most similar to X.
3. X replaces Y if X is better than Y; otherwise, X is discarded.

A robust rule of thumb is to set $w = \min\{n, N/20\}$ where n is the number of decision variables in the problem and N is the population size [29, 26].

The run is terminated when a good enough solution has been found, when the population has not improved for a long time, or when the number of generations has exceeded a given upper bound. Figure 1 shows the pseudocode of hBOA. The procedure of hBOA is illustrated in figure 2. For a more detailed description of hBOA, see [28, 25, 26].

3 Parameter-less hBOA

The parameter-less genetic algorithm (GA) [11] eliminates the need for setting parameters—such as the population size, selection pressure, crossover rate, and mutation rate—in genetic algorithms.The crossover rate and selection pressure are

```
Hierarchical BOA (hBOA)
  t := 0;
  generate initial population P(0);
  while (not done) {
    select population of promising solutions S(t);
    build Bayesian network B(t) with local struct. for S(t);
    sample B(t) to generate offspring O(t);
    incorporate O(t) into P(t) using RTR yielding P(t+1);
    t := t+1;
  };
```

Fig. 1. The pseudocode of the hierarchical BOA (hBOA).

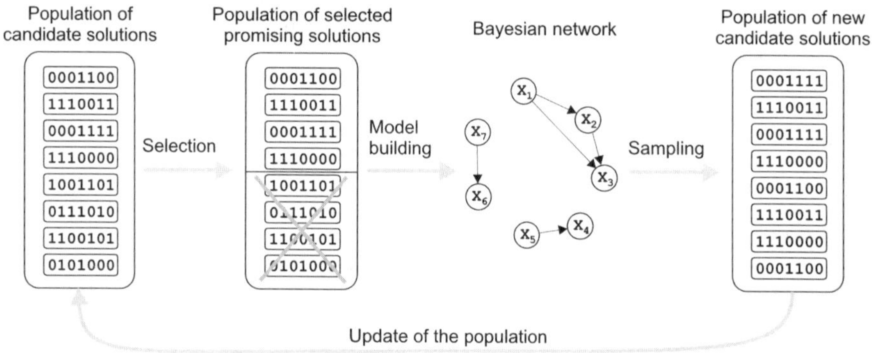

Fig. 2. hBOA starts by generating the initial population of candidate solutions at random. In each iteration, hBOA generates new solutions by sampling a Bayesian network built for the selected population of candidate solutions. The iteration concludes by incorporating new solutions into the original population using restricted tournament replacement. The new iteration is executed if the termination criteria are not met.

set to ensure consistent growth of building blocks based on the schema theorem. The mutation rate can be eliminated in a similar manner. The population size is eliminated by simulating a collection of populations of different sizes. In the context of hBOA, eliminating the population size is the most important part of the parameter-less GA, because the selection pressure influences performance of hBOA by only a constant factor and there are no crossover or mutation rates in hBOA.

The parameter-less hBOA incorporates the population-sizing technique of the parameter-less GA into hBOA by simulating a collection of populations

$$P = \{P_0, P_1, \ldots\}.$$

The size of the first population P_0 is denoted by N_0 and it is called the base population size. The size of the population P_i is denoted by N_i and it can be obtained by multiplying the base population size by a factor of 2^i:

```
Parameter-less hBOA
  initialize P[0];
  generation[0]=0;
  max_initialized=0;
  i=0;
  while (not done) {
    simulate one generation of P[i];
    generation[i] = generation[i]+1;
    if (generation[i] mod k = 0) {
      i = i + 1;
      if (i>max_initialized) {
        initialize P[i];
        max_initialized=i;
      }
    }
    else
      i = 0;
  };
```

Fig. 3. The pseudocode of the parameter-less hBOA.

$$N_i = N_0 2^i.$$

For all $i \in \{1, 2, \ldots\}$, one generation of P_i is executed after executing $k \geq 2$ generations of P_{i-1}. Here we use $k = 2$ so that all populations proceed at the same speed with respect to the number of evaluations. Each population is initialized just before its first iteration is executed. The pseudocode that can be used to simulate the collection of populations described here is shown in Figure 3. This implementation is slightly different from the one based on a k-ary counter described in the first parameter-less GA study [11].

3.1 When to Terminate a Population?

To terminate populations in the collection, the parameter-less hBOA can use the same termination criteria as the parameter-less GA [11]. However, since hBOA uses niching, no population in the collection can be expected to converge for a long time. To improve performance of the parameter-less hBOA, we upper-bound the number of generations for each population using the estimates predicted by theory [23, 38, 10, 39, 35, 26] so that the number of iterations is at least several times greater than necessary to solve each problem.

Nonetheless, similarly as in the case of the parameter-less GA [34, 33], the parameter-less hBOA can run all populations indefinitely long without increasing the logarithmic growth of the factor by which the computational complexity of hBOA increases compared to the case with the optimal population size.

4 Experiments

This section describes experiments, presents experimental results, and discusses the results.

4.1 Description of Experiments

Both hBOA and the parameter-less hBOA have been applied to artificial hierarchical and nearly decomposable problems and 2D $\pm J$ spin glasses with nearest neighbor interactions and periodic boundary conditions. Standard hBOA uses the minimum population size required for reliable convergence to the global optimum, whereas the parameter-less hBOA uses the parameter-less population sizing scheme with the base population size $N_0 = 10$.

For all problems, problem size was varied to investigate the scalability of both algorithms. For spin glasses, systems of different size were tested and 1000 random instances were examined for each problem size to ensure that the results would provide an insight into hBOA performance on a wide range of spin glass instances. To enhance efficiency of hBOA on spin glasses, cluster exact approximation (CEA) is used to improve each candidate solution in the population [15, 32].

For each problem instance and problem size, the parameter-less hBOA is first run to find the optimum in 100 independent runs and the total number of evaluations in every run is recorded. The average number of function evaluations is then displayed. These results are compared to the results for hBOA with the minimum population size that ensures that 30 independent runs converge to the optimum. The minimum population size is determined using bisection until the width of the resulting interval is at most 10% of the lower bound [25, 26].

In all experiments, binary tournament selection without replacement is used to select promising solutions, and RTR is used as a replacement strategy where the window size is $w = \min\{n, N/20\}$.

The remainder of this section discusses test problems and experimental results.

4.2 Concatenated Trap of Order 5 (trap-5)

In concatenated traps of order 5 (trap-5) [1, 8], the number of bits in solution strings must be divisible by 5. The input string is first partitioned into independent groups of 5 bits each. This partitioning is unknown to the algorithm, but it does not change during the run. A 5-bit trap function is applied to each group of 5 bits and the contributions of all traps are added together to form the fitness.

The 5-bit trap used to compute the contribution of each group of 5 bits is defined as

$$trap_5(u) = \begin{cases} 5 & \text{if } u = 5 \\ 4 - u & \text{if } u < 5 \end{cases}, \tag{1}$$

where u is the number of 1s in the input string of 5 bits. Trap-5 has one global optimum in the string of all ones and $(2^{n/5} - 1)$ other local optima. Trap-5 is fully deceptive [7]. Consequently, variation operators should not break interactions between the bits in each group because all statistics of lower order lead away from the optimum.

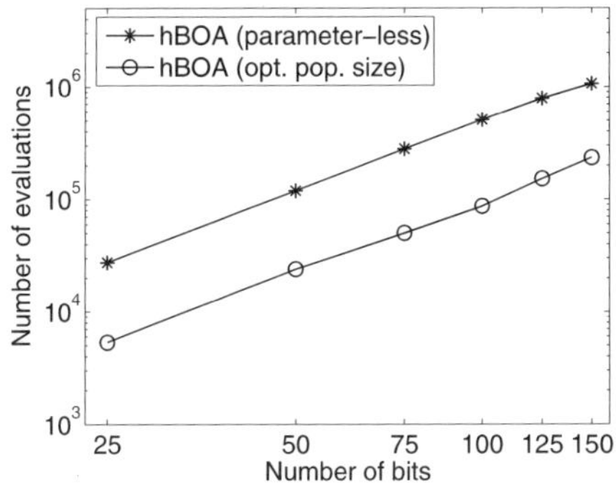

Fig. 4. Parameter-less hBOA and hBOA with the optimal population size on trap-5 of $n = 25$ to $n = 150$ bits.

Because trap-5 is a fully deceptive decomposable problem of bounded difficulty, this function provides a good test for optimization algorithms that address decomposable problems of bounded difficulty. If an algorithm is not capable of identifying appropriate problem decomposition, it should be mislead away from the optimum. That is why standard crossover and mutation operators fail to solve trap-5 scalably [10, 37, 22].

Figure 4 shows the number of evaluations of hBOA with the optimal population size determined by the bisection method and the parameter-less hBOA on trap-5 of $n = 25$ to $n = 150$ bits. The results indicate that the number of function evaluations for decomposable problems increases only by a nearly constant factor, and eliminating the parameters of hBOA thus does not seem to affect hBOA scalability on decomposable problems qualitatively; the same result is expected due to the existing worst-case theory of the parameter-less population sizing scheme [34].

4.3 Hierarchical Traps of Order 3 (hTrap)

In an order-3 hierarchical trap (hTrap) with L levels [28, 27], the total number of bits in a candidate solution is $n = 3^L$. A candidate solution is evaluated on multiple levels and the overall value of the hierarchical trap is computed as the sum of contributions on all levels.

In the variant of hTrap used in this work, on the lowest level, groups of 3 bits contribute to the overall fitness using a generalized 3-bit trap

$$trap_3(u) = \begin{cases} f_{high} & \text{if } u = 3 \\ f_{low} - u\frac{f_{low}}{2} & \text{otherwise} \end{cases}, \tag{2}$$

where $f_{high} = f_{low} = 1$.

Each group of 3 bits corresponding to one of the traps is then mapped to a single symbol on the next level; a 000 is mapped to a 0, a 111 is mapped to a 1, and everything else is mapped to the null symbol '-'. The bits on the next level again

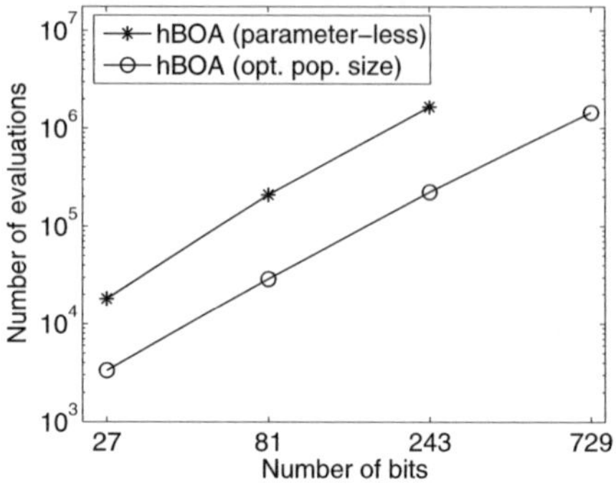

Fig. 5. Parameter-less hBOA and hBOA with the optimal population size on the hierarchical trap of $n = 27$ to $n = 729$ bits.

contribute to the overall fitness using 3-bit traps defined above, and the groups are mapped to an even higher level. This continues until the top level is evaluated that contains 3 bits total. However, on the top level, a trap with $f_{high} = 1$ and $f_{low} = 0.9$ is applied. Any group of bits containing the null symbol does not contribute to the overall fitness.

To make the overall contribution at each level of the same magnitude, the contributions of traps on ith level from the bottom are multiplied by 3^i.

hTraps have many local optima, but only one global optimum in the string of all ones. Nonetheless, any single-level decomposition into subproblems of bounded order will lead away from the global optimum. That is why hTraps necessitate an optimizer that can build solutions hierarchically by juxtaposing good partial solutions over multiple levels of difficulty until the global optimum if found.

Figure 5 shows the number of evaluations of the hierarchical BOA with optimal population size determined by the bisection method and the parameter-less hBOA on hierarchical traps of $n = 27, 81, 243, \text{and } 729$ bits. The results indicate that the number of function evaluations for hierarchically decomposable problems increases only by a constant factor, and eliminating the parameters of hBOA does not affect scalability of hBOA on this class of problems qualitatively.

4.4 Ising Spin Glasses

Ising spin glasses provide a class of prototypical models for disordered systems and have played a central role in statistical physics during the last three decades [2, 20, 9, 40]. Here we discuss a simple two-dimensional Ising spin glass model, which provides a rich class of difficult optimization problems that are intractable with most standard crossover and mutation operators [26] as well as state-of-the-art Markov chain Monte Carlo methods [6]. There are three primary sources of problem difficulty of Ising spin glasses: (1) Ising spin glasses contain a large number of local optima, which often grows exponentially with problem size, (2) local optima in the objective-function landscape are surrounded by low-quality points, and (3) effective exploration of the space of candidate solutions (spin configurations) necessitates that the exploration operator considers interactions between a large number of spins (problem variables).

A 2D spin-glass system consists of a regular 2D grid containing n nodes where each node i corresponds to a spin s_i and each edge $\langle i, j \rangle$ corresponds to a coupling between two spins s_i and s_j. Each edge has a real value associated with it that defines the relationship between the two connected spins. To approximate the behavior of the large-scale system, periodic boundary conditions are often used that introduce a coupling between the first and the last element along each dimension.

For the classical Ising model, each spin s_i can be in one of two states: $s_i = +1$ or $s_i = -1$. Note that this simplification corresponds to highly anisotropic systems, which do indeed exist in some experimental situations. Nevertheless, the two-state Ising model comprises all basic effects also found in models with more degrees of freedom. A specific set of coupling constants define a spin glass instance. Each possible setting of the states of all spins is called a spin configuration.

Given a set of coupling constants $J_{i,j}$, and a configuration of spins $C = \{s_i\}$, the energy can be computed as

$$E(C) = \sum_{\langle i,j \rangle} s_i J_{i,j} s_j \ , \tag{3}$$

where $i, j \in \{0, 1, \ldots, n-1\}$ and $\langle i, j \rangle$ denote the nearest neighbors in the underlying grid (allowed edges).

Here we consider the task of finding *ground states* of a 2D spin glass with specified coupling constants, where a ground state is a configuration of spins that minimizes the energy of the system given by equation 3.

Each spin glass configuration is represented by a string of bits where each bit corresponds to one spin: a 0 represents a spin -1, and a 1 represents a spin $+1$. We created 1000 random spin glasses of sizes 20×20 (400 spins) to 50×50 (2500 spins) by generating coupling constants to be $+1$ or -1 with equal probabilities. Spin glasses with coupling constants restricted to $+1$ and -1 are called $\pm J$ spin glasses. The optimum for most 2D instances was verified with the branch-and-cut algorithm provided at the Spin Glass Ground State Server at the University of Köln [36].

To enhance hBOA efficiency, cluster exact approximation (CEA) [15] is incorporated into hBOA to improve all solutions in the population similarly as in [32]. CEA proceeds by selecting a cluster of non-frustrated spins and changes the cluster optimally given the fixed values of all remaining spins. This can be done efficiently using the maximum flow algorithm. CEA can be repeated a number of times because the clusters of selected spins tend to differ and so do the effects of the corresponding updates.

In the hybrid algorithm hBOA+CEA created by combining hBOA and CEA, before evaluating each spin glass configuration in hBOA, CEA is applied to improve the configuration for a number of steps. The application of CEA to each configuration is terminated once the configuration cannot be improved for a certain number of iterations; more specifically, the bound on the number of failures is $\sqrt{n}$ where n is the number of spins. Since one iteration of CEA can be expected to be at least as expensive as one evaluation of a configuration [21], instead of counting the number of evaluated candidate solutions, we consider the number of CEA iterations. The overall number of CEA iterations ran on candidate solutions during the entire run represents the time complexity of this run.

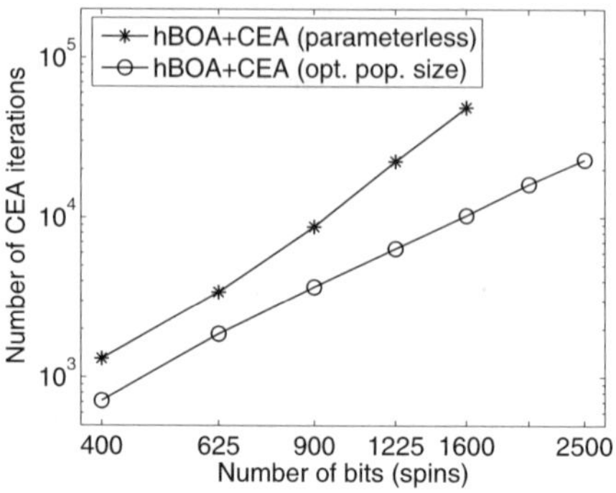

Fig. 6. Parameter-less hBOA and hBOA with the optimal population size on 2D Ising spin glasses with random $\pm J$ couplings and periodic boundary conditions. Both methods are combined with cluster exact approximation (CEA).

Figure 5 shows the number of CEA iterations for hBOA+CEA with the optimal population size determined by the bisection method and the parameter-less hBOA+CEA on 2D $\pm J$ spin glasses. The good news is that the number of CEA iterations for the parameter-less hBOA still grows as a low-order polynomial with respect to the number of decision variables. The bad news is that, unlike for single-level and hierarchical traps, in this case the parameter-less population sizing does influence the asymptotic complexity of hBOA. Although the number of CEA iterations seems to be still upper-bounded by a low-order polynomial, the order of the polynomial that approximates the number of CEA iterations increases slightly by the factor bounded by $O(n^{0.75})$.

The increase in the order of the polynomial that approximates the asymptotic growth of the number of CEA steps seems to disagree with the existing theory, which claims that the factor for the number of evaluations should be at most logarithmic [34]. One factor that leads to these results is that for hybrid variants of hBOA, the time complexity of evaluating a candidate solution depends on the solution itself. For any population, early in the run, it can be expected to be rather easy to improve most configurations with CEA and each solution will most likely lead to a number of CEA iterations because configurations are random or nearly random. On the other hand, later in the run, most configurations should be near a local optimum and CEA would usually lead to no improvement or only little improvement at a time. In the parameter-less hBOA, the proportion of solutions that were processed only for a few generations to the overall number of evaluated candidate solutions increases faster than in the standard hBOA with the optimum population size. Consequently, the number of CEA iterations per evaluation of a candidate solution is expected to increase, leading to the increase of the asymptotic complexity of hBOA+CEA using the parameter-less population sizing. However, this factor still does not fully explain the results because based on the analysis of experimental results, this leads to the factor of only about $O(n^{0.05})$, leaving the overall increase in the asymptotic time complexity unexplained.

Another hypothesis is that due to the change in the structure of the global population that consists of all evaluated solutions, the division of labor between the local and global searcher may change, and more bias is given to one of the searchers at the expense of the other one. This hypothesis can be supported by the experiments with the parameter-less variant of hBOA hybridized with other local search methods [33], where hBOA was combined with a simple discrete hill climber and the parameter-less version of this hybrid algorithm also led to an increase of the asymptotic growth of the number of evaluations.

5 Eliminating Population Size in Other EDAs

Eliminating the parameters of hBOA is important on its own right, but an important question is whether the proposed scheme can be used to eliminate parameters of other EDAs. Intuitively, the same approach can be used to eliminate the need for setting the population size in most other EDAs. We present two examples that should apply to many EDAs.

5.1 Eliminating Population Size in Population-Based EDAs

Any EDA that maintains a population of candidate solutions can be modified by incorporating the parameter-less population-sizing scheme of the parameter-less genetic algorithm analogically to the parameter-less hBOA. For example, the parameter-less population-sizing can be used in the extended compact genetic algorithm (ECGA) [19, 17]. Nonetheless, if there are other parameters that crucially depend on the problem, such as the standard deviation of Gaussian kernel distributions, more parameters would have to be eliminated to create a true parameter-less EDA.

5.2 Parameter-less Compact Genetic Algorithm

Although some EDAs do not maintain an explicit population of candidate solutions, they represent a population implicitly using a probabilistic model; this probabilistic model is updated based on the results of tournaments between candidate solutions generated according to the model. Despite the absence of the population, there is usually a parameter that can be seen as a counterpart of the population size and it is just as important as the population size in population-based EDAs. One of the algorithms that use incremental updates of the probabilistic model is the compact genetic algorithm [13, 14].

In the compact genetic algorithm (cGA) [13, 14], the population is replaced by a probability vector that stores the estimated probability of a 1 in each string position. The probability-vector entries are initialized to 0.5 (uniform distribution). In each iteration, the probability-vector entries are updated by performing a binary tournament between two solutions generated according to the probability vector and modifying vector entries for the bits that differ in the two competing solutions.

N=10, generation=4, learning rate=1/10	→	0.40	0.50	0.70	0.50	0.60	0.60
N=20, generation=2, learning rate=1/20	→	0.55	0.55	0.60	0.50	0.50	0.50
N=40, generation=1, learning rate=1/40	→	0.53	0.50	0.53	0.50	0.50	0.53

Fig. 7. Probability vectors of the parameter-less compact genetic algorithm after running the first iteration on the third probability vector for a 6-bit onemax. Since the optimum of onemax is in the string of all ones, it is expected that the proportions of ones increase over time.

The update rule depends on the learning rate, which corresponds to the effective population size.

The parameter-less cGA can be created by replacing the multiple populations of the parameter-less genetic algorithm by probability vectors; each population is represented by one probability vector [18]. In each probability vector, the learning rate $1/N$ is used to represent the population size N. Using the population sizes that grow with an integer power of two, this leads to a series of exponentially decreasing learning rates. See figure 7 for an illustration of this concept.

All probability vectors can be simulated indefinitely, similarly as in the parameter-less hBOA. Another approach is to terminate each probability vector when its fitness average over a specific period of time becomes lower than the fitness average for the probability vector representing a larger population size.

Since the learning rate is the only parameter of cGA, the parameter-less cGA is a true parameter-less algorithm. A similar approach could be used to update the learning rate in other similar EDAs, such as PBIL, although the update of the appropriate parameters may not be as straightforward as in cGA.

6 Summary and Conclusions

This chapter described, implemented and tested the parameter-less hierarchical Bayesian optimization algorithm (hBOA). The parameter-less hBOA enables the practitioner to simply plug the problem into hBOA without requiring the practitioner to first estimate an adequate population size or other problem-specific parameters. Despite the parameter-less scheme, low-order polynomial time complexity of hBOA on broad classes of optimization problems is retained. The parameter-less hBOA is thus a true black-box optimization algorithm, which can be applied to hierarchical and nearly decomposable problems without setting any parameters whatsoever.

The parameter-less population-sizing scheme can be used in other estimation of distribution algorithms, including the algorithms that replace the population by a probabilistic model and use incremental updates of the probabilistic model.

Acknowledgments

This project was sponsored by the National Science Foundation under CAREER grant ECS-0547013, by the Air Force Office of Scientific Research, Air Force Materiel

Command, USAF, under grant FA9550-06-1-0096, by the University of Missouri in St. Louis through the High Performance Computing Collaboratory sponsored by Information Technology Services, and the Research Award and Research Board programs, and by the *VolkswagenStiftung* (Germany) within the program "Nachwuchsgruppen an Universitäten". Most experiments were done using the hBOA software developed by Martin Pelikan and David E. Goldberg at the University of Illinois at Urbana-Champaign. The experiments presented in this paper were completed at the Beowulf cluster at the University of Missouri in St. Louis.

References

1. D. H. Ackley. An empirical study of bit vector function optimization. *Genetic Algorithms and Simulated Annealing*, pages 170–204, 1987.
2. K. Binder and A.P. Young. Spin-glasses: Experimental facts, theoretical concepts and open questions. *Rev. Mod. Phys.*, 58:801, 1986.
3. Peter A. N. Bosman and Dirk Thierens. Continuous iterated density estimation evolutionary algorithms within the IDEA framework. *Workshop Proceedings of the Genetic and Evolutionary Computation Conference (GECCO-2000)*, pages 197–200, 2000.
4. Erick Cantú-Paz, Martin Pelikan, and Kumara Sastry, editors. *Scalable optimization via probabilistic modeling: From algorithms to applications.* Springer-Verlag, 2006.
5. David M. Chickering, David Heckerman, and Christopher Meek. A Bayesian approach to learning Bayesian networks with local structure. Technical Report MSR-TR-97-07, Microsoft Research, Redmond, WA, 1997.
6. P. Dayal, S. Trebst, S. Wessel, D. ürtz, M. Troyer, S. Sabhapandit, and S. Coppersmith. Performance limitations of flat histogram methods and optimality of Wang-Langdau sampling. *Physical Review Letters*, 92(9):097201, 2004.
7. K. Deb and D. E. Goldberg. Analyzing deception in trap functions. IlliGAL Report No. 91009, University of Illinois at Urbana-Champaign, Illinois Genetic Algorithms Laboratory, Urbana, IL, 1991.
8. K. Deb and D. E. Goldberg. Sufficient conditions for deceptive and easy binary functions. *Annals of Mathematics and Artificial Intelligence*, 10:385–408, 1994.
9. K.H. Fischer and J.A. Hertz. *Spin Glasses.* Cambridge University Press, Cambridge, 1991.
10. David E. Goldberg. *The design of innovation: Lessons from and for competent genetic algorithms*, volume 7 of *Genetic Algorithms and Evolutionary Computation.* Kluwer Academic Publishers, 2002.
11. Georges Harik and Fernando Lobo. A parameter-less genetic algorithm. *Proceedings of the Genetic and Evolutionary Computation Conference (GECCO-99)*, I:258–265, 1999.
12. Georges R. Harik. Finding multimodal solutions using restricted tournament selection. *Proceedings of the International Conference on Genetic Algorithms (ICGA-95)*, pages 24–31, 1995.
13. Georges R. Harik, Fernando G. Lobo, and David E. Goldberg. The compact genetic algorithm. IlliGAL Report No. 97006, University of Illinois at Urbana-Champaign, Illinois Genetic Algorithms Laboratory, Urbana, IL, 1997.

14. Georges R. Harik, Fernando G. Lobo, and David E. Goldberg. The compact genetic algorithm. *Proceedings of the International Conference on Evolutionary Computation (ICEC-98)*, pages 523–528, 1998.
15. A. K. Hartmann. Cluster-exact approximation of spin glass ground states. *Physica A*, 224:480, 1996.
16. Pedro Larrañaga and Jose A. Lozano, editors. *Estimation of Distribution Algorithms: A New Tool for Evolutionary Computation.* Kluwer, Boston, MA, 2002.
17. Claudio F. Lima and Fernando G. Lobo. Parameter-less optimization with the extended compact genetic algorithm and iterated local search. *Proceedings of the Genetic and Evolutionary Computation Conference (GECCO-2004)*, pages 1328–1339, 2004.
18. Tz-Kai Lin and Martin Pelikan. Parameter-less compact genetic algorithm. Unpublished Technical Report, Department of Mathematics and Computer Science, University of Missouri, St. Lous, 2004.
19. Fernando G. Lobo. *The parameter-less genetic algorithm: Rational and automated parameter selection for simplified genetic algorithm operation.* PhD thesis, Faculdade de Ciencias e Tecnologia da Universidade Nova de Lisboa, Lisbon, Portugal, 2000.
20. M. Mezard, G. Parisi, and M.A. Virasoro. *Spin glass theory and beyond.* World Scientific, Singapore, 1987.
21. Alan Middleton and Daniel S. Fisher. The three-dimensional random field Ising magnet: Interfaces, scaling, and the nature of states. *Phys. Rev. B*, 65:134411, 2002.
22. H. Mühlenbein. How genetic algorithms really work: I.Mutation and Hillclimbing. In R. Männer and B. Manderick, editors, *Parallel Problem Solving from Nature*, pages 15–25, Amsterdam, Netherlands, 1992. Elsevier Science.
23. H. Mühlenbein and D. Schlierkamp-Voosen. Predictive models for the breeder genetic algorithm: I. Continuous parameter optimization. *Evolutionary Computation*, 1(1):25–49, 1993.
24. Heinz Mühlenbein and Gerhard Paaß. From recombination of genes to the estimation of distributions I. Binary parameters. *Parallel Problem Solving from Nature*, pages 178–187, 1996.
25. Martin Pelikan. *Bayesian optimization algorithm: From single level to hierarchy.* PhD thesis, University of Illinois at Urbana-Champaign, Urbana, IL, 2002. Also IlliGAL Report No. 2002023.
26. Martin Pelikan. *Hierarchical Bayesian optimization algorithm: Toward a new generation of evolutionary algorithms.* Springer-Verlag, 2005.
27. Martin Pelikan and David E. Goldberg. Hierarchical problem solving by the Bayesian optimization algorithm. *Proceedings of the Genetic and Evolutionary Computation Conference (GECCO-2000)*, pages 275–282, 2000. Also IlliGAL Report No. 2000002.
28. Martin Pelikan and David E. Goldberg. Escaping hierarchical traps with competent genetic algorithms. *Proceedings of the Genetic and Evolutionary Computation Conference (GECCO-2001)*, pages 511–518, 2001. Also IlliGAL Report No. 2000020.
29. Martin Pelikan and David E. Goldberg. Hierarchical boa solves ising spin glasses and maxsat. *Proceedings of the Genetic and Evolutionary Computation Conference (GECCO-2003)*, II:1275–1286, 2003. Also IlliGAL Report No. 2003001.

30. Martin Pelikan and David E. Goldberg. A hierarchy machine: Learning to optimize from nature and humans. *Complexity*, 8(5):36–45, 2003.
31. Martin Pelikan, David E. Goldberg, and Fernando Lobo. A survey of optimization by building and using probabilistic models. *Computational Optimization and Applications*, 21(1):5–20, 2002. Also IlliGAL Report No. 99018.
32. Martin Pelikan and Alexander Hartmann. Searching for ground states of Ising spin glasses with hierarchical BOA and cluster exact approximation. In E. Cantú-Paz, M. Pelikan, and K. Sastry, editors, *Scalable optimization via probabilistic modeling: From algorithms to applications*. Springer, 2006.
33. Martin Pelikan and Tz-Kai Lin. Parameter-less hierarchical boa. *Proceedings of the Genetic and Evolutionary Computation Conference (GECCO-2004)*, 2:24–35, 2004.
34. Martin Pelikan and Fernando G. Lobo. Parameter-less genetic algorithm: A worst-case time and space complexity analysis. IlliGAL Report No. 99014, University of Illinois at Urbana-Champaign, Illinois Genetic Algorithms Laboratory, Urbana, IL, 1999.
35. Martin Pelikan, Kumara Sastry, and David E. Goldberg. Scalability of the Bayesian optimization algorithm. *International Journal of Approximate Reasoning*, 31(3):221–258, 2002. Also IlliGAL Report No. 2001029.
36. Spin Glass Ground State Server. `http://www.informatik.uni-koeln.de/ls_juenger/research/sgs/sgs.html`, 2004. University of Köln, Germany.
37. D. Thierens. *Analysis and design of genetic algorithms*. PhD thesis, Katholieke Universiteit Leuven, Leuven, Belgium, 1995.
38. D. Thierens and D.E. Goldberg. Convergence models of genetic algorithm selection schemes. *Parallel Problem Solving from Nature*, pages 116–121, 1994.
39. Dirk Thierens, David E. Goldberg, and Angela G. Pereira. Domino convergence, drift, and the temporal-salience structure of problems. *Proceedings of the International Conference on Evolutionary Computation (ICEC-98)*, pages 535–540, 1998.
40. A.P. Young, editor. *Spin glasses and random fields*. World Scientific, Singapore, 1998.

Evolutionary Multi-Objective Optimization Without Additional Parameters

Kalyanmoy Deb

Department of Mechanical Engineering
Indian Institute of Technology Kanpur
Kanpur, PIN 208016, India
deb@iitk.ac.in

Summary. The present-day evolutionary multi-objective optimization (EMO) algorithms had a demonstrated history of evolution over the years. The initial EMO methodologies involved additional niching parameters which made them somewhat subjective to the user. Fortunately, soon enough parameter-less EMO methodologies have been suggested thereby making the earlier EMO algorithms unpopular and obsolete. In this paper, we present a functional decomposition of a viable EMO methodology and discuss the critical components which require special attention for making the complete algorithm free from any additional parameter. A critical evaluation of existing EMO methodologies suggest that the elitist non-dominated sorting GA (NSGA-II) is one of EMO algorithms which does not require any additional implicit or explicit parameters other than the standard EA parameters, such as population size, operator probabilities, etc. This parameter-less property of NSGA-II is probably the reason for its popularity to most EMO studies thus far.

1 Introduction

Evolutionary algorithms (EAs) came out as serious contenders in optimization studies during the past two decades. EAs have certain features (such as, flexible operators, no need for using gradients, ease in tackling mixed-integer problems and combinatorial problems, etc.) which are rare to be found in any other single optimization method, including classical methods. However, the flexibilities in their usage come with some onus on the part of the user. To put different flexible EA operators together to make an effective search, they have to be tuned properly as to make a proper balance [22, 19] of two conflicting aspects needed in a successful optimization algorithm: (i) exploitation of available resource and (ii) exploration of search space. The former aspect deals with the selection operator and is related to the selection probabilities assigned to better population members for them to be used as parents. The more the assigned selection probability, the larger is the exploitation of available resource. But since a population early on need not have the optimal solution, too much exploitation may lead to a premature convergence of the

K. Deb: *Evolutionary Multi-Objective Optimization Without Additional Parameters*, Studies in Computational Intelligence (SCI) **54**, 241–257 (2007)
www.springerlink.com

overall algorithm. Whether a premature convergence actually takes place or not is also closely related to the associated explorative search power of the algorithm. This aspect is related to the variation (recombination and mutation) operators used in an EA. If these operators are designed in a way to have a large search power (large ergodicity, as defined by Radcliffe [32]), a large exploitation may be allowed to constitute a successful search [22]. Thus, it is clear that a successful application of an EA is far from being simple and a proper balancing act of exploitation-exploration issue is essential in an EA. The balance between these issues is controlled by a proper choice of different EA parameters (such as, population size, selection pressure parameter, EA operators and their probabilities, etc.). Unfortunately, different optimization problems require different values of some or all of these parameters and a successful application of an EA often requires a proper tuning of these parameters from problem to problem. Since these parameters are essential to be fixed in any form of an EA, it is not surprising that these parameters must have to be set properly in a successful application of a multi-objective EA.

As the name suggests, multi-objective optimization deals with multiple conflicting objectives and usually the optimal solution of one of the objectives is not necessarily the optimum for any of the other objectives. In such a scenario, instead of one optimal solution, a number of solutions are optimal. These solutions are called Pareto-optimal solutions. In the growing literature on evolutionary multi-objective optimization (EMO) [5], the task is to first find a well-distributed set of Pareto-optimal solutions and then, based on the trade-off information about the solutions, choose one specific solution for implementation [10].

Intuitively, EMO algorithms have a more stringent task to be achieve than the single-objective EAs. While in most single-objective EAs, the task is to find a single optimal solution, an EMO is expected to find a set of Pareto-optimal solutions causing an optimal trade-off among multiple conflicting objectives. Besides the requirement of these optimized solutions to be as close to the true Pareto-optimal front as possible, they also have an additional requirement of being well distributed over the entire Pareto-optimal region. The task of finding multiple, well spread-out optimal solutions resembles the task in a niching EA, in which the target is to find multiple optimal solutions in a multi-modal, single-objective optimization problem [23, 11]. Besides, an EMO algorithm must have another important consideration – the need of assigning an appropriate implicit or explicit fitness measure so that solutions can progress towards the true Pareto-optimal frontier. All these operations seem demanding and it is not obvious whether an EMO algorithm can be devised without the need of any additional parameter.

In this paper, we first discuss the nature of optimal solutions in a multi-objective optimization problem and then outline some standard EA parameters which are usually associated with any single-objective EA. Thereafter, we attempt to decompose different tasks necessary to a multi-objective EA functionally into three major components and discuss the difficulties which one can face in trying to devise them as parameter-free procedures. Finally, we consider a number of existing EMO methodologies and show that a specific algorithm has a systematic implementation which allowed the complete algorithm to have no additional parameter. This specific algorithm (the elitist non-dominated sorting GA or NSGA-II [7, 8]) is by far the most popular EMO methodology used and cited in the EMO literature. Through a comparison of NSGA-II with other contemporary EMO algorithms according to their requirement of additional parameters, we argue that the parameter-less approach of NSGA-II is one of the main reasons for its success and popularity among researchers and practitioners. The systematic discussion for achieving parameter-less algorithms presented in this paper should encourage readers to develop more effective parameter-less EMO or EA approaches.

2 Multi-Objective Optimization

A multi-objective optimization problem involves a number of objective functions which are to be either minimized or maximized. As in the single-objective optimization problem, the multi-objective optimization problem usually has a number of constraints which any feasible solution (including the optimal solution) must satisfy. In the following, we state the multi-objective optimization problem (MOOP) in its general form:

$$\left.\begin{array}{lll} \text{Minimize/Maximize} & f_m(\mathbf{x}), & m = 1, 2, \ldots, M; \\ \text{subject to} & g_j(\mathbf{x}) \geq 0, & j = 1, 2, \ldots, J; \\ & h_k(\mathbf{x}) = 0, & k = 1, 2, \ldots, K; \\ & x_i^{(L)} \leq x_i \leq x_i^{(U)}, & i = 1, 2, \ldots, n. \end{array}\right\} \quad (1)$$

A solution $\mathbf{x}$ is a vector of n decision variables: $\mathbf{x} = (x_1, x_2, \ldots, x_n)^T$. The solutions satisfying the constraints and variable bounds constitute a *feasible decision variable space* $\mathcal{S}$, or simply the variable space. One of the striking differences between single-objective and multi-objective optimization is that in multi-objective optimization the objective functions constitute a multi-dimensional space, in addition to the usual variable space. For each solution $\mathbf{x}$ in the variable space, there exists a point in the objective space, denoted by $\mathbf{f}(\mathbf{x}) = \mathbf{z} = (z_1, z_2, \ldots, z_M)^T$. A mapping exists between an n-dimensional solution vector and an M-dimensional objective vector through the objective function, constraints, and variable bounds. Figure 1 illustrates these two spaces and a mapping between them.

Although the fundamental difference between single and multiple objective optimization lies in the cardinality in the optimal set, from a practical standpoint a user needs only one solution, no matter whether the associated optimization problem is single-objective or multi-objective. In the case of multi-objective optimization, the user is now in a dilemma. Which of these optimal solutions must one choose? This is not an easy question to answer. It involves many higher-level information which are often non-technical, qualitative and experience-driven. However, if a set of many trade-off solutions are already worked out or available, one can evaluate the pros and cons of each of these solutions based on all such non-technical and qualitative, yet still important, considerations and compare them to make a choice. Thus, in a multi-objective optimization, ideally the effort must be made in finding the set of trade-off optimal solutions by considering all objectives to be important. After a set of such trade-off solutions are found, a user can then use higher-level qualitative considerations to make a choice. In view of these discussions, the following principle was suggested as an *ideal multi-objective optimization procedure*:

Step 1 Find multiple trade-off optimal solutions with a wide range of values for objectives.
Step 2 Choose one of the obtained solutions using higher-level information.

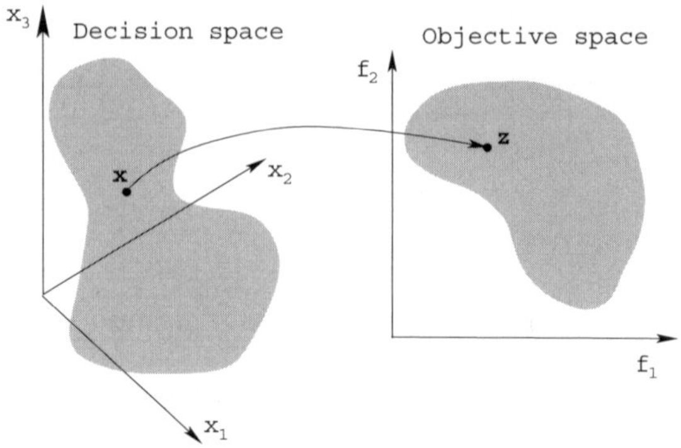

Fig. 1. Representation of the decision variable space and the corresponding objective space.

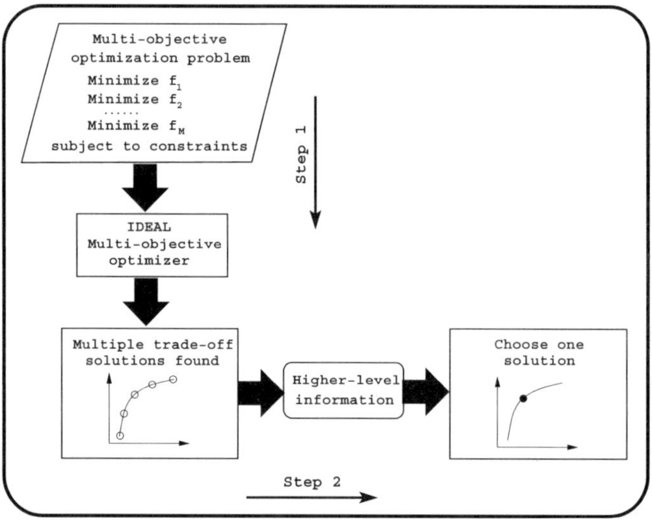

Fig. 2. Schematic of an ideal multi-objective optimization procedure.

Figure 2 shows schematically the principles in an ideal multi-objective optimization procedure. In Step 1 (vertically downwards), multiple trade-off solutions are found. Thereafter, in Step 2 (horizontally, towards the right), higher-level information is used to choose one of the trade-off solutions. With this procedure in mind, it is easy to realize that single-objective optimization is a degenerate case of multi-objective optimization. In the case of single-objective optimization with only one global optimal solution, Step 1 will find only one solution, thereby not requiring us to proceed to Step 2. In the case of single-objective optimization with multiple global optima, both steps are necessary to first find all or many of the global optima and then to choose one from them by using the higher-level information about the problem. Realizing this degeneracy between single and multi-objective optimization, we have designed an *omni-optimizer*, which is capable of solving different types of optimization problems [14]. We shall discuss this generic procedure in Section 7, but first enumerate the necessary components of an evolutionary multi-objective optimization.

2.1 Evolutionary Principles

Since an evolutionary algorithm deals with a number of population members in each generation, an EA is an ideal candidate for finding multiple Pareto-optimal solutions in a multi-objective optimization problem. In Sections 4 to 7, we shall discuss a few modified EAs for doing this task, but all of these methods perform the following tasks:

1. Emphasize *non-dominated* solutions for progressing towards the Pareto-optimal front.
2. Emphasize *less-crowded* solutions for maintaining a good diversity among obtained solutions.
3. Emphasize *elites* to provide a faster and reliable convergence near the Pareto-optimal front.

Different algorithms implement the above three features differently, but they are mostly implemented in the *selection* operator of an EA. They can be used while choosing parent solutions for recombination and they can also be used while deciding to accept an offspring solution into the population. Potentially, a naive implementation may involve setting additional parameters for each of the above features, but intuitively the latter two tasks (determining less-crowded solutions and determining elite population members) may become difficult to implement without any user-defined parameter. We discuss these two matters in the following paragraphs.

The less-crowded solutions are those which are not surrounded by too many solutions. First of all, the neighborhood can be defined in either of the two spaces: (i) objective space and (ii) variable space. Secondly, how many neighboring solutions are too many? The answer to this question may require a parameter to resolve. The elite solutions are those which are best in the population. In a multi-objective problem, how does a solution define to be the best, when there are multiple conflicting objectives? This question can be answered somewhat by declaring all non-dominated solutions to be the best. As shown elsewhere [5], problems having a large number of objectives, almost all population members belong to the non-dominated front. It is also natural to understand that not all solutions can be declared as elites and be passed on to the next population. This leaves us with a question of which or what proportion of non-dominated solutions are to be redefined as elites? This involves setting another parameter.

Thus, it is easy to think of an evolutionary multi-objective optimization algorithm which will have components to take care of the above three essential tasks, but such an algorithm may involve a number of additional parameters which the user must have to set. It becomes a challenging task to come up with an EMO algorithm which would perform all three tasks efficiently but without any additional parameter. We shall discuss how parameter-less implementation of such salient tasks is achieved in a few popular EMO methodologies later in this study. But before we do this, we would like to mention a few essential parameters which a single-objective EA run usually demands. These parameters are also required to be supplied in running most multi-objective EA algorithms.

3 Essential Parameters in an Evolutionary Algorithm

Many chapters of this book have dealt with the essential parameters needed in a single-objective evolutionary algorithm (EA) and suggested various procedures of setting up the adequate values of these parameters. Here, we briefly mention the essential parameters often needed to be fixed in an EA:

1. Population size (N),
2. Number of generations (T),
3. Parameters related to Selection (selection pressure (s), selection operator, etc.),
4. Parameters related to recombination (crossover probability (p_c), crossover operator, etc.),
5. Parameters related to mutation (mutation probability (p_m), mutation operator, etc.),
6. Generation gap parameter (G),
7. Number of independent runs of an EA.

Population size is a crucial parameter in the successful application of an EA. For an application without any limit in the number of overall function evaluations, there exists a *critical* population size below which the EA is not expected to perform to its best and on or beyond of which the EA performs at its best. However, additional population members demand more function evaluations compared to that needed by the EA which uses the critical population size. However, if an EA is applied for a limited number of function evaluations, there exists a critical range of population sizes below and above of which the performance of an EA gets degraded. Thus, in such cases, it is essential to choose a population size as close to the critical population size as possible. Some statistics-based estimation of the size of a random initial population exist [21, 24, 33] and some numerical procedures do exist as well [1, 35].

Even with an adequate population size and other EA operators, the EA must be run for a *critical* number of generations for a convergence near the optimal solution. For some problems (mostly test problems), a bound on number of generations is computed from a complexity analysis [40]. However, such an analysis is usually problem-specific and cannot be generalized for any arbitrary problem [41].

An EA is a flexible optimization algorithm, because it allows the user to choose any suitable EA operators. Since it is flexible, it also puts an onus on the part of the user to choose *proper* operators. There are mainly two different EA operators – a selection operator mimicking the Darwin's survival of the fittest principle and one or more variation operators causing the generation of a new offspring population. Usually, two variation operators are used in most applications: (i) recombination or crossover operator and (ii) mutation operator. Each of these operators have many variations and each variation may involve different additional user-defined parameters. For example, the following is a list of selection operators popularly used:

1. s-ary tournament selection, in which s solutions are picked with or without replacement and the best solution (in terms of their fitness values) is selected.
2. Ranking selection, in which population members are ranked according to the fitness values and a higher rank (better solution) solution is selected proportionately more often (by assigning s copies to the best solution).

In these selection operators, the parameter s is loosely called the selection pressure and the performance of an EA procedure depends on the choice of this parameter s. Goldberg and Deb [20] computed the selection pressure for the above two operators and a few others.

Similarly, different recombination operators and their associated parameters affect the performance of an EA and must be chosen properly. One common parameter among different crossover operators is the 'probability of crossover' (p_c), which denotes the expected proportion of the population which participates in the crossover operation. However, more operator-specific parameters exist and must also be set by the user. For example, the binary-coded crossover operator requires the user to specify the number of cross-sites. For real-parameter recombination operators such as SBX [6], BLX [16] and FR [39] require a parameter, controlling the extent of the search, whereas some recombination operators such as PCX [9], UNDX [29], SPX [25], differential evolution (DE) [38], operator in particle swarm optimization (PSO) [28] require more than one parameters to be set by the user.

Mutation operator also involves a probability of mutation p_m to be set. In addition, at least one mutation-specific parameter is often required to be supplied by the user.

Besides, the generation gap is an important matter which controls the overall selection pressure per function evaluation. On one extreme (called the 'generational EA'), the generation gap can be equal to the population size, meaning that all N offspring solutions must be created and evaluated before any of them is put back into the population. On the other extreme (called the 'steady-state EA'), the generation gap is one, meaning that after every new offspring solution is created, it is considered for its inclusion to the population. The latter causes a larger selection pressure and the EA is often termed as an 'greedy EA'. Other intermediate propositions are certainly possible [34, 27].

3.1 Desired and Undesired Parameters

In most optimization studies, an algorithm involving parameters which the users are required to be set is considered negatively, simply because of the obtained optimized solution may be sensitive to the parameter values. In fact, more the number of parameters involved in the algorithm, the worse is the situation. Often, good optimization studies make a parametric analysis and present the sensitivities of each parameter on the obtained result. Often researchers suggest ways to estimate parameters which are found to have worked well on known test problems.

However, there may exist certain parameters which are *desired* in an algorithm and provide the user (a designer or an applicationist) flexibility and control in the obtained solution. Although the outcome of the optimization depends on these tunable parameters, instead they being controlled by the developers of the algorithm, the users prefer to choose them on their own. In multi-objective optimization algorithms, such tunable parameters (for example, ϵ in the ϵ-MOEA [12] or a fixation of one or more reference points in a reference-point based EMO [13]) allows a decision-maker to find solutions closer to the desired region on the Pareto-optimal frontier. We call these parameters as 'desired parameters' for a problem solving task.

4 NSGA, Contemporary Algorithms, and Additional Parameters

In this section, we investigate the non-elitist EMO methodologies for their requirement of any additional parameter for achieving two tasks: (i) in determining non-dominated solutions and (ii) in determining less-crowded solutions. Since these methods did not use any elite-preservation operator, we ignore the third task in these first-generation EMO algorithms. A detail description of these algorithms can be found in [5].

Non-dominated sorting GA or NSGA [37] and multi-objective GA or MOGA [18] emphasized non-dominated solutions without any additional parameter, but the niched-Pareto GA or NPGA [26] introduced a subpopulation size parameter for defining the non-dominated solutions. This parameter is required to be set by the user.

Next, we discuss their requirement for additional parameter in implementing the niching operator. The non-elitist EMO methodologies, which were able to find a set well-distributed near-Pareto-optimal solutions, all used an explicit niche-preserving operator involving a niching parameter. Thus, these methodologies demanded an additional parameter on top of EA parameters mentioned above. For example, the non-dominated sorting GA or NSGA [37] required the niching parameter σ_{share}, which defined the maximum Euclidean distance among two solutions in the variable space for them to qualify as solutions from the same niche. Although a subsequent study [4] has suggested a procedure of estimating this parameter, the performance of NSGA was shown to depend on the choice of this parameter [37]. The purpose of the niching parameter is to define the maximum distance between any two solutions to remain in a single niche. If a small niching parameter is used, fewer solutions will qualify to remain in a single niche, thereby clustering all solutions to more number of niches. The effect of choosing a small niching parameter in an EMO is, therefore, to converge to many Pareto-optimal solutions. There are some difficulties in using a fixed niching parameter throughout the run. In most problems, the search region where an algorithm is started is often wider than the Pareto-optimal region in which the algorithm is expected to converge at the end. By keeping the same niching parameter throughout, many different niches must have to be encountered early on to keep a good distribution of solutions, thereby requiring a larger population size. However, near the Pareto-optimal region, a large population may not be required and with a large population, multiple solutions will exist in each niche. Although the use of population-best and population-worst function values can be used to normalize the objective values and an identical niching parameter can be used throughout, this idea causes another difficulty. Early on, the effective niche size will be large for a similar argument given above and there may not be enough representative solutions to search the region effectively. Thus, for both of the above cases (with or without normalization), there is a need of a variable population size to make an effective search. Although an adaptive change of population size with generation is in order, most EMO methodologies use a fixed population size.

Similarly, MOGA [18] required an identical niching parameter as in NSGA, but the distance measure was computed in the objective space. The study suggested a clever procedure for estimating the niching parameter and a later study [5] suggested a normalized version of the procedure and presented a table showing the σ_{share} value as a function of a fixed population size and number of objectives. However, these estimation procedures are not free from approximations and some tuning may be necessary for their proper working in an arbitrary problem.

NPGA [26] required two parameters: σ_{share} and an additional parameter t_{dom} – the number of population members chosen for counting the number of similar solutions to two competing solutions. The original study did not provide any specific guidelines for estimating these parameters.

The need for these extra parameters for multi-objective optimization and their demonstrated dependence on the performance of the corresponding EMO procedure made researchers think of parameter-less EMO procedures.

5 NSGA-II and No Additional Parameters

Next, we consider the elitist EMO methodologies and begin our discussion with the elitist non-dominated sorting or NSGA-II [8] algorithm.

The requirement of the additional parameter setting (σ_{share}) in NSGA is avoided in the elitist non-dominated sorting GA or NSGA-II [8]. Although the non-dominated sorting approach is used, the NSGA-II procedure is somewhat different from NSGA in the following aspects:

1. NSGA-II eliminates the need of the additional parameter, which NSGA required.
2. NSGA-II uses an elite-preserving strategy, whereas NSGA did not have any elite-preservation mechanism.
3. NSGA-II uses a computationally fast niching strategy which is applied with objective values, instead of decision parameter values used in NSGA.
4. NSGA-II allows a flexible platform to develop different strategies, such as a steady-state EMO, a preference-based EMO, and others, which were not possible with the proportionate selection framework of NSGA.

The determination of the non-dominated solutions does not require any additional parameter. The clever method of combining offspring and parent populations together and choosing the best half of the combined population according to a non-dominated sorting and crowding scheme did not require any additional parameter in implementing the overall elite-preservation scheme. This way, if a parent solution is better than all offspring solutions, it naturally has a higher chance of getting included in the next population, thereby implementing the elite-preservation principle without any extra parameter. Such a scheme has also been used in other single-objective EA methodologies, such as CHC [15] and evolution strategy studies [36].

Next, we discuss how the niching strategy is implemented without requiring any additional niching parameter in NSGA-II. The original NSGA procedure degraded the *rank* assigned to each solution (based on its non-dominated ranking) by the niche count – a measure of crowdedness near the solution in the decision space. This procedure required a niching parameter. The NSGA-II procedure used a completely different niching strategy. Instead of degrading the fitness or assigning a rank to each solution, a two-stage procedure is adopted here. First, the solutions are chosen based on the non-domination ranking. Second, solutions, belonging to a maximum ranking which could not be completely accepted to keep the size of the population constant, are evaluated for their crowdedness. This two-stage procedure allows a niching procedure to be applied to a non-dominated set of population members. The niching task remains as the one of finding a subset of population members having as few neighbors as possible. Such a niching task is possible to achieve without requiring any niching parameter.

There are two ways such a task can be achieved: (i) a block reduction strategy or (ii) a one-at-a-time pruning strategy. In both approaches, every population member can be assigned a crowding metric value based on the extent of neighboring solutions in the objective space or in the variable space. Thereafter, in the block reduction strategy, the required number of solutions having the largest crowding metric value (meaning more crowding) can be eliminated at once. In the one-at-a-time pruning strategy [31], only one solution having the largest crowding metric value is eliminated and the crowding metric value is recomputed for the remaining solutions. One solution is eliminated again and the procedure is continued till the required number of solutions are eliminated. It is clear that the latter strategy would constitute a better niching operation than the former one, but at the expense of more computations.

Thus it is clear that a two-phase task of fitness assignment and niching operation allow no additional parameters for achieving both the tasks. NSGA-II procedure only requires the standard EA parameters (population size, operator probabilities etc.) to be supplied.

We argue that requirement of no further parameters in NSGA-II is one of the main reasons of its popularity in EMO studies in the recent past. We observe that the recent EMO conference proceedings (EMO-2005 [2]) had a total of 59 papers on multi-objective optimization and 33 papers (about 56%) used NSGA-II directly or used as basis for comparison. Similarly, 19 of 56 papers (about 34%) of EMO-2003 conference proceedings [17] used NSGA-II. A recent recognition of a high rate of citation of the article describing NSGA-II by the ISI Web of Science [3] is another testimony to its success as an efficient multi-objective optimization algorithm.

6 Other EMO Methodologies and Additional Parameters

The strength Pareto EA or SPEA [43] uses a parameter-less fitness assignment scheme based on domination level of each solution. However, the niching operator and the elite-preserving operators involve additional parameters.

SPEA uses a clustering technique for niching to make a one-at-a-time deletion of crowded solutions. In this approach, every non-dominated solution is assumed to reside in a separate cluster. Thereafter, two clusters having the shortest distance between them are merged together and the solution closest to the centroid of the merged cluster is retained and other solutions in the merged cluster are not considered. This procedure is continued as many times as the required attrition in population members. This way more-crowded solutions are deleted and less-crowded solutions get emphasized. Although this niching strategy apparently does not require any additional niching parameter, the extended SPEA or SPEA2 [42] used the distance to the k-th nearest neighbor to cluster solutions and suggested $k = \sqrt{N + N'}$

(N is the EA population size and N' is the archive size). In this sense, the original SPEA procedure used an additional niching parameter but authors suggested a fixed value of $k = 1$ (in the original SPEA) or $k = \sqrt{N + N'}$ in SPEA2. Besides, the one-at-a-time deletion strategy causes SPEA or SPEA2 to have a large computational burden [12], to have a much better distribution of solutions particularly for problems having more than two objectives.

SPEA and SPEA2 also require an archive, the size of which must be carefully chosen to provide an adequate selection pressure to the better solutions (another parameter!). This is because the archive contains the best non-dominated solutions (elites) found thus far in any generation and a combination of archive and EA population is used for choosing potential parent solutions. Thus, the ratio between the archive size and EA population size sets up a critical selection pressure for the elite solutions. The developers of SPEA suggested to use an archive size which is about 1/5-th to the size of the EA population. Potentially, the archive size is another additional parameter which SPEA or SPEA2 requires the user to set. It is interesting to note both these parameters (k and archive size N') in associated in the proper working of SPEA or SPEA2 and are not the 'desired parameters' from the point of view of a user.

The Pareto-archived evolution strategy or PAES [30] used a pairwise domination-based selection scheme, which does not require any additional parameter. However, the niching approach compartmentalizes the objective space into a fixed number of hyper-boxes, which required an additional niching parameter. In PAES, the number of divisions in each objective or the size of hyper-boxes in each objective is used for this purpose. The PAES procedure attempts to have only one solution in each hyper-box, thereby ensuring a diversity among population members. Thus, a smaller box-size is, in effect, capable of finding more optimized solutions. Since this parameter directly controls the number of optimized solutions found at the end of the optimization run, it can be called as a 'desired parameter'. By fixing the box-size (niching parameter), the user can control the number of optimized solutions. However, the user does not have a direct control of the exact number of non-dominated solutions to be found, as the the box-size parameter is pre-specified and it is not known beforehand the extent of the Pareto-optimal set. The elite-preserving operator compared the newly-created child solution with the parent which created it and used a parameter-less acceptance rule based on domination level of two solutions and the number of solutions resident to hyper-boxes of child and parent.

7 Omni-Optimizer and No Additional Parameters

Recently, the author, with the help of his student, suggested an omni-optimizer which is capable of solving four different optimization tasks without any intervention and additional parameter [14]:

1. Problems having single-objective, single optimum,
2. Problems having single-objective, multiple optima,
3. Problems having multi-objective, single optimum for each Pareto-optimal point, and
4. Problems having multi-objective, multiple optimal for each Pareto-optimal point.

The advantage of such an optimization algorithm is that the algorithm degenerates itself to find the necessary optimal solutions (one or more) depending on the supplied objective(s). This way the user needs only to know a single optimization algorithm and may be able to solve different kinds of optimization problems offered by the description of the problem (objectives and constraints).

The algorithm has a generic structure of a multi-objective optimization algorithm (NSGA-II). The domination criterion of comparing two solutions for superiority degenerates to making a real-valued comparison of two solutions in the case of single-objective optimization. Thus, this first feature of the omni-optimizer is parameter-free. For single-objective optimization, the algorithm also degenerates to an elite-preserving operation by first combining both parent and offspring populations and then systematically choosing half the population. Thus, the elite-preserving operator is also parameter-free.

Since this algorithm claims to find multiple optimal solutions simultaneously, the omni-optimizer has an explicit niching operator. Since the algorithm also claims also to solve both single and multi-objective optimization problems, the niching operator is also expected to perform niching in both spaces: (i) variable space and (ii) objective space. The difficulty with such a requirement is that the niching information in one space can be completely different from that in the other space. Thus, it may be difficult to come up with a combined niching metric to define the extent of crowding in both spaces. If a sharing function based approach [23] is to be used, two different niching parameters must have to be introduced, thereby requiring the user to set two niching parameters. However, the omni-optimizer uses a parameter-less approach which make a hierarchical comparison of the two niching quantities and constitute a viable algorithm. We give a brief description of the procedure below.

First, every solution in a front is compared with its neighbors objective-wise and an objective-space crowding distance (say D_o) (similar to that in NSGA-II [8]) is computed. The extreme solutions are assigned a very large crowding distance to ensure their presence in the population. Thereafter, a variable-space crowding distance (say D_v) is computed by using a similar procedure, but all computations are performed in the variable space. Here, the extreme solutions in the variable space are not assigned an arbitrary large crowding distance, instead twice the crowding distance from their nearest solutions are assigned. This way, every solution in the population is assigned two niching quantities computed without using any extra niching parameter: the crowding distance in the objective space and crowding distance in the variable space. To choose a subset of solutions from a non-dominated set, the next task is to use these two measures and decide which all solutions to keep and which all solutions to delete. One way to do this would be to choose a threshold for each space and keep all solutions having the crowding distance greater than the chosen threshold value. But this simple-minded procedure will require two new niching parameters to be set by the user. The omni-optimizer uses a parameter-less procedure, which we discuss next.

First, we compute the front-average of objective-wise and variable-wise crowding distance values. Thereafter, instead of comparing the individual crowding distance values with a threshold of some sort, we compare them with these front-wise average values. If a solution does not have above-front-average crowding distance in both spaces, we de-emphasize the solution by assigning the smallest of the two crowding distance values as its overall crowding distance measure. On the other hand, if a solution has above-average crowding distance value in any of the two spaces, we encourage it by assigning it the larger of the two crowding distance values. This procedure is free from any niching parameter and emphasizes any solution which has above-average crowding distance in any of the two spaces.

In the event of a single-objective problem having multiple, identical optimal function value, all high-performing solutions will have very similar objective values, thereby making the objective-wise crowding distance value small (and more or less identical) for each solution. In this case, solutions are chosen mainly based on the variable-wise crowding distance value. The effect is that solutions residing near different and well-separated optima in the variable space get emphasized in the population, thereby allowing multiple optimal solutions to be remain in the population. Most niching procedures work using a similar principle, but unfortunately using a niching parameter. For example, the sharing function approach can find multiple optimal solutions simultaneously, but requires a sharing parameter to be set properly. By comparing a crowding measure with the population-average crowding measure, solutions can be given differential preference, thereby encouraging niche-formation without the need of any additional niching parameter.

Similarly, if in a multi-objective optimization problem, each Pareto-optimal solution corresponds to multiple variable vectors, it would be desirable to locate all or as many such optimal variable vectors as possible. The above two-space crowding distance consideration allows a nice way to find such multiple optimal solutions as well. To illustrate, we consider the following two-objective minimization problem:

$$\begin{aligned} \text{Minimize } f_1(x) &= \sum_{i=1}^{n} \sin(\pi x_i), \\ \text{Minimize } f_2(x) &= \sum_{i=1}^{n} \cos(\pi x_i), \\ 0 \leq x_i &\leq 6, \quad i = 1, 2, \ldots, n. \end{aligned} \tag{2}$$

Here, $n = 5$ and both objectives are periodic functions with period of 2. The Pareto-optimal solutions are $x_i \in [2m + 1, 2m + 3/2]$, where m is an integer. In the above bound, there are exactly three solutions (in [1.0,1.5], [3.0,3.5] and [5.0,5,5]) which produce the same objective values. For every efficient point in the objective space there are in general $3^5 = 243$ Pareto-optimal solutions in the variable space. We choose a population of size 1,000 and run the algorithm for 500 generations to capture as many Pareto-optimal solutions as possible. It is interesting to note that both omni-optimizer and the original NSGA-II find the entire range of efficient points in the objective space, as shown in Figures 3 and 4. However, the variable space plots show a different scenario. The lower diagonal plots in Figure 5 show the performance of the omni-optimizer and the upper diagonal plots show that of the original NSGA-II. It is clear that in a pair-wise plot of variables, all 3^2 or 9 optimal regions are found by the omni-optimizer, whereas since no variable-space crowding is considered in the original NSGA-II, not all optimal combinations are found.

8 Conclusions

In this paper, we have functionally decomposed the tasks needed in an evolutionary multi-objective optimization (EMO) algorithm and identified the critical components which are vulnerable to be designed for additional parameters. These are the niching operator and the elite-preserving operator. By systematically analyzing a number of existing EMO methodologies, we have shown that an EMO methodology can be designed without introducing any additional parameter to those needed in a standard single-objective EA. The elitist non-dominated sorting GA or NSGA-II is one such EMO algorithm and a major reason for its popularity can be contributed to the fact that it does not demand setting any further parameters than those required by single-objective EAs. Often, such parameter-less implementations are innovative and attempt to design such procedures may spur novel ideas in other similar algorithm developmental tasks.

Acknowledgments

The author wishes to thank Santosh Tiwari for implementing the idea of omni-optimizer. The author also acknowledges the support provided by STMicroelectronics for performing this study.

References

1. T. P. Bagchi and K. Deb. Calibration of GA parameters: The design of experiments approach. *Computer Science and Informatics*, 26(4):46–56, 1996.
2. C. A. Coello Coello, A. H. Aguirre, and E. Zitzler, editors. *Evolutionary Multi-Criterion Optimization: Third International Conference*. Berlin, Germany: Springer, 2005. LNCS 3410.
3. K. Deb. IEEE TEC Paper (2002, vol. 6, pp. 182–197) on NSGA-II: Fast breaking paper in the field of engineering. ISI Web of Science, http://esi-topics.com/fbp/2004/february04-KalyanmoyDeb.html.
4. K. Deb. Multi-objective evolutionary optimization: Past, present and future. In I. C. Parmee, editor, *Evolutionary Design and Manufacture*, pages 225–236. London: Springer, 2000.
5. K. Deb. *Multi-objective optimization using evolutionary algorithms*. Chichester, UK: Wiley, 2001.
6. K. Deb and R. B. Agrawal. Simulated binary crossover for continuous search space. *Complex Systems*, 9(2):115–148, 1995.
7. K. Deb, S. Agrawal, A. Pratap, and T. Meyarivan. A fast elitist non-dominated sorting genetic algorithm for multi-objective optimization: NSGA-II. In *Proceedings of the Parallel Problem Solving from Nature VI (PPSN-VI)*, pages 849–858, 2000.
8. K. Deb, S. Agrawal, A. Pratap, and T. Meyarivan. A fast and elitist multi-objective genetic algorithm: NSGA-II. *IEEE Transactions on Evolutionary Computation*, 6(2):182–197, 2002.
9. K. Deb, A. Anand, and D. Joshi. A computationally efficient evolutionary algorithm for real-parameter optimization. *Evolutionary Computation Journal*, 10(4):371–395, 2002.
10. K. Deb, S. Chaudhuri, and K. Meitinnen. I-EMO: An interactive evolutionary multi-objective optimization tool. Technical Report KanGAL Report Number 2005005, Department of Mechanical Engineering, IIT Kanpur, India, 2005.

Also in Proceedings of Pattern Recognition in Machine Intelligence (PReMI'05), Springer.

11. K. Deb and D. E. Goldberg. An investigation of niche and species formation in genetic function optimization. In *Proceedings of the Third International Conference on Genetic Algorithms*, pages 42–50, 1989.
12. K. Deb, M. Mohan, and S. Mishra. Towards a quick computation of well-spread pareto-optimal solutions. In *Proceedings of the Second Evolutionary Multi-Criterion Optimization (EMO-03) Conference (LNCS 2632)*, pages 222–236, 2003.
13. K. Deb and J. Sundar. Reference point based multi-objective optimization using evolutionary algorithms. In *Proceeedings of the Genetic and Evolutionary Computation Conference (GECCO-2006)*, in press.
14. K. Deb and S. Tiwari. Omni-optimizer: A generic evolutionary algorithm for global optimization. *European Journal of Operations Research (EJOR)*, in press.
15. L. J. Eshelman. The CHC adaptive search algorithm: How to have safe search when engaging in nontraditional genetic recombination. In *Foundations of Genetic Algorithms 1 (FOGA-1)*, pages 265–283, 1991.
16. L. J. Eshelman and J. D. Schaffer. Real-coded genetic algorithms and interval-schemata. In *Foundations of Genetic Algorithms 2 (FOGA-2)*, pages 187–202, 1993.
17. C. Fonseca, P. Fleming, E. Zitzler, K. Deb, and L. Thiele. *Proceedings of the Second Evolutionary Multi-Criterion Optimization (EMO-03) Conference (Lecture Notes in Computer Science (LNCS) 2632)*. Heidelberg: Springer, 2003.
18. C. M. Fonseca and P. J. Fleming. An overview of evolutionary algorithms in multi-objective optimization. *Evolutionary Computation Journal*, 3(1):1–16, 1995.
19. D. E. Goldberg. *Genetic Algorithms for Search, Optimization, and Machine Learning*. Reading, MA: Addison-Wesley, 1989.
20. D. E. Goldberg and K. Deb. A comparison of selection schemes used in genetic algorithms. In *Foundations of Genetic Algorithms 1 (FOGA-1)*, pages 69–93, 1991.
21. D. E. Goldberg, K. Deb, and J. H. Clark. Genetic algorithms, noise, and the sizing of populations. *Complex Systems*, 6(4):333–362, 1992.
22. D. E. Goldberg, K. Deb, and D. Thierens. Toward a better understanding of mixing in genetic algorithms. *Journal of the Society of Instruments and Control Engineers (SICE)*, 32(1):10–16, 1993.
23. D. E. Goldberg and J. Richardson. Genetic algorithms with sharing for multimodal function optimization. In *Proceedings of the First International Conference on Genetic Algorithms and Their Applications*, pages 41–49, 1987.
24. G. Harik, E. Cantú-Paz, D. E. Goldberg, and B. L. Miller. The gambler's ruin problem, genetic algorithms, and the sizing of populations. *Evolutionary Computation Journal*, 7(3):231–254, 1999.
25. T. Higuchi, S. Tsutsui, and M. Yamamura. Theoretical analysis of simplex crossover for real-coded genetic algorithms. In *Parallel Problem Solving from Nature (PPSN-VI)*, pages 365–374, 2000.
26. J. Horn, N. Nafploitis, and D.E. Goldberg. A niched Pareto genetic algorithm for multi-objective optimization. In *Proceedings of the First IEEE Conference on Evolutionary Computation*, pages 82–87, 1994.
27. K.A. De Jong and J. Sharma. Generation gap revisited. In *Foundations of Genetic Algorithms (FOGA-II)*, pages 19–28, 1992.

28. James Kennedy and Russell C. Eberhart. *Swarm intelligence.* Morgan Kaufmann, 2001.
29. H. Kita, I. Ono, and S. Kobayashi. Multi-parental extension of the unimodal normal distribution crossover for real-coded genetic algorithms. In *Proceedings of the 1999 Congress on Evolutionary Computation*, pages 1581–1587, 1999.
30. Joshua D. Knowles and David W. Corne. Approximating the non-dominated front using the Pareto archived evolution strategy. *Evolutionary Computation Journal*, 8(2):149–172, 2000.
31. S. Kukkonen and K. Deb. Improved pruning of non-dominated solutions based on crowding distance for bi-objective optimization problems. In *Proceedings of the Congress on Evolutionary Computation (CEC-06)*, in press.
32. N. J. Radcliffe. Forma analysis and random respectful recombination. In *Proceedings of the Fourth International Conference on Genetic Algorithms*, pages 222–229, 1991.
33. C. R. Reeves. Using genetic algorithms with small populations. In *Proceedings of the Fifth International Conference on Genetic Algorithms*, pages 92–99, 1993.
34. H. Satoh, M. Yamamura, and S. Kobayashi. Minimal generation gap model for gas considering both exploration and exploitation. In *Proceedings of the IIZUKA: Methodologies for the Conception, Design, and Application of Intelligent Systems*, pages 494–497, 1996.
35. J. D. Schaffer, R. A. Caruana, L. J. Eshelman, and R. Das. A study of control parameters affecting online performance of genetic algorithms. In *Proceedings of the International Conference on Genetic Algorithms*, pages 51–60, 1989.
36. H.-P. Schwefel. *Evolution and Optimum Seeking.* New York: Wiley, 1995.
37. N. Srinivas and K. Deb. Multi-objective function optimization using non-dominated sorting genetic algorithms. *Evolutionary Computation Journal*, 2(3):221–248, 1994.
38. R. Storn and K. Price. Differential evolution – A fast and efficient heuristic for global optimization over continuous spaces. *Journal of Global Optimization*, 11:341–359, 1997.
39. H.-M. Voigt, H. Mühlenbein, and D. Cvetković. Fuzzy recombination for the Breeder Genetic Algorithm. In *Proceedings of the Sixth International Conference on Genetic Algorithms*, pages 104–111, 1995.
40. I. Wegener and T. Jansen. Real royal road functions - where crossover provably is essential. *Discrete Applied Mathematics*, 149:111–125, 2005.
41. D. H. Wolpert and W. G. Macready. No free lunch theorems for optimization. *IEEE Transactions on Evolutionary Computation*, 1(1):67–82, 1977.
42. E. Zitzler, M. Laumanns, and L. Thiele. SPEA2: Improving the strength pareto evolutionary algorithm for multiobjective optimization. In K. C. Giannakoglou, D. T. Tsahalis, J. Périaux, K. D. Papailiou, and T. Fogarty, editors, *Evolutionary Methods for Design Optimization and Control with Applications to Industrial Problems*, pages 95–100, Athens, Greece, 2001. International Center for Numerical Methods in Engineering (Cmine).
43. E. Zitzler and L. Thiele. Multiobjective evolutionary algorithms: A comparative case study and the strength pareto approach. *IEEE Transactions on Evolutionary Computation*, 3(4):257–271, 1999.

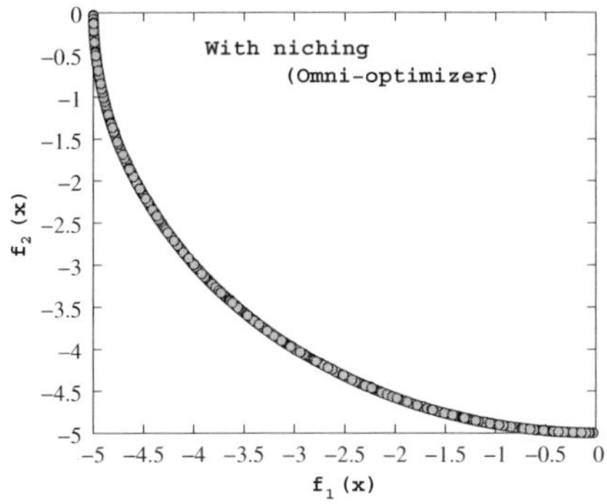

Fig. 3. Efficient points using omni-optimizer.

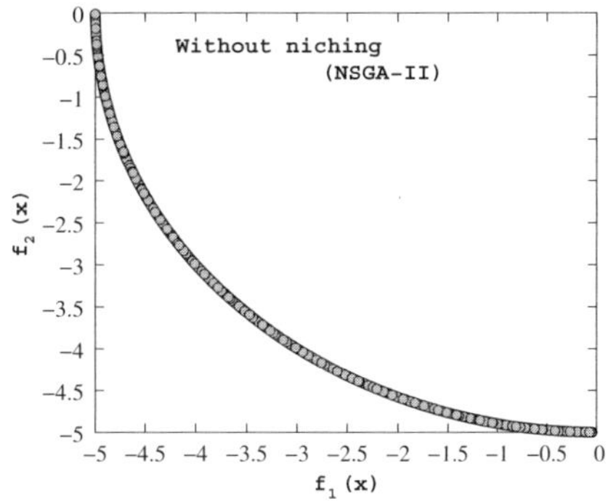

Fig. 4. Efficient points using NSGA-II.

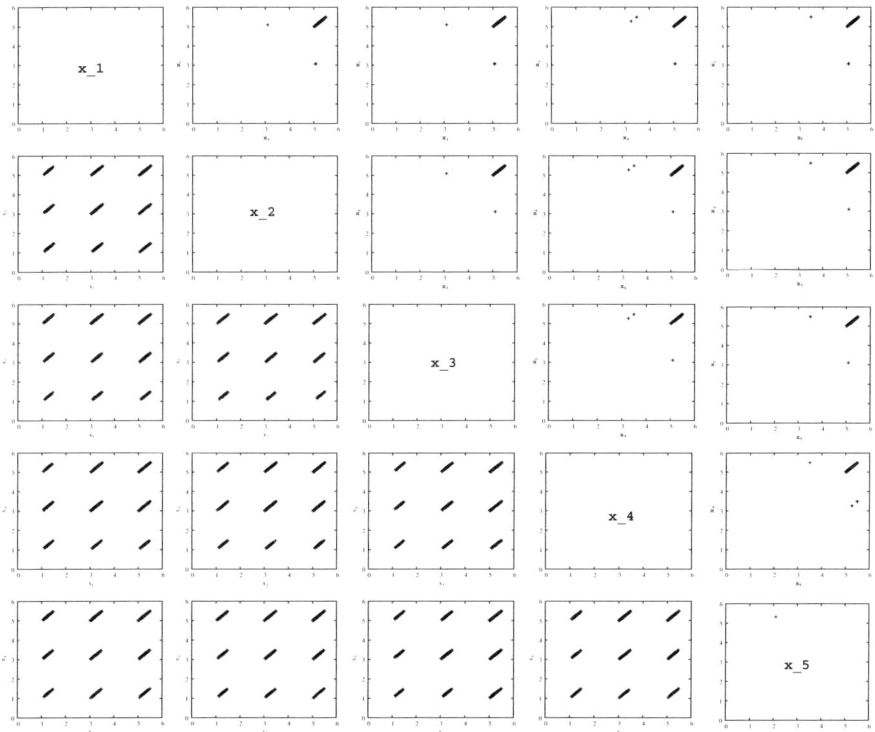

Fig. 5. Pareto-optimal solutions with omni-optimizer (left) and NSGA-II (right). The axes in a (i, j)-plot correspond to variables x_i and x_j.

Parameter Setting in Parallel Genetic Algorithms

Erick Cantú-Paz

Yahoo!, Inc.
701 First Avenue
Sunnyvale, CA 94089
cantupaz@acm.org

Summary. Parallel genetic algorithms (GAs) have numerous parameters that affect their efficiency and accuracy. Traditionally, these parameters have been studied using empirical studies whose generality and limitations are difficult to assess. This chapter reviews existing theoretical models that predict the effects of the parameters. The models are used to examine the effect of communication topologies, migration rates, population sizing, and the choice of migrants and the individuals they replace in the receiving populations. The models should help practitioners make informed decisions about the setting of parameters of parallel GAs.

1 Introduction

One of the most promising alternatives to improve the efficiency of genetic algorithms (GAs) is to use parallel implementations [1, 40, 11, 4, 60]. While it is relatively easy to implement efficient parallel versions of GAs, parallel GAs are complex non-linear algorithms that are controlled by many parameters that affect their efficiency and the quality of their search. For example, one must decide whether to use a single or multiple populations. In either case, the size of the populations must be determined carefully, and for multiple populations, one must decide how many to use. In addition, the populations may remain isolated or they may communicate during the run by exchanging individuals or some other information. Communication involves extra costs and additional decisions on the pattern of communications, on the number of individuals to be exchanged, and on the frequency of communications. This chapter reviews the existing theory that explains the effect of the numerous parameters of these algorithms on their accuracy and efficiency.

The chapter deals with two common types of parallel GAs. Single-population master-slave GAs have a master node that stores the population and executes the GA operations (selection, crossover, and mutation). The evaluation of fitness is distributed among several slave processors (see Figure 1). Despite being very simple algorithms, master-slave implementations can be very efficient as will be shown in section 2.

E. Cantú-Paz: *Parameter Setting in Parallel Genetic Algorithms*, Studies in Computational Intelligence (SCI) **54**, 259–276 (2007)
www.springerlink.com

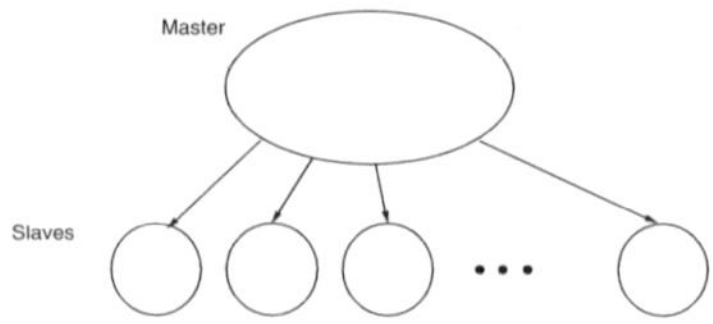

Fig. 1. A schematic of a master-slave parallel GA. The master stores the population, executes the GA operations, and distributes individuals to the slaves. The slaves evaluate the fitness of the individuals.

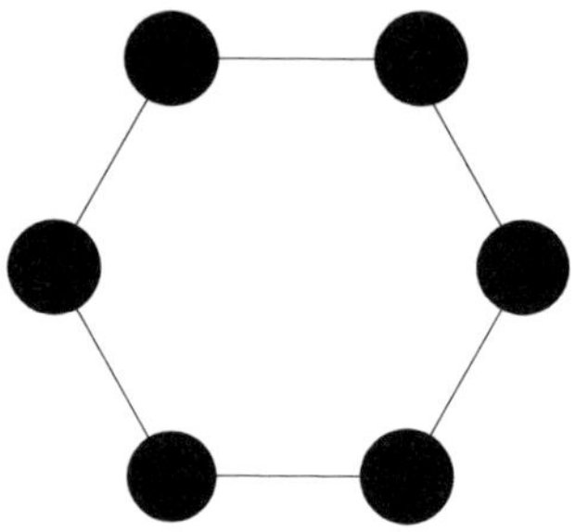

Fig. 2. A schematic of a multiple-population parallel GA. Each circle represents a simple GA, and there is (infrequent) communication between the populations. In this example, the populations are arranged in a ring, but many other communication topologies have been used.

Multiple-population GAs are the most sophisticated and popular type of parallel GAs. They consist of several subpopulations that exchange individuals (Figure 2 has a schematic). This exchange of individuals is called migration, and is controlled by several parameters such as the frequency of migration, the number and destination of migrants, and the method used to select which individuals migrate. The effect of these parameters on the search is complex and these algorithms are the focus of most of the research on parallel GAs. Section 3 reviews the theory relevant to these algorithms.

2 Master-Slave Parallel GAs

Probably the easiest way to implement GAs on parallel computers is to distribute the evaluation of fitness among several slave processors while one master executes the GA operations (selection, crossover, and mutation). Master-slave parallel GAs are interesting and important for several reasons: (1) they explore the search space in exactly the same manner as serial GAs, and therefore the parameter settings used for simple GAs are directly applicable; (2) they are very easy to implement, which

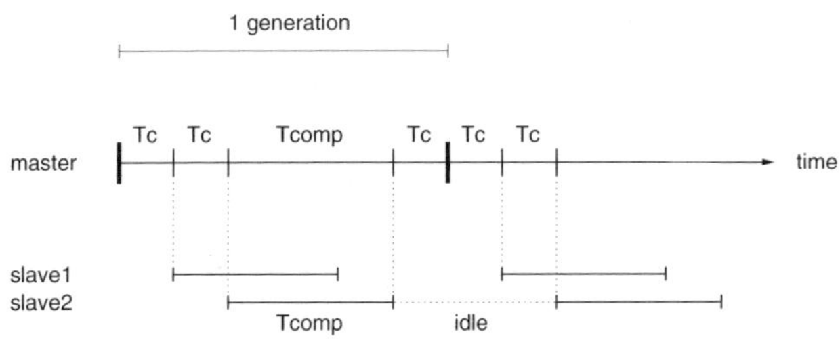

Fig. 3. One generation in a master-slave parallel GA when the master evaluates a fraction of the population.

makes them popular with practitioners; and (3) in many cases master-slave GAs result in significant improvements in performance.

The only parameter introduced by this mode of parallelization is the number of processors. Although at first it may seem best to use all available processors, the communication time might increase to the point of obliterating the gains obtained by distributing the computing load between the processors.

Centering our attention on the master processor, Figure 3 depicts the sequence of events in every generation. First, the master sends a fraction of the population to each of the slaves, using time T_c to communicate with each. Next, the master evaluates a fraction of the population using time $\frac{nT_f}{\mathcal{P}}$, where T_f is the time required to evaluate one individual, n is the size of the population, and $\mathcal{P}$ is the number of processors used.

The slaves start evaluating their portion of the population as soon as they receive it, and return the evaluations to the master as soon as they finish. We ignore the time consumed by selection, crossover, and mutation because it is usually much shorter than the time used to evaluate and to communicate individuals. In addition, we assume that the same number of individuals are assigned to each slave and that evaluation time is the same for all individuals. With these assumptions, the elapsed time for one generation of the parallel GA may be estimated as the sum of elapsed time spent in computations and time used to communicate information among the processors:

$$T_p = \mathcal{P}T_c + \frac{nT_f}{\mathcal{P}}. \tag{1}$$

As more slaves are used, the computation time decreases as desired, but the communication time increases. Making $\frac{\partial T_p}{\partial \mathcal{P}} = 0$ and solving for $\mathcal{P}$, we obtain the optimal number of processors that minimizes the execution time:

$$\mathcal{P}^* = \sqrt{\frac{nT_f}{T_c}}. \tag{2}$$

Defining the parallel speedup as

$$\frac{T_s}{T_p} = \frac{nT_f}{\frac{nT_f}{\mathcal{P}} + \mathcal{P}T_c} \tag{3}$$

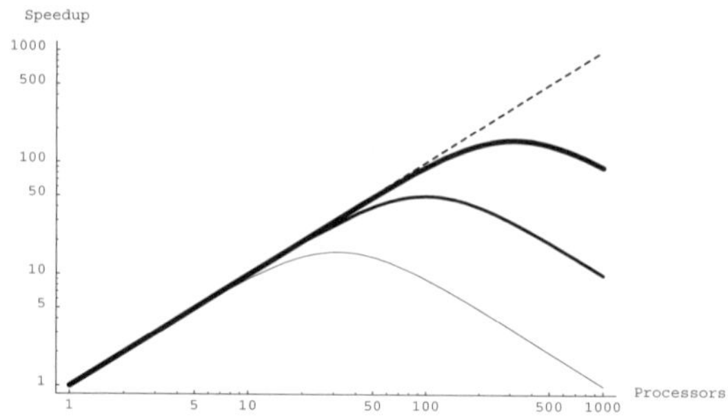

Fig. 4. Theoretical speedups of a master-slave GA varying the value of γ. The thinest line corresponds to $\gamma = 1$, the intermediate to $\gamma = 10$, and the thickest to $\gamma = 100$. The dotted line is the ideal (linear) speedup.

and substituting $\mathcal{P}^*$, gives the maximum speedup possible as

$$S_p^* = \frac{1}{2}\mathcal{P}^*. \tag{4}$$

Figure 4 shows the theoretical speedups of a master-slave GA varying the ratio $\gamma = T_f/T_c$. As γ increases, more processors may be used effectively to reduce the execution time. The figure considers a master-slave GA with a population of $n = 1000$ individuals, and the speedups are plotted for $\gamma = 1, 10, 100$. In many practical problems, the function evaluation time is much greater than the time of communications, $T_f \gg T_c$ ($\gamma \gg 1$), and master-slave parallel GAs can deliver near-linear speedups for a large range of processors.

However, equation 4 clearly points out that at their optimal configuration, the simple master-slave GAs have an efficiency of 50%: half of the time the processors are idle or communicating. A major cause of this inefficiency is that the master waits for *all* the slaves to finish the fitness evaluations before selecting the parents to the next generation. This synchronization ensures that the master-slave produces exactly the same results as a simple serial GA, but it is easy to avoid if we are willing to accept a different behavior of the algorithm.

In asynchronous master-slave GAs, the master generates individuals and sends them to the slaves to be evaluated. As soon as a slave finishes its share of the evaluations, the master inserts the evaluated individual(s) into the population. Then, the master generates new individuals, and sends them to any available slaves. Asynchronous master-slaves have the potential to be more efficient than the synchronous algorithm (e.g., [63, 55]), especially if the evaluation times are not constant for all individuals, but they introduce additional parameters.

In particular, there are several options about how many individuals are generated at a time, and about how they are incorporated into the population. Some of these choices may increase the selection pressure and require larger population sizes [13, 15]. In addition, in asynchronous master-slaves, individuals may return from the slaves in a different order in which they were created, because some slaves will finish their evaluations faster than others. Davison et al. [23] identified this problem and studied experimentally the effect of allowing random returns while varying the number of slaves. They found that by replacing a low-fit individual that is similar to the new one, the effect of accepting individuals in random order was very small.

Asynchronous master-slave GAs are only slightly more difficult to implement than the synchronous, and the gains in performance may easily offset the extra cost of development. However, their mathematical analysis is significantly more difficult.

Until this point, the calculations have assumed that T_c is constant, and that each slave received a single message from the master with all the individuals that the slave must evaluate. In reality, a better expression for T_c may be $T_c = Bx + L$, where B is the inverse of the bandwidth of the network, x is the amount of information transmitted, and L is the latency of communications. The latency is the overhead per message that depends on the operating system, the programming environment, and on the particular hardware, and can easily dominate the communication time. Cantu-Paz [13] contains calculations that indicate that equation 2 is a good estimator of the optimal number of slaves.

Gagné et al. [24] perform similar calculations and include a term to account for possible failures of the slave nodes. Their calculations also take into account that the master can use multiple messages to send the individuals that each slave will evaluate.

Of course, the master-slave method has been used to parallelize other evolutionary algorithms, such as evolution strategies [5], evolutionary programming [37], and genetic programming [45]. The models presented in this section are applicable to those algorithms as well as to GAs.

3 Multi-Population Parallel GAs

The existing theory of multi-population parallel GAs deals mainly with (1) the sizing of each population and (2) with the effects of migration. These two factors are related, but we will look first at the effects of population sizing, since this is extremely important for the quality and efficiency of the algorithm.

3.1 Models of Population Sizing

The existing population sizing models for multi-population parallel GAs are derived from the gambler's ruin model of population sizing for simple GAs [34]. This is a very simple model that originates from considering the number of correct building blocks in a partition of the population as a random quantity (see Figure 5). A partition is given by a template of k fixed symbols F and "don't care" symbols $*$ that match any character in the alphabet chosen to represent solutions. Assuming binary alphabets, a partition specified by k F symbols contains 2^k schemata. A schema represents the class of individuals that match each of the fixed positions. The schema with the highest average fitness and that matches the global optimum in the fixed positions is called the correct building block.

The gambler's ruin model views the GA search as a series of competitions that ends either when the entire population contains the building block or when the building block has completely disappeared from the population. For some problems, we can calculate the probability p that selection will add one copy of the correct building block to the population [27]. Given p, the probability that the population

Fig. 5. The bounded one-dimensional space of the gambler's ruin problem. The absorbing barrier on the right represents a population full of representatives of the correct building block. The probability of reaching that barrier depends on the starting point and on the probability of adding one building block in each step, given by p.

will be eventually full of copies of the correct building block can be approximated as [34]:

$$P_{bb} \approx 1 - \left(\frac{1-p}{p}\right)^{n/2^k}, \tag{5}$$

where n is the population size and k is the order of the building blocks we are considering. The exponent of the equation above is the number of correct building blocks of order k we expect in a randomly initialized population of size n if we are using binary encodings. For a given desired probability of success $\hat{P}_1$, we can solve the equation above for n and obtain a population-sizing equation:

$$n = \frac{2^k \ln(1-\hat{P}_1)}{\ln\left(\frac{1-p}{p}\right)}. \tag{6}$$

This simple yet accurate model for simple GAs is the basis for modeling the effect of different parameters on the solution quality of multipopulation parallel GAs. An extremely simple case of a multi-population parallel GA is to run r independent GAs and report the answer of the best of the runs to the user. Intuitively, this strategy makes sense: Since GAs are stochastic and there will be some variance in the quality of the results, reporting the best of r runs should result in a better solution than running the GA only once. An alternative view is that if we keep the desired quality constant, we could reduce the population size, execute the independent (reduced size) GAs in parallel, and reach the desired solution faster than with a single GA. However, at least for linearly decomposable functions, this strategy is not very efficient [17, 18]. To reduce the population size and still reach the same solution quality, we would need to substitute $\hat{P}_1$ in our population sizing equation by

$$\hat{P}_r = \hat{P}_1 - \frac{\sqrt{\ln r}}{\sqrt{2m}}, \tag{7}$$

where m is the number of partitions in the problem. This new probability of success does not differ much from $\hat{P}_1$ as more independent runs r as used. Therefore, the population size required by multiple independent runs is not much smaller than the size required by a single GA to reach the same solution, and there are almost no time savings.

Empirical studies [58, 18] confirm that using isolated populations is not a very effective or efficient approach. Most multi-population parallel GAs exchange a number of individuals between the populations. Empirically, this exchange has been known for a long time to improve the effectiveness and efficiency of the algorithms, and early studies [58] suggest that using "moderate" values for the migration parameters results in a good tradeoff between efficiency and accuracy.

The modeling of multi-population GAs concentrates on one population and finding out the effect of the migration parameters on the solution quality. Examining the gambler's ruin model (equation 5) we can see that the probability of success depends on the value of p and the initial position of the imaginary particle on the random walk space. The value of p depends only on the problem to be solved, not on the algorithm or any of its parameters. On the other hand, the initial position of the particle corresponds to the number of BBs present at the beginning of a GA run, and is the focus of the analysis.

To use of the gambler's ruin model, we consider an extreme case where migration occurs only after the populations converge [17, 19]. After convergence, the populations communicate, each sending a number of individuals to some of the other populations. The populations incorporate the migrants and restart. If we can compute the expected number of BBs contained on the exchanged individuals, we can predict the quality of solutions after the populations restart simply by changing the exponent of Equation 5 to reflect the expected number of correct building blocks after the first migration.

Following the idea of changing the initial state for the gambler's ruin model, Cantú-Paz and Goldberg [19] analyzed the effect of arbitrary migration rates and sparse topologies. Essentially, their calculations assumed that to reach a particular solution quality, the populations must start with some initial number of correct building blocks x_1 (replacing this number in the exponent of the gambler's ruin model gives the required success probability). For a given a migration rate, it is easy to compute the probability that a population receives at least x_1 building blocks from its neighbors. Actually, there are several combinations of migration rate, number of neighbors, and population size that result in the same probability of success: large populations with few neighbors have the same chance of contributing x_1 copies of the BB as small populations with many neighbors. Since the size of the populations mainly determines the computation time and the number of neighbors largely determines the communications time, there is a tradeoff between communications and computation similar to the one for the master-slaves. This tradeoff can be optimized to find the optimal number of neighbors and corresponding population size that minimizes the parallel execution time. Interestingly, the optimal number of neighbors is $O\left(\sqrt{nT_f/T_c}\right)$, which is the same as the optimal number of processors in master-slaves.

With multiple restarts, the populations are influenced not only by their direct neighbors, but by the neighbors of their neighbors and so on. The multiple restarts in effect create an "extended neighborhood" composed of all the populations that contribute migrants to a population. The gambler's ruin model again gives us a way to calculate the probability of success. For some topologies of communications where the populations have δ neighbors, after τ epochs the extended neighborhood grows linearly reaching $\delta(\tau-1)+1 = \delta\tau'+1$ populations. The number of individuals in these extended neighborhood is

$$n_\tau = (\sqrt{\delta}\tau' + 1)n_d, \tag{8}$$

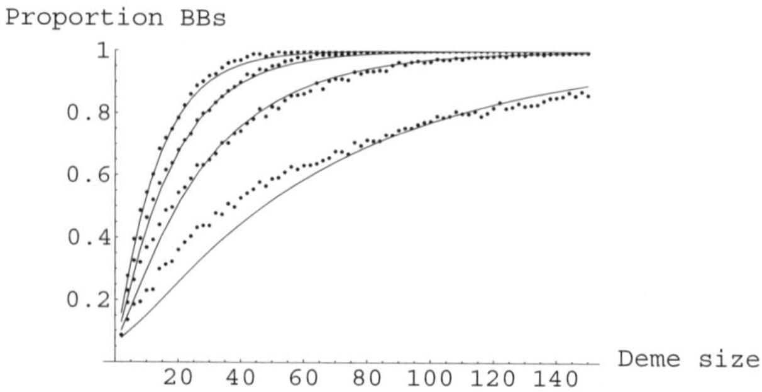

Fig. 6. Theoretical predictions (line) and experimental results (dots) of the average quality per deme after 1, 2, 3, and 4 epochs (from right to left) using eight demes connected by a +1+2 topology.

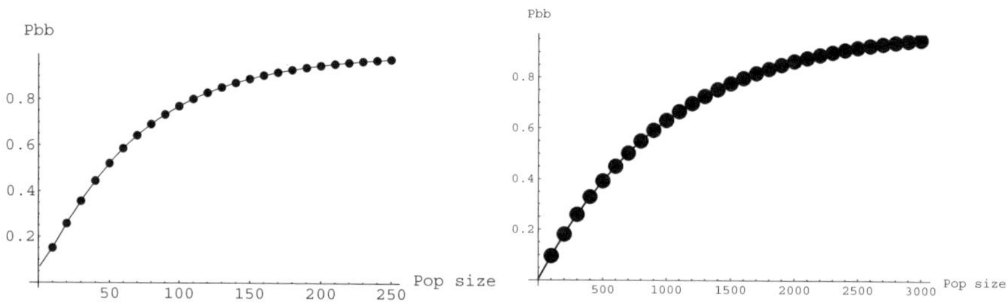

(a) 20-BB 4-bit trap.

(b) 10-BB 8-bit trap.

Fig. 7. In the limit, a parallel GA with r fully connected populations using a maximal migration rate (dots) has the same chance of finding the solution as a simple GA with an aggregate population (continuous line).

where n_d is the number of individuals in each population. Simply substituting this extended population size into the gambler's ruin model (Equation 5) gives very accurate predictions of the probability of success, as evidenced in figure 6. Empirical evidence suggests that even when the extended neighborhood does not grow linearly, using n_τ in the gambler's ruin model is very accurate [13]. These results imply that the degree of the connectivity is a crucial factor in the solution quality and that the specific details of how the populations are connected is not as important. This observation is corroborated by other empirical studies [20] as well as by a more exact analysis with Markov chains [14].

Markov chains allow us to predict the solution quality after each epoch and also in the long run, when all the populations converge to the same solution and no further improvement is possible. Figure 7 presents plots of the long-run behavior of four demes with $0 \leq n_d \leq 100$ individuals each, and of P_{bb} (Equation 5) using a population size of $n = 4n_d$. The plots overlap perfectly, suggesting that the probability that r fully-connected demes of size n_d converge correctly in the long run is

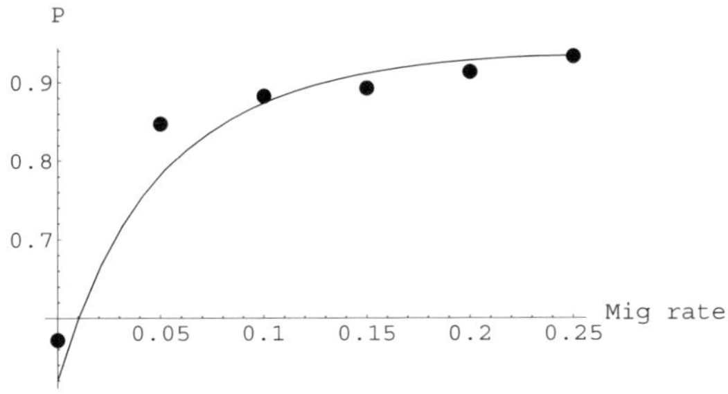

Fig. 8. The probability of converging to the correct BB increases with higher migration rates. The theoretical predictions (continuous line) are compared against experimental results.

the same as the probability of success of a GA with a single population with rn_d individuals. This is important because it suggests that, *in the long run*, the solution's quality does not degrade or improve when a population is partitioned into smaller fully-connected demes that communicate with the maximum migration rate possible.

Figure 8 illustrates the probability of reaching the correct solution as a function of the migration rate. The example uses four fully-connected demes with 50 individuals each; the test function is a 20-BB 4-bit trap problem; and only the first two epochs are considered (i.e., the populations run independently, converge, exchange individuals, restart, and converge again). Note that the probability of success increases rapidly with higher migration rates.

Figure 9 shows the probability that, in the long run, the parallel GA will converge to the correct solution as a function of the migration rate. The plot suggests that only a moderate migration rate is sufficient to reach the same solution as a simple GA with a aggregate population. Note that since all the individuals are the same when migration occurs, there are no cost penalties associated with higher rates: only one individual needs to be sent and it can be replicated any number of times at the receiving deme.

The next set of results compares the quality of the solutions reached using different topologies. Figure 10 shows plots of the quality versus the number of epochs for a fully-connected topology and three topologies frequently used by practitioners: a uni-directional ring, a bi-directional ring, and a hypercube. Each of the eight demes has 30 individuals, and the migration rate was set to the maximum value possible for each topology: 50% for the uni-directional ring, 33% for the bi-directional ring, and 25% for the hypercube.

In all cases, the quality increases as more epochs are used, but the rate of increase depends on each topology. The uni-directional ring needs the most epochs to reach the highest quality possible—which is the same quality that simple GA with an aggregate population would reach—while the fully-connected topology realizes its full potential using the fewest epochs.

The results above suggest that, at any given time, topologies where the demes have more neighbors reach solutions of higher quality than sparse topologies: at any epoch the fully-connected topology reached the best solutions, while the uni-directional ring reached the worst. To visualize the effect of the degree more clearly,

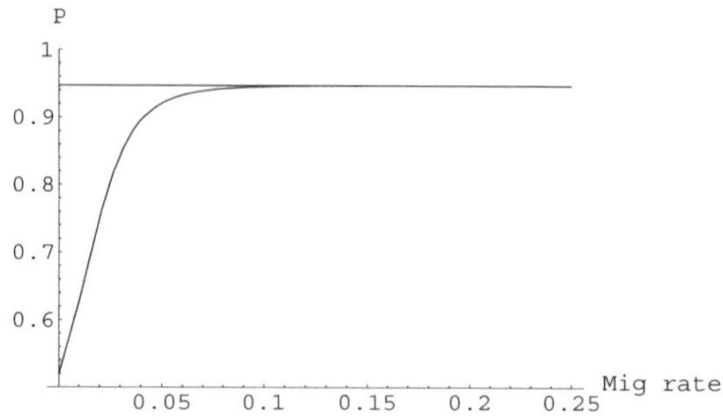

Fig. 9. The long-run probability of converging to the correct BB increases rapidly with higher migration rates. The example considers four fully-connected demes, with 50 individuals each, working on a 20-BB 4-bit trap function. The horizontal line is the probability that four fully-connected demes with maximal migration (or a simple GA with 200 individuals) will eventually converge to the right BB.

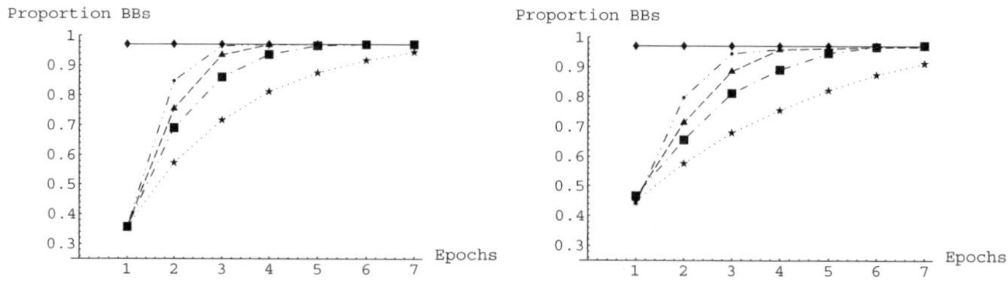

(a) Theory. (b) Experiments.

Fig. 10. Solution quality after several epochs. The graphs include data for uni- and bi-directional rings, a hypercube, and a fully-connected topology (from bottom to top, respectively). The horizontal line in both graphs is the prediction of the quality that would be reached by a simple GA with an aggregate population.

Figure 11 shows experiments where the degree varies while the deme size and migration rate are constant. The experiments used the four topologies used previously (with degrees 1, 2, 3, and 7). The experiments considered eight demes with 40 individuals each, and the quality was measured at the end of the second epoch. In the first experiment, the migration rate was set to 10% per deme (so the number of migrants received by a deme increases with the degree), and in the second experiment the total number of migrants was set to 20.

The analyses presented in these section allow us to make some observations relating the parameters with the quality of solutions. First, we observed that the greatest gain in quality always occurs after the second epoch. Second, the results suggest that—in the long run—the parallel GAs reach solutions of the same quality as a simple GA with a population equivalent to the aggregate of the demes. Partitioning the population does not degrade or improve the solution quality, as long as

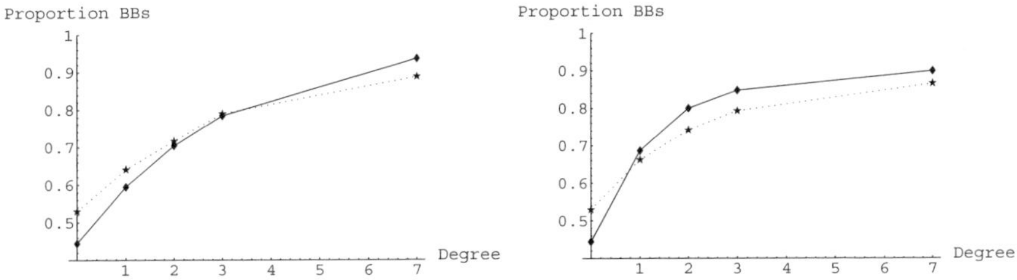

(a) 10% migration per deme.

(b) 20 migrants total.

Fig. 11. Solution quality as a function of the degree of the topology. The continuous lines are the theoretical predictions, and the dashed lines are the experimental results.

the migration rate is not very low. Furthermore, it appears that only a few epochs are necessary to reach the same solution as the simple GA. Another important observation is that the quality improves with higher migration rates and more neighbors per deme. of the migration rate.

3.2 Other Effects of Migration

The previous section showed that migration greatly affects the solution quality and that extensions of the gambler's ruin model can predict the quality very accurately, given certain assumptions. In particular, in the previous section we assumed that migration occurred after the populations had converged to a unique solution. In that case, there is no question about which individuals should migrate, because all the individuals are identical. This section reviews theory that considers other effects of migration and pays particular attention to the choice of migrants and how they are incorporated into the receiving population.

All evolutionary algorithms have a selection mechanism that identifies which individuals in the population will survive to generate new solutions. The choice of selection method influences the speed of convergence of the algorithm, and it has been demonstrated that an excessively slow or fast convergence may cause the algorithm to fail [28, 59]. In addition, the selection pressure is related to the optimal mutation rate and to the population size [43, 6]. This is relevant to us because migration can cause changes in the selection pressure.

The speed of convergence of different selection methods was first studied by Goldberg et al. [26], who introduced the concept of takeover time.

We consider that individuals are selected to migrate either randomly or based on their fitness. Likewise, existing individuals can be replaced by incoming migrants randomly or based on their fitness.

One way to study selection methods—and therefore migration policies—is to examine the selection intensity, which is the difference of the average fitness of the selected individuals and of the original population, normalized by the standard deviation of the fitnesses of the original population:

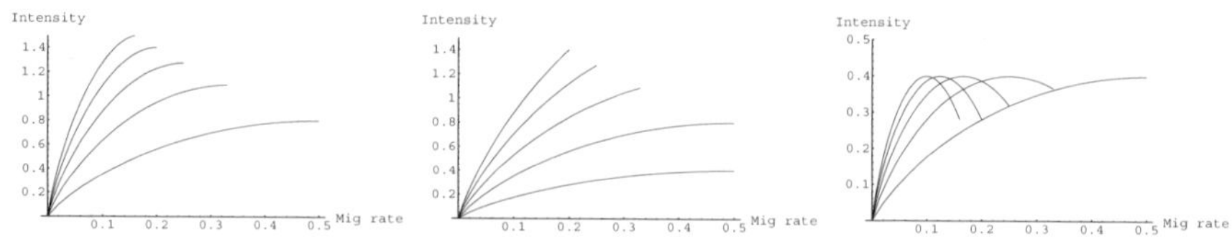

(a) Best replace worst. (b) Best replace random. (c) Random replace worst.

Fig. 12. Selection intensity for different migration policies varying the migration rate and the number of neighbors (δ=1,2,..,5, from bottom to top in each graph).

$$I = \frac{\bar{f}_s - \bar{f}}{\sigma}.$$

Cantu-Paz [16] derived equations for the additional selection intensity caused by selecting migrants and their replacements based on their fitnesses. The calculations work for arbitrary migration rates and any regular communication topology where each population communicates with δ neighbors. If the fitness is distributed normally, we can derive closed-form equations for the increases in selection intensity caused by migration. When a fraction ρ of the top individuals in a population migrate, the added selection intensity is

$$I_e \approx \delta\phi(\Phi^{-1}(1-\rho)), \tag{9}$$

where $\phi(z) = \exp(-z^2/2)/\sqrt{2\pi}$ is the density of a normal distribution and $\Phi^{-1}(x)$ is the value of z where $\int_{-\infty}^{z} \phi(y)dy = x$. If the migrants are chosen randomly, there is no increase in the selection pressure and $I_e = 0$.

Replacing the worst individuals by incoming migrants causes an increase in the selection pressure given by

$$I_r =\approx \phi(\Phi^{-1}(1-\delta\rho)). \tag{10}$$

The maximum of I_r is $\phi(0) = 1/\sqrt{2\pi} = 0.3989$, which is a fairly low value, but it is not negligible (consider that the intensity of pairwise tournament selection is 0.5642). When individuals are replaced randomly in the receiving population, $I_r = 0$.

The overall selection intensity caused by migration is simply $I_m = I_e + I_r$. Figure 12 presents plots of I_m for different migration policies considering topologies with different degrees and varying the migration rate. The plots show that the migration policy with the highest intensity is when the best individuals migrate and replace the worst.

A higher selection pressure causes the algorithms to converge faster. Faster convergence can be desirable, but converging too quickly without allowing enough time for crossover to explore and combine useful building blocks may cause the algorithm to converge prematurely to a solution of poor quality.

The additional selection pressure can also be the cause of some controversial claims of superlinear parallel speedups. Despite the attempts of careful experimenters to set up "fair" experiments (using the same parameters for mutation and crossover rates, keeping the total population sizes equal, using the same selection method, and comparing algorithms when they reach the same solutions, etc.), the migration and replacement of individuals is usually done according to their fitness. As we have seen, these choices increase the selection pressure and cause the parallel algorithm to converge faster than it would if migrants and replacements were chosen randomly. Clearly, comparing parallel and serial algorithms with different selection intensities is not fair. Superlinear speedups can also be caused by other reasons. For example, the smaller populations might fit completely in cache memory [8] or the parallel code that executes unnecessary communications is used to measure the serial performance [7]. More details and discussions about fair comparisons can be found elsewhere [13, 3].

4 Conclusions

Parallel implementations of genetic and other evolutionary algorithms have long been recognized as a promising alternative to improve the efficiency of these algorithms. Most of the research in this field has been empirical, and while important observations were made even in very early studies, it is difficult to assess their generality or limitations. The existing theory summarized in this chapter explains many of the observations and provides guidelines that should be useful to practitioners.

There is, however, much that remains to be done. All theoretical models make assumptions about the problem, the algorithm, or both. Relaxing those assumptions suggests opportunities to extend the theory in several directions. For example, extending models for multiple populations to consider frequent migrations or communication behavior that adapts to the progress of the search. In addition, there is still relatively little theory related to cellular GAs. Continuing the study of these algorithms seems to be a promising avenue of research.

References

1. P. Adamidis. Review of parallel genetic algorithms bibliography. Tech. rep. version 1, Aristotle University of Thessaloniki, Thessaloniki, Greece, 1994.
2. E. Alba, M. Giacobini, M. Tomassini, and S. Romero. Comparing synchronous and asynchronous cellular genetic algorithms. In J.J. Merelo et al., editor, *Parallel Problem Solving from Nature VII*, pages 601–610, Berlin Heidelberg, 2002. Springer Verlag.
3. E. Alba and M. Tomassini. Parallelism and evolutionary algorithms. *IEEE Transactions on Evolutionary Computation*, 6(5):443–462, 2002.
4. E. Alba and J. M. Troya. A survey of parallel distributed genetic algorithms. *Complexity*, 4(4):31–52, 1999.
5. T. Bäck. Parallel optimization of evolutionary algorithms. In Y. Davidor, H.-P. Schwefel, and R. Männer, editors, *Parallel Problem Solving fron Nature, PPSN III*, pages 418–427, Berlin, 1994. Springer-Verlag.
6. T. Bäck. *Evolutionary algorithms in theory and practice.* Oxford University Press, New York, 1996.
7. D.H. Bailey. Misleading performance reporting in the supercomputing field. *Scientific Programming*, 1(2):141–151, 1992.

8. T. C. Belding. The distributed genetic algorithm revisited. In L. Eschelman, editor, *Proceedings of the Sixth International Conference on Genetic Algorithms*, pages 114–121, San Francisco, CA, 1995. Morgan Kaufmann.
9. F. H. Bennett III, J. R. Koza, J. Shipman, and O. Stiffelman. Building a parallel computer system for $18,000 that performs a half peta-flob per day. In W. Banzhaf, J. Daida, A. E. Eiben, M. H. Garzon, V. Honavar, M. Jakiela, and R. E. Smith, editors, *Proceedings of the Genetic and Evolutionary Computation Conference 1999: Volume 2*, pages 1484–1490, San Francisco, CA, 1999. Morgan Kaufmann Publishers.
10. P. R. Calégari. *Parallelization of Population-Based Evolutionary Algorithms for Combinatorial Optimization Problems.* Unpublished doctoral dissertation, École Polytechnique Fédérale de Lausanne (EPFL), 1999.
11. E. Cantú-Paz. Using Markov chains to analyze a bounding case of parallel genetic algorithms. *Genetic Programming 98*, pages 456–462, 1998.
12. E. Cantú-Paz. Migration policies and takeover times in parallel genetic algorithms. In W. Banzhaf, J. Daida, A. E. Eiben, M. H. Garzon, V. Honavar, M. Jakiela, and R. E. Smith, editors, *GECCO-99: Proceedings of the 1999 Genetic and Evolutionary Computation Conference*, page 775, San Francisco, CA, 1999. Morgan Kaufmann Publishers.
13. E. Cantú-Paz. *Efficient and Accurate Parallel Genetic Algorithms.* Kluwer Academic Publishers, Boston, MA, 2000.
14. E. Cantú-Paz. Markov chain models of parallel genetic algorithms. *IEEE Transactions on Evolutionary Computation*, 4(3), 2000. Also available as IlliGAL Report No. 98010.
15. E. Cantú-Paz. Selection intensity in genetic algorithms with generation gaps. In D. Whitley, D. E. Goldberg, E. Cantú-Paz, L. Spector, I. Parmee, and H.-G. Beyer, editors, *GECCO-2000: Proceedings of the Genetic and Evolutionary Computation Conference*, pages 911–918, San Francisco, CA, 2000. Morgan Kaufmann.
16. E. Cantú-Paz. Migration policies, selection pressure, and parallel evolutionary algorithms. *Journal of Heuristics*, 7(4):311–334, 2001.
17. E. Cantú-Paz and D. E. Goldberg. Modeling idealized bounding cases of parallel genetic algorithms. In J. Koza, K. Deb, M. Dorigo, D. Fogel, M. Garzon, H. Iba, and R. Riolo, editors, *Genetic Programming 1997: Proceedings of the Second Annual Conference*, pages 353–361, San Francisco, CA, 1997. Morgan Kaufmann Publishers.
18. E. Cantú-Paz and D. E. Goldberg. Predicting speedups of idealized bounding cases of parallel genetic algorithms. In T. Bäck, editor, *Proceedings of the Seventh International Conference on Genetic Algorithms*, pages 113–121, San Francisco, 1997. Morgan Kaufmann.
19. E. Cantú-Paz and D. E. Goldberg. Efficient parallel genetic algorithms: theory and practice. *Computer Methods in Applied Mechanics and Engineering*, 1999.
20. E. Cantú-Paz and M. Mejía-Olvera. Experimental results in distributed genetic algorithms. In *International Symposium on Applied Corporate Computing*, pages 99–108, Monterrey, Mexico, 1994.
21. Fuey-Sian Chong. Java-based distributed genetic programming on the internet. In A. Wu, editor, *Evolutionary Computation and Parallel Processing Workshop, Proceedings of the 1999 GECCO Workshops*, pages 163–166, July 1999.
22. Y. Davidor. The ECOlogical framework II: Improving GA performance at virtually zero cost. In S. Forrest, editor,*Proceedings of the Fifth International*

Conference on Genetic Algorithms, pages 171–176, San Mateo, CA, 1993. Morgan Kaufmann.

23. Brian D. Davison and Khaled Rasheed. Effect of global parallelism on a steady state ga. In A. Wu, editor, *Evolutionary Computation and Parallel Processing Workshop, Proceedings of the 1999 GECCO Workshops*, pages 167–170, July 1999.
24. C. Gagne, M. Parizeau, and M. Dubreuil. The master-slave architecture for evolutionary computations revisited. In E. Cantú-Paz et al., editor, *Genetic and Evolutionary Computation—GECCO-2003*, pages 1578–1579, Berlin, 2003. Springer-Verlag.
25. M. Giacobini, E. Alba, A. Tettamanzi, and M. Tomassini. Modeling selection intensity for toroidal cellular evolutionary algorithms. In K. Deb et al., editor, *Genetic and Evolutionary Computation—GECCO 2004*, pages 1138–1149, Berlin Heidelberg, 2004. Springer Verlag.
26. D. E. Goldberg and K. Deb. A comparative analysis of selection schemes used in genetic algorithms. *Foundations of Genetic Algorithms*, 1:69–93, 1991. (Also TCGA Report 90007).
27. D. E. Goldberg, K. Deb, and J. H. Clark. Genetic algorithms, noise, and the sizing of populations. *Complex Systems*, 6:333–362, 1992.
28. D. E. Goldberg, K. Deb, and D. Thierens. Toward a better understanding of mixing in genetic algorithms. *Journal of the Society of Instrument and Control Engineers*, 32(1):10–16, 1993.
29. V. S. Gordon and D. Whitley. Serial and parallel genetic algorithms as function optimizers. In S. Forrest, editor, *Proceedings of the Fifth International Conference on Genetic Algorithms*, pages 177–183, San Mateo, CA, 1993. Morgan Kaufmann.
30. M. Gorges-Schleuter. Explicit parallelism of genetic algorithms through population structures. In H.-P. Schwefel and R. Männer, editors, *Parallel Problem Solving from Nature*, pages 150–159, Berlin, 1991. Springer-Verlag.
31. M. Gorges-Schleuter. An analysis of local selection in evolution strategies. In W. Banzhaf, J. Daida, A. E. Eiben, M. H. Garzon, V. Honavar, M. Jakiela, and R. E. Smith, editors, *Proceedings of the Genetic and Evolutionary Computation Conference 1999: Volume 1*, pages 847–854, San Francisco, CA, 1999. Morgan Kaufmann Publishers.
32. J. J. Grefenstette. Parallel adaptive algorithms for function optimization. Tech. Rep. No. CS-81-19, Vanderbilt University, Computer Science Department, Nashville, TN, 1981.
33. J.-P. Gwo, F.M. Hoffman, and W.W Hargrove. Mechanistic-based genetic algorithm search on a Beowulf cluster of Linux PCs. In *Proceedings of the High Performance Computing Conference*, Washington, DC, 2000.
34. G. Harik, E. Cantú-Paz, D. E. Goldberg, and B. L. Miller. The gambler's ruin problem, genetic algorithms, and the sizing of populations. In *Proceedings of 1997 IEEE International Conference on Evolutionary Computation*, pages 7–12, Piscataway, NJ, 1997. IEEE.
35. W. E. Hart. *Adaptive Global Optimization with Local Search.* PhD thesis, University of California, San Diego, 1994.
36. W. E. Hart, S. Baden, R. K. Belew, and S. Kohn. Analysis of the numerical effects of parallelism on a parallel genetic algorithm. In *Proceedings of the 10th International Parallel Processing Symposium*, pages 606–612. IEEE Press, 1996.

37. J.E. Hirsh and D.K. Young. Evolutionary programming strategies with self adaptation applied to the design of rotorcraft using parallel processing. In V. W. Porto, N. Saravanan, D. Waagen, and A.E. Eiben, editors, *Proceedings of the Seventh Annual Conference on Evolutionary Programming*, pages 147–156, Berlin, 1998. Springer Verlag.
38. M. Kirley. An empirical investigation of optimisation in dynamic environments using the cellular genetic algorithm. In D. Whitley, D. E. Goldberg, E. Cantú-Paz, L. Spector, I. Parmee, and H.-G. Beyer, editors, *GECCO-2000: Proceedings of the Genetic and Evolutionary Computation Conference*, pages 11–18, San Francisco, CA, 2000. Morgan Kaufmann.
39. S.-C. Lin, E. D. Goodman, and W. F. Punch III. Investigating parallel genetic algorithms on job shop scheduling problems. In P. J. Angeline, R. G. Reynolds, J. R. McDonnell, and R. Eberhart, editors, *Evolutionary Programming VI*, pages 383–393, Berlin, 1997. Springer.
40. S-C Lin, W. Punch, and E. Goodman. Coarse-grain parallel genetic algorithms: Categorization and new approach. In *Sixth IEEE Symposium on Parallel and Distributed Processing*, Los Alamitos, CA, October 1994. IEEE Computer Society Press.
41. B. Manderick and P. Spiessens. Fine-grained parallel genetic algorithms. In J. D. Schaffer, editor, *Proceedings of the Third International Conference on Genetic Algorithms*, pages 428–433, San Mateo, CA, 1989. Morgan Kaufmann.
42. H. Mühlenbein. Evolution in time and space-The parallel genetic algorithm. In G. J. E. Rawlins, editor, *Foundations of Genetic Algorithms*, pages 316–337, San Mateo, CA, 1991. Morgan Kaufmann.
43. H. Mühlenbein and D. Schlierkamp-Voosen. The science of breeding and its application to the breeder genetic algorithm (BGA). *Evolutionary Computation*, 1(4):335–360, 1994.
44. P. Nangsue and S. E. Conry. An agent-oriented, massively distributed parallelization model of evolutionary algorithms. In J. R. Koza, editor, *Late Breaking Papers at the Genetic Programming 1998 Conference*, pages 160–168, Madison, WI, 1998. Omni Press.
45. M. Oussaidène. *Genetic programming methodology, parallelization and applications.* Unpublished doctoral dissertation, Université de Genève, Genève, 1997.
46. W. F. Punch. How effective are multiple programs in genetic programming. In J. R. Koza, W. Banzhaf, K. Chellapilla, K. Deb, M. Dorigo, D. B. Fogel, M. H. Garzon, D. E. Goldberg, H. Iba, and R. L. Riolo, editors, *Genetic Programming 98*, pages 308–313, San Francisco, 1998. Morgan Kaufmann Publishers.
47. G. G. Robertson. Parallel implementation of genetic algorithms in a classifier system. In J. J. Grefenstette, editor, *Proceedings of the Second International Conference on Genetic Algorithms*, pages 140–147, Hillsdale, NJ, 1987. Lawrence Erlbaum Associates.
48. G. Rudolph. On takeover times in spatially structured populations: Array and ring. In K.K. Lai, O. Katai, M. Gen, and B. Lin, editors, *Proceedings of the Second Asia-Pacific Conference on Genetic Algorithms and Applications*, pages 144–151, Hong Kong, 2000. Global Link Publishing Company.
49. Günter Rudolph and Joachim Sprave. A cellular genetic algorithm with self-adjusting acceptance threshold. In *Proceedings of the First IEE/IEEE International Conference on Genetic Algorithms in Engineering Systems: Innovations and Applications*, pages 365–372, London, 1995. Institution of Electrical Engineers and Institute for Electrical and Electronics Engineers.

50. J. Sarma and K. De Jong. An analysis of the effects of neighborhood size and shape on local selection algorithms. In H.-M. Voigt, W. Ebeling, I. Rechenberg, and H.-P. Schwefel, editors, *Parallel Problem Solving from Nature, PPSN IV*, pages 236–244, Berlin, 1996. Springer-Verlag.
51. J. Sarma and K. De Jong. An analysis of local selection algorithms in a spatially structured evolutionary algorithm. In T. Bäck, editor, *Proceedings of the Seventh International Conference on Genetic Algorithms*, pages 181–187, San Francisco, 1997. Morgan Kaufmann.
52. J. Sarma and K. De Jong. The behavior of spatially distributed evolutionary algorithms in non-stationary environments. In W. Banzhaf, J. Daida, A. E. Eiben, M. H. Garzon, V. Honavar, M. Jakiela, and R. E. Smith, editors, *Proceedings of the Genetic and Evolutionary Computation Conference 1999: Volume 1*, pages 572–578, San Francisco, CA, 1999. Morgan Kaufmann Publishers.
53. P. Spiessens and B. Manderick. A massively parallel genetic algorithm: Implementation and first analysis. In R. K. Belew and L. B. Booker, editors, *Proceedings of the Fourth International Conference on Genetic Algorithms*, pages 279–286, San Mateo, CA, 1991. Morgan Kaufmann.
54. J. Sprave. A unified model of non-panmictic population structures in evolutionary algorithms. In Peter J. Angeline, Zbyszek Michalewicz, Marc Schoenauer, Xin Yao, and Ali Zalzala, editors, *Proceedings of the Congress on Evolutionary Computation*, volume 2, pages 1384–1391. IEEE Press, 1999.
55. T. J. Stanley and T. Mudge. A parallel genetic algorithm for multiobjective microprocessor design. In L. Eschelman, editor, *Proceedings of the Sixth International Conference on Genetic Algorithms*, pages 597–604, San Francisco, CA, 1995. Morgan Kaufmann.
56. T. Sterling. Beowulf-class clustered computing: Harnessing the power of parallelism in a pile of PCs. In J. R. Koza, W. Banzhaf, K. Chellapilla, K. Deb, M. Dorigo, D. B. Fogel, M. H. Garzon, D. E. Goldberg, H. Iba, and R. L. Riolo, editors, *Genetic Programming 98*, pages 883–887, San Francisco, 1998. Morgan Kaufmann Publishers.
57. R. Tanese. Parallel genetic algorithm for a hypercube. In J. J. Grefenstette, editor, *Proceedings of the Second International Conference on Genetic Algorithms*, pages 177–183, Hillsdale, NJ, 1987. Lawrence Erlbaum Associates.
58. R. Tanese. Distributed genetic algorithms. In J. D. Schaffer, editor, *Proceedings of the Third International Conference on Genetic Algorithms*, pages 434–439, San Mateo, CA, 1989. Morgan Kaufmann.
59. D. Thierens and D. E. Goldberg. Mixing in genetic algorithms. In S. Forrest, editor, *Proceedings of the Fifth International Conference on Genetic Algorithms*, pages 38–45, San Mateo, CA, 1993. Morgan Kaufmann.
60. M. Tomassini. Parallel and distributed evolutionary algorithms: A review. In K. Miettinen, M. Mäkelä, P. Neittaanmäki, and J. Periaux, editors, *Evolutionary Algorithms in Engineering and Computer Science*, pages 113–133. J. Wiley and Sons, Chichester, UK, 1999.
61. D. Whitley. A free lunch proof for gray versus binary encodings. *Proceedings of the Genetic and Evolutionary Computation Conference 1999: Volume 1*, pages 726–733, 1999.
62. D. Whitley and T. Starkweather. Genitor II: A distributed genetic algorithm. *Journal of Experimental and Theoretical Artificial Intelligence*, 2:189–214, 1990.

63. B. P. Zeigler and J. Kim. Asynchronous genetic algorithms on parallel computers. In S. Forrest, editor, *Proceedings of the Fifth International Conference on Genetic Algorithms*, page 660, San Mateo, CA, 1993. Morgan Kaufmann.

Parameter Control in Practice

Zbigniew Michalewicz[1] and Martin Schmidt[2]

[1] School of Computer Science, University of Adelaide, Adelaide, SA 5005, Australia, also with Institute of Computer Science, Polish Academy of Sciences and Polish-Japanese School of Information Technology, Warsaw, Poland
zbyszek@cs.adelaide.edu.au

[2] SolveIT Software, PO Box 3161, Adelaide, SA 5000, Australia
martin.schmidt@solveitsoftware.com

Summary. In this chapter we summarize our experience of tuning and/or controlling various parameters of evolutionary algorithms from working on a variety of complex real word problems. We illustrate some issues on one particular case study (car distribution system). This chapter contains also general discussion on the prediction and optimization issues present in dynamic environments, and explains the ideas behind Adaptive Business Intelligence.

1 Introduction

The two major steps in applying any heuristic search algorithm (in particular: evolutionary algorithm) to any problem are: (1) the specification of the representation and (2) the evaluation function. These two items form the bridge between the original problem context and the problem-solving framework. When defining an evolutionary algorithm one needs to choose its components, such as variation operators (mutation and recombination) that suit the representation, selection mechanisms for selecting parents and survivors, and an initial population. Each of these components may have parameters, for instance: the probability of mutation, the tournament size of selection, or the population size. The values of these parameters greatly determine whether the algorithm will find a near-optimum solution, and whether it will find such a solution efficiently.

Choosing the right parameter values, however, is a time-consuming task and considerable effort has gone into developing good heuristics for it. As discussed in Chapter 2, we distinguish two major forms of setting parame-ter values: parameter *tuning* and parameter *control*. By parameter tuning we mean the commonly practiced approach that amounts to finding good values for the parameters *before* the run of the algorithm and then running the algorithm using these values, which remain fixed during the run. Pa-rameter control, on the other hand, forms an alternative, as it amounts to starting a run with initial parameter values which are changed *during* the run.

Z. Michalewicz and M. Schmidt: *Parameter Control in Practice*, Studies in Computational Intelligence (SCI) **54**, 277–294 (2007)
www.springerlink.com

There are many theoretical and experimental results on parameter tuning and parameter control. However, very often real world problems (here we mean problems for which a production system is required to optimize some aspects of company's operations in real time) call for shortcuts and/or some *ad hoc* solutions. Some of them work very well, some others often result in disappointing performance.

In this chapter we summarize our experience of tuning and/or controlling various parameters of evolutionary algorithms from working on a variety of complex real word problems. We illustrate some issues on one particular case study (car distribution system). The chapter is organized as follows. The next section presents the problem of recommending the best distribution of used cars among many available auction sites. Section 3 provides a brief overview of some issues connected with parameter control methods. Section 4 discusses "good practices that work," and section 5 discusses "uncertain environments" — typical scenarios for real world problems. Section 6 introduces the concept of *Adaptive Business Intelligence* and section 7 concludes this chapter.

2 A Case Study

The problem is to recommend the best distribution of used cars (which are coming back to the company after the term of the lease or rental is over) among many available auction sites [7]. By "best distribution," we mean a distribution, which maximizes the net total proceeds from all these sales. Many issues have to be considered in the process of making optimal recommendation, which vary from price prediction for various types of cars at different locations, to price depreciation and volume effect, to transportation issues. We discuss these in this section.

The leased/rented car is owned by a manufacturing company; the company gets the car back when it is returned. Each car is different, and these differences include make and model, mileage, model year, color, type of transmission, body style and options, wear and tear on the vehicle (i.e., the damage level), etc. There are hundreds of auctions around the US, and each returned car has to be sold at one of them. The characteristics listed above (plus some others) influence the sales price of a car at each particular location. The central question is: Where should each car be sent to maximize the total proceeding from all these sales?

The total annual number of returned leased/rental cars for just one large US-based car manufacturer is more than 1,000,000. This would correspond to approximately 4,000 cars per day. In other words, every day a remarketing team has to make 4,000 decisions: they have to assign each car to the best possible auction site that would maximize the sales price. Nevertheless, due to volume effects, the assignment of cars to auctions is highly interrelated and hence the number of possible decisions becomes astronomical! If a company operates on, say, 50 auction sites and processes "only" 1,000 cars a day, there are $50^{1,000}$ distribution choices. This gives roughly 10,000,000,000,...,000,000 possible choices — a number where 1 at the front is followed by 1,700 zeros! This is a mind-boggling number and no computer can check out all combinations in a human lifetime. Never-theless, it is necessary to make decisions on all of the cars today!

Again, the task is to distribute cars to auction sites. For example, we can look at four-door silver 2002 Toyota Corolla with 34,983 miles on its odometer, sunroof, automatic transmission, power windows, power seats, etc., which, at the moment, sits at a dealership in North Carolina, and decide where to send it. At first glance, this looks easy. We might be tempted just take one car at a time, look up in some file (or a book)[3] the average sale price for this particular type of car (remember to adjust the price for mileage, options, etc.) at each auction, and decide to send the car to the auction with the highest current average sales price. Of course, we should also estimate the transportation cost (usually, the longer distance, the higher cost), but all calculations seem manageable. So, what is the problem? Well, the problem is that the above approach would not work very well. When you ship a whole truckload of cars (say, 10 cars) from one place to another, then you will get a cheaper rate per car than if you send only one car (or a few cars) at a time. A typical transportation cost structure usually looks as follow (again, we assume fix locations "from" and "to"):

- Transporting 1 - 6 cars costs $120 per car.
- Transporting 7 - 10 cars costs $95 per car.
- Transporting 11 - 14 cars costs $85 per car.
- The cost of transporting more than 14 cars usually is calculated as follows: You pay for transporting 14 cars (on one truck, it would cost $85 per car, as indicated above) and you repeat it as many times as possible (to get the cheapest rate). Once the number of remaining cars is less than 14, you use other rates. For example, transporting 20 cars would cost you $85 per car for 14 cars and $120 per car for 6 cars, making 14 x $85 + 6 x $120 = $1,910.

On top of the transportation price, you also have to deal with additional issues, such as risk factors (cars can fall down from trucks and get damaged, cars might get stolen, etc.), insurance, depreciation, and calendars of auctions (auctions are not organized every day). We will discuss some of these issues later in this section.

To complicate matters further, if you have, say, 45 red cars of the same make/model and you decide to send all of them to the same auction site, which might seem reasonable since the selected auction site offers the best net price. However, if you do this then the so-called *volume effect* kicks in, which results in a reduced price per car.

Figure 1 illustrates the volume effect phenomena: you get more money per car when you sell a few cars than when you sell many similar cars. In other words, the volume effect would adjust the price of a car on the basis of the number of cars. For example, the current average sale price for a particular car on a particular auction site is $10,400. You can get the same price if you ship up to seven cars to that location. However, if you transport 30 similar[4] cars there, the average price per car will drop to $9,450 per car.

To complicate the matter even further, every auction has a typical sales day (e.g., every second Friday at 11 am.) Assume that at one particular location you

---

[3] There are plenty of such books available, e.g., *Black Book*, *Kelly Blue Book*, *Manheim Auction Report*, etc.

[4] Note that the term "similar" can mean much more than just same make/model or color. For example, you could have many white compact cars of different makes and models, which would still compete for the same buyers, so they would press down the average sales price per car. Note also, that the volume effect might be different for different types of cars, e.g., the curve of the volume effect is different for Honda Accords and for Porsche 911s.

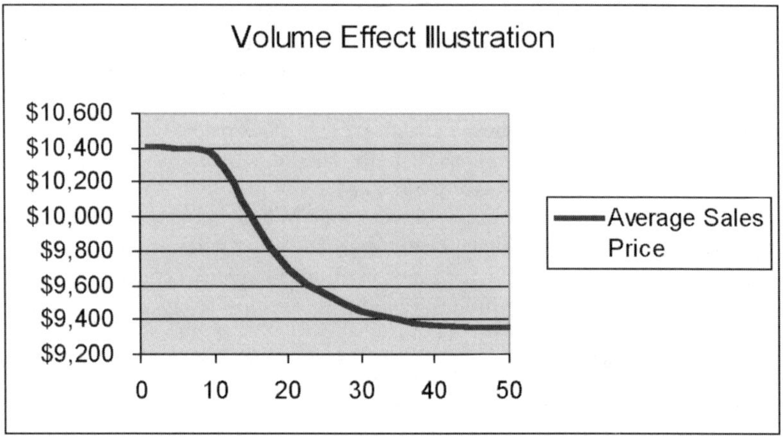

Fig. 1. Volume effect

have 20 cars that you would like to ship to an auction site. Usually the transport between these two locations takes 10 days. The closest auction is 11 days away; if there is a slight delay in delivery of these 20 cars, then they might miss the auction by one day! If that happens then the cars remaining on the auction's parking lot will stay there for almost two weeks. This is bad for the company because it would like to sell the cars as quickly as possible. Apart from other reasons (money is received only after the cars are sold), the cars lose value each day. This phenomenon is called *price depreciation* and is often (for an average car) around $10 per day.

Another thing that makes your job difficult is that the market prices for cars change very frequently — sometimes slightly and sometimes dramatically (e.g., after the terrorist attack on September 11th, 2001, or after a major hurricane storm in Florida). So, on top of all the other issues that you have to consider, you also have to stay up-to-date with prices that may change from one day to the next. Further, remember that car prices may change in different regions in different ways. For example, it is not easy to sell a convertible Corvette in Boston in the fall, but it might be an excellent idea to sell the car in Florida since the temperature is becoming just right for a convertible. This is referred to as the *seasonality effect*. Hence, you have to stay on top of car prices at each auction each day. This is difficult, to say the least.

At the end of the day, you are ready with all your decisions. Whether they are "optimal" or not, whether they maximize the profit for the leasing company or not, whether there is another distribution that would increase net sales by $150 per car (if you distribute 4,000 cars per day, this would be equivalent increasing total net sales by $600,000 on that day) or not, you can visualize the overall picture for that day. In Figure 2, the darker circles illustrate areas where the leased cars were returned on a particular day. The larger the circle, the more cars were returned in that area (clearly, most cars were returned in the east of the US). Note that the distribution of cars would look different on the following day since different people and organizations might return their cars at different locations. The lighter circles, on the other hand, illustrate 50 auction sites that are available for sending selected the cars to. The locations of these auction sites are fixed: they are the same every day for any number of returned cars.[5]

Fig. 2. Cars to be distributed (darker circles) and 50 auction sites (lighter circles)

Clearly, this is not a trivial problem to deal with. In order for you to make your decision you have to deal with the details of each car, many different possible auctions with their timetables and inventories, complex transportation costs, volume effects, country-wide inventory of vehicles currently on the way to auctions, price depreciations, dynamic price changes at each auction, being able to predict sale prices anywhere from one week to two months into the future, etc. Every leasing company therefore has a whole remarketing team dedicated to distributing cars that are returned from leases and rentals. These teams face the formidable task of recommending the best possible distribution of the daily load of cars, day after day, week after week, month after month. A small mistake or an infe-rior recommendation may result in a net loss of "only" $150 per car, but this can equate into hundreds of thousands of dollars lost in a single day if many cars are distributed! On the other hand, if a smart decision-support system is used to improve the daily distribution of cars in such a way that the net sales lift is around $200 per car (note here, that this corresponds to only 1.33% increase in the price of an average off-lease car valued at

[5] It may happen, however, that from time to time (usually, once a year) a set of auction sites that the company does business with is changed. For example, a few current auction sites are dropped, and several new auction sites are added. This may raise additional questions: how to evaluate such business decision? Do we in-crease or decrease profits by swapping some auction sites?

$15,000), then the leasing company would increase annual profits by hun-dreds of million of dollars! Definitely, it is worthwhile to examine this possibility further.

3 Some Issues

During the 1980s, a standard genetic algorithm (GA) based on bit-representation, one-point crossover, bit-flip mutation and roulette wheel selection (with or without elitism) was widely applied (see chapter 1). Algorithm design was thus limited to choosing the so-called control parameters, or strategy parameters (i.e., the parameters of the algorithm, not those of the problem), such as mutation rate, crossover rate, and population size. Many researchers based their choices on tuning the control parameters "by hand," that is experimenting with different values and selecting the ones that gave the best results. Later, they reported their results of applying a particular algorithm to a particular problem, paraphrasing here:

> "*...for these experiments, we have used the following parameters: population size of 100, probability of crossover equal to 0.85, etc.*"

without much justification of the choice made.

Two main approaches were tried to improve GA design in the past. First, De Jong [1] put a considerable effort into finding parameter values (for a traditional GA), which were good for a number of numeric test problems. Later, Grefenstette [3] used a GA as a meta-algorithm to optimize values for the same parameters for both on-line and off-line performance. Note, however, that in both of these approaches, an attempt was made to find the optimal and general set of parameters; in this context, the word "general" means that the recommended values can be applied to a wide range of optimization problems. Formerly, genetic algorithms were seen as robust problem solvers that exhibit approximately the same performance over a wide range of problems [2]. The contemporary view on EAs, however, acknowledges that specific problems (problem types) require specific EA settings for satisfactory performance [6]. Thus, the scope of "optimal" parameter settings is necessarily narrow. Any quest for generally (near-)optimal parameter settings is lost *a priori* [12]. This stresses the need for efficient techniques that help finding good parameter settings for a given problem, in other words, the need for good parameter tuning methods.

As an alternative to tuning parameters before running the algorithm (see chapter 2), controlling them during a run was realized quite early (e.g., mutation step sizes in the evolution strategy (ES) community). Analysis of the simple corridor and sphere problems in large dimensions led to Rechenberg's 1/5 success rule [9], where feedback was used to control the mutation step size. Later, self-adaptation of mutation was used, where the mutation step size and the preferred direction of mutation were controlled without any direct feedback. For certain types of problems, self-adaptive mutation was very successful and its use spread to other branches of evolutionary computation (EC).

As mentioned earlier, parameter tuning by hand is a common practice in evolutionary computation. Typically one parameter is tuned at a time, which may cause some sub-optimal choices, since parameters often interact in a complex way. Simultaneous tuning of more parameters, however, leads to an enormous amount of experiments. The technical drawbacks to parameter tuning based on experimentation can be summarized as follows:

- Parameters are not independent, but trying all different combinations systematically is practically impossible.
- The process of parameter tuning is time consuming, even if parameters are optimized one by one, regardless to their interactions.
- For a given problem the selected parameter values are not necessarily optimal, even if the effort made for setting them was significant.

Other options for designing a good set of static parameters for an evolutionary method to solve a particular problem include *parameter setting by analogy* and the use of theoretical analysis. Parameter setting by analogy amounts to the use of parameter settings that have been proved successful for "similar" problems. However, it is not clear whether similarity between problems as perceived by the user implies that the optimal set of EA parameters is also similar. As for the theoretical approach, the complexities of evolutionary processes and characteristics of interesting problems allow theoretical analysis only after significant simplifications in either the algorithm or the problem model. Therefore, the practical value of the current theoretical results on parameter settings is unclear.

A general drawback of the parameter tuning approach, regardless of how the parameters are tuned, is based on the observation that a run of an EA is an intrinsically dynamic, adaptive process. The use of rigid parameters that do not change their values is thus in contrast to this spirit. Additionally, it is intuitively obvious that different values of parameters might be optimal at different stages of the evolutionary process. For instance, large mutation steps can be good in the early generations helping the exploration of the search space and small mutation steps might be needed in the late generations to help fine tuning the suboptimal chromosomes. This implies that the use of static parameters itself can lead to inferior algorithm performance. The straightforward way to treat this problem is by using parameters that may change over time, that is, by replacing a parameter p by a function $p(t)$, where t is the generation counter. However, as indicated earlier, the problem of finding optimal static parameters for a particular problem can be quite difficult, and the optimal values may depend on many other factors (like the applied recombination operator, the selection mechanism, etc.). Hence designing an optimal function $p(t)$ may be even more difficult. Another possible drawback to this approach is that the parameter value $p(t)$ changes are caused by a deterministic rule triggered by the progress of time t, without taking any notion of the actual progress in solving the problem, i.e., without taking into account the current state of the search. Yet researchers have improved their evolutionary algorithms, i.e., they improved the quality of results returned by their algorithms while working on particular problems, by using such simple deterministic rules. This can be explained simply by superiority of changing parameter values: suboptimal choice of $p(t)$ often leads to better results than a suboptimal choice of p.

To this end, recall that finding good parameter values for an evolutionary algorithm is a poorly structured, ill-defined, complex problem. But on this kind of problem, EAs are often considered to perform better than other methods! It is thus seemingly natural to use an evolutionary algorithm not only for finding solutions to a problem, but also for tuning the (same) algorithm to the particular problem. Technically speaking, this amounts to modifying the values of parameters during the run of the algorithm by taking the actual search process into account. Basically, there are two ways to do this. Either one can use some heuristic rule which takes feedback from the current state of the search and modifies the parameter values accordingly, or incorporate parameters into the chromosomes, thereby making them subject to evolution. The first option, using a heuristic feedback mechanism, allows one to base changes on triggers different from elapsing time, such as population diversity measures, relative improvements, absolute solution quality, etc. The second option, incorporating parameters into the chromosomes, leaves changes entirely based on the evolution mechanism. In particular, natural selection acting on solutions (chromosomes) will drive changes in parameter values associated with these solutions.

In general, methods for changing the value of a parameter can be classified into one of three categories:

- *Deterministic* parameter control. This takes place when the value of a strategy parameter is altered by some deterministic rule. This rule modifies the strategy parameter deterministically without using any feedback from the search. Usually, a time-varying schedule is used, i.e., the rule will be used when a set number of generations have elapsed since the last time the rule was activated.
- *Adaptive* parameter control. This takes place when there is some form of feedback from the search that is used to determine the direction and/or magnitude of the change to the strategy parameter. The assignment of the value of the strategy parameter may involve credit assignment, and the action of the EA may determine whether or not the new value persists or propagates throughout the population.
- *Self-adaptive* parameter control. The idea of the evolution of evolution can be used to implement the self-adaptation of parameters. Here the parameters to be adapted are encoded into the chromosomes and undergo mutation and recombination. The better values of these encoded parameters lead to better individuals, which in turn are more likely to survive and produce offspring and hence propagate these better parameter values.

The main criteria for classifying methods that change the values of the strategy parameters of an algorithm during its execution are: (1) What is changed? and (2) How is the change made? These two questions address two issues: the component of the evolutionary algorithm (which incorporates the parameter) and the type of control. The component and type entries are orthogonal and encompass typical forms of parameter control within EAs. The component of parameters' change consists of six categories: representation, evaluation function, variation operators (mutation and recombination), selection, replacement, and population. The type of parameters' change consists of three (discussed earlier) categories: deterministic, adaptive, and self-adaptive mechanisms. In the following section, we give a few practical observations on some good practices regarding some components of EAs and types of control that worked well on a variety of commercial projects we executed over the last decade.

4 Good Practices That Work

Section 2 introduced an example of a complicated real-life problem and addressed some problem issues. Let us have a look at some easy to implement and yet very efficient good practices for solving complex real-life problems. A general strategy is to apply forms of adaptation. However, very much like the *Ockham's Razor Principle* for prediction and classification systems, *simplicity* is more often than not the key to quick, robust and stable implementations that perform well. Let us discuss different parameter settings and controls that follow our "simplicity principle"; in the following discussion we would refer to a solution we developed for the car distribution problem introduced in section 2.

Initial population size. Often the initial population size is determined by measuring the runtime length and quality of the solutions found by the optimizer on a few test cases. While this might work for simple problems, a better (and still very simple approach) is to make the size of an initial population linearly dependent on the size of the search space.[6] Obviously, this is not perfect as the convergence of an evolutionary algorithm is not linear in the population size, but it is simple and robust. Apart from the obvious imperfection, the authors have repeatedly used this simple adaptation/control of the initial population size with good results. Often minimum and maximum initial population sizes should also be enforced. The minimum population size usually is set to 10–100, which assures a reasonable level of performance on very small problems (a very small population more or less performs like a population-based hill-climbing). The maximum population size usually is determined by the user requirements and the available computer resources, i.e., by how fast the computer is and how much memory it has. The reason for a maximum initial population size is that the end-user often wants 1) an answer within a certain time frame, and 2) has a limited amount of available computer resources. If we set the maximum initial population size too large then 1) the evolutionary process would take too long time and 2) the population would require too much memory (which results in swapping memory content, which, in turn, would reduce the performance dramatically). Generic guidance for the maximum population size can not be given as it depends on the specific problem, the available hardware, and the user requirements.

In the system developed for the car distribution example (section 2) the length of the chromosome is equal to the number of cars to be distributed. For twice as many cars, the length of chromosomes is doubled, and we also double the initial population size. We use the minimum population size of 100 and the maximum of 5,000.

Operator probability adaptation. Quite often problem-specific operators are developed (or local search methods are added) to improve the performance of evolutionary algorithm. It is not uncommon to work with an EA that incorporates five or more operators. Most practitioners still identify the operators' probabilities just by testing (see section 3 of this chapter); finally some static probabilities are chosen and stay fixed during the whole run. However, in our commercial applications we have been using a very simple and effective parameter control, which falls into "adaptive parameter control" methods (section 3). The idea is to adapt the probability of each operator during the evolutionary process based on the operator's performance. Initially each operator has a given probability for being applied (e.g., each operator could initially have the same probability). For a number of iterations

[6] If the size of the search space is very difficult or time consuming to calculate then one can often create an initial population size that is linearly depending on the chromosome length.

I we simply count how often each operator creates an offspring that is better than all of its parents and we calculate the success ratio SR for operator i by dividing the number of improved offspring[7] created by this operator by the number of all improved offspring (whichever operator was responsible for an improvement) in the last N iterations. Hence, for operator i we would find

$$SR[i] = improvement[i]/all\_improvements,$$

where $SR[i]$ is the success ratio of operator i, $improvement[i]$ is the number of improvements operator i created, and $all\_improvements$ is the total number of improved offspring. If $all\_improvements = 0$ then $SR[i]$ is undefined and we don't update any operators' probabilities. Otherwise, the probability of each operator is afterwards updated:

$$probability[i] = probability[i] + k \cdot SR[i],$$

and normalized such that the sum of all $probability[i]$'s adds up to 1.0 (where k is an adjustment constant $0 \leq k \leq 1$). Also, just before the normalization step is executed, it is often a good idea to make sure no operator $probability[i]$ becomes smaller than some threshold l (and a reasonable value for l could be $l = 0.01$). Hence, if $probability[i] < l$ then $probability[i]$ is set to l.

With these adaptation steps the operators that do relatively well (i.e., have a relatively high success ratios $SR[i]$) will slowly increase their probabilities (assuming $k > 0$) and those that have a below average performance will decrease their probabilities. Nevertheless, just like in Machine Learning one has to pay attention to adapt carefully and avoid too large adjustments. The adjustments are controlled by the constants k and I, and while good values for k and I are problem dependent, often a good starting point is $k = 0.001$ and I is set in such a way, that the operators are updated every time 1,000 offspring are generated. However, for different problems and different types of evolutionary algorithms the values of k and I have to be adjusted accordingly.

The benefits of adaptation of the operator probabilities are multi-fold:

1. Time consuming tuning of operators is not needed anymore.
2. If an operator is added or removed at some stage of the development process, the probabilities of all operators are re-adjusted.
3. When instances (of the same problem type) with very different sizes are executed (e.g., distribution of 2,000 cars versus 6,000 cars), or the characteristics of different instances are quite different (e.g., one instance includes a majority of similar cars — many white Fords returned by a large government organization in one location, whereas the other instance has many cars of different types returned in many locations all over the nation), then the adaptation process sometimes creates very different adaptations of the operator probabilities to compensate for problem specific details.[8]

[7] An improved offspring is an offspring that is better than all its parents.

[8] This was experienced many times by the authors. Often to our initial surprise but further investigation revealed that a particular problem input had certain characteristics that favored one type of operators over the others, e.g. mutation versus crossover based operators.

4. Many studies were and are still focusing on self-adaptive *mutation* operators. However, very often practitioners simply implement *multiple problem-specific* mutation operators and assign them some static probabilities. Some of these mutation operators may act like large macro-mutations while others may act more like small local hill-climbing mutations. With the described simple adaptation scheme of the operator probabilities, no time-consuming process of assigning operator probabilities is required: whenever large or small mutations are appropriate they are more likely to be applied.
5. Finally, it is very important to understand that the operator probability adaptation has its clear limits. One of the most important (and often easy to avoid) mistakes is to implement operators that have very different computational complexities. For example, if one compares the performance of uniform mutation with an exhaustive neighborhood search operator then, not surprisingly, the neighborhood search operator will be more successful than the blind uniform mutation. However, this is due to the simple fact that the neighborhood search operator takes many more computational cycles. A simple way to avoid this problem is to "limit" the neighborhood search operator to look at a very limited neighborhood, e.g., just one neighbor. In other words, each operator is allowed to make a simple iteration using its own logic. A simple way to think about this is that each operator should (roughly) take the same amount of computational time. If this is the case, then the operator probability adaptation most often works surprisingly well.

In the system developed for the car distribution example (section 2) we have used six problem-specific adaptive operators ranging from "large" mutation via "small" mutation and permutations to different crossovers. Initially all operator probabilities were the same: $1/6$. In the system we used adaptive operators with the control parameters $I = 1,000$ and $k = 0.001$.

Diversity preservation. As indicated earlier, most real-world applications of evolutionary algorithms require several problem-specific operators, decoders, fine-tuning, etc. However, many practitioners neglect the issue of diversity. Our experience indicates that it is very important to deal with this issue and it is usually a mistake not to address it. Diversity preservation need not be a hassle to implement and tune; at least some aspects of diversity are often easy and computationally fast to implement. Different schemes for diversity preservation have been proposed over the years (e.g., sharing, crowding), but these are time consuming. Instead of trying to ensure that most genotypes and/or phenotypes are different one can simply ensure that the fitness of each individual is different. If the fitness value of a new offspring already exists in the population, one can either drop the offspring (death penalty, i.e., not insert the offspring into the population) or continue to mutate the offspring until its fitness is different from any other individual (this can take a significant amount of extra computational cycles). Such diversity preservation approach typically works well for fitness functions that result in floating point numbers; it does ensure that all solutions in the population are different from each other and hence represent a diverse set of solutions. However, this approach obviously has its limitations, when the number of all possible fitness values is very small (e.g., a limited range of integers).

In the system developed for the car distribution example (section 2) the fitness of each solution is a floating point value[9] and it turned out to be very beneficial to apply the above diversity preservation scheme.

Termination issues. A simple and popular stopping criteria is to quit after a number of steps L without any improvement of the best individual (again, the parameter L is usually found by manual tweaking). However, often large-scale complex optimization problems can take a very long time to fine-tune at the end of the evolutionary optimization process. Hence, it often makes more sense to terminate the evolutionary algorithm based on a percentage of fitness improvement of the best individual. In other words, if the percentage of the fitness improvement of the best individual over the last L iterations is below a certain limit (e.g. 0.1%) then the evolutionary optimization process terminates. Obviously the population could consist of individuals that potentially might be further fine-tuned.

To address this issue, we found it beneficial to run a post-processing phase (after the termination of the basic iterations of evolutionary algorithm) during which we continue with a final phase of evolutionary optimization using only the N best individuals (N is much smaller than the initial population size). This approach combined with the earlier described adaptive operator probabilities and diversity preservation creates N fine-tuned and different solutions. This approach focuses the computational cycles on the N best individuals instead of the whole population. One could also apply an explicit hill-climber on each of the N solutions, but often this can be implemented as a hill-climbing operator (typically a mutation based operator). Of course, one could try to have intermediate steps where the population is gradually reduced to focus more and more on the N best individuals. However, this creates additional complexity which might not be worthwhile.

In the system developed for the car distribution example (section 2) we applied such simple 2-phased approach where we terminate the first EA phase after $L = 50,000$ iterations, using the termination limit of 0.01%. This is followed by a much smaller population ($N = 20$) performing local fine tuning. Interestingly, many mutation operators adaptively had their probabilities significantly increased during this second EA phase, since they tended to be better than the crossover-based operators.

5 Uncertain Environments

Let us address one additional issue we experienced while solving many real world problems. Real world problems are very often set in uncertain (and possibly changing) environments, which are usually categorized into four categories: (1) noise, (2) robustness, (3) approximation, and (4) time-varying environments [4]. Before we present and discuss the fifth category, and argue that this fifth category is the most common in real word situations, let's first discuss the main features of these four categories.

[9] The fitness of each solution is the sum of the net profit for each car, and each allele assigns an auction site to a car.

Noise. Sometimes evaluation functions are subject to noise. This happens when evaluation functions return sensory measurements or results of randomised simulations. In other words, the evaluation procedure for the same solution (i.e., the solution defined as a vector of some design variables) may return different values. The common approach in such scenarios is to approximate a noisy evaluation function eval by an averaged sum of several evaluations:

$$eval(\mathbf{x}) = 1/n \sum_{i=1}^{n} (f(\mathbf{x}) + z_i), \quad (1)$$

where $\mathbf{x}$ is a vector of design variables (i.e., variables controlled by a method), $f(\mathbf{x})$ is the evaluation function, z_i represents additive noise, and n is the sample size. Note that the only measurable (returned) values are $f(\mathbf{x}) + z$.

Robustness. Sometimes design variables, other variables, or constraints of the problem are subject to perturbations after the solution is determined. The general idea is that such (slightly modified) solutions should have quality evaluations (thus making the original solution robust). This is important in scenarios involving manufacturing tolerances, or when it is necessary to modify the original solution because of employee illness or machine failure. The common approach to such scenarios is to use evaluation function eval based on the probability distribution of possible disturbances δ, which is approximated by Monte Carlo integration:

$$eval(\mathbf{x}) = 1/n \sum_{i=1}^{n} f(\mathbf{x} + \delta_i). \quad (2)$$

Note that $eval(\mathbf{x})$ depends on the shape of $f(\mathbf{x})$ at point $\mathbf{x}$; in other words, the neighbourhood of $\mathbf{x}$ determines the value of $eval(\mathbf{x})$.

Approximation. Sometimes it is too expensive to evaluate a candidate solution. In such scenarios, evaluation functions are often approximated based on experimental or simulation data (the approximated evaluation function is often called the meta-model). In such cases, evaluation function eval becomes:

$$eval(\mathbf{x}) = f(\mathbf{x}) + E(\mathbf{x}), \quad (3)$$

where $E(\mathbf{x})$ is the approximation error of the meta-model. Note that the approximation error is quite different than noise, as it is usually deterministic and systematic.

Time-varying environments. Sometimes evaluation functions depend on an additional variable: time. In such cases, evaluation function eval becomes:

$$eval(\mathbf{x}) = f(\mathbf{x}, t), \quad (4)$$

where t represents the time variable. Clearly, the landscape defined by the function f changes over time; consequently, the best solution may change its location over time. There are two main approaches for handling such scenarios: (1) to restart the method after a change, or (2) require that the method is capable of chasing the changing optimum.

However, it seems the largest class of real world problems is not included in the above four categories. From our business/industry experience of the last decade, it is clear that in many real world problems the evaluation functions are based on the predicted future values of some variables. In other words, evaluation function eval is expressed as:

$$eval(\mathbf{x}) = f(\mathbf{x}, P(\mathbf{x}, \mathbf{y}, t)), \tag{5}$$

where $P(\mathbf{x}, \mathbf{y}, t)$ represents an outcome of some prediction for solution vector $\mathbf{x}$ and additional (environmental, beyond our control) variables $\mathbf{y}$ at time t. Let's compare this category with the four categories defined earlier to see the differences between them.

First of all, noise may or may not be involved. If the prediction model is deterministic, then there is no noise in the scenario: every solution vector $\mathbf{x}$ is evaluated to the same value. On the other hand, if the prediction model involves simulations, noise might be present. Second, the meaning of robustness is quite different. Unexpected disturbances (e.g., delays) influence the outcomes of the prediction model, and should be handled accordingly. Third, the concept of approximation is different. Note, that we can evaluate a candidate solution precisely (i.e., the evaluation function is not expensive), however, approximation is connected with uncertainties of the predictions. Finally, the time-changing environment also has a different meaning. As the real world changes, the prediction model needs constant updates and/or parameter adjustments, thus changing the problem landscape in an implicit way.

Such real world problems call for so-called *Adaptive Business Intelligence* solutions, i.e., intelligent decision support systems, which combine prediction, optimization, and adaptiveness [7], [8]. Clearly, the car distribution problem falls into this category, as it requires prediction of car prices at different auction sites at different times and the recommended distributions — outcomes of optimization process — are based on predicted prices. We discuss the basic concepts of such systems in the following section.

6 Adaptive Business Intelligence

Adaptive Business Intelligence systems assume the existence of a few key modules that interact with each other. We discuss them in turn.

The prediction module, in general, generates a predicted output based upon some input (see Figure 3).

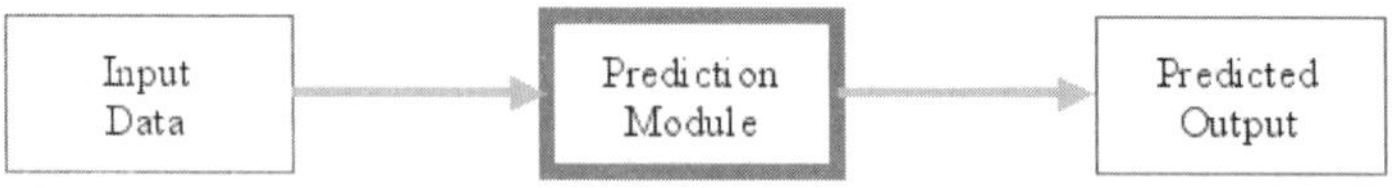

Fig. 3. Prediction module

There are many techniques for constructing a prediction model, from classic forecasting methods [5] to modern heuristic methods, such as neural networks, evolutionary programming, and genetic programming [11], [6]. A recent study [10] investigated the development of a new dynamic genetic programming model specifically tailored for prediction in time-changing environments. The model incorporated methods to automatically adapt to changes in the environment, as well as retain knowledge learned from previously encountered environments. A similar approach was proved to be effective in the price prediction module of the car distribution system.

Next, the optimisation module has to be capable of recommending the best answer. Note, that the optimisation module's recommendation is based on the prediction module's output, so there is a strong relationship between the prediction and optimization modules. The overall concept is displayed in Figure 4.

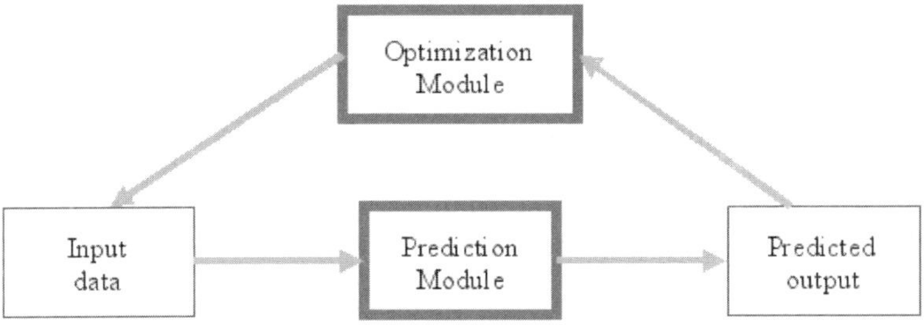

Fig. 4. Prediction and optimization modules

What happens here is that the optimization module generates some possible solutions to the problem at hand (e.g., creates a possible distribution of cars to the auction sites), which serves as *input data* for the prediction module. The predicted *output data* is then used evaluate the generated answers. In other words, the optimisation module tries different *input data* combinations in order to find the best *predicted output.*

A prediction and optimisation module are necessary components of an intelligent software system, but by themselves they are not sufficient in today's constantly changing environment. The prediction module has to be adaptive and learn from changes in the environment. Today's accurate prediction might not be that accurate tomorrow! Adaptability can be accomplished by slightly changing the learned relationship between input and output each time it is needed. This could be every minute, hour, day, week, month, or for any other time period. The update frequency depends on how fast the environment changes. Some classic forecasting methods (e.g., exponential smoothing methods) [5] approach this problem by putting more emphasis on more recent data. However, a really good adaptive solution can make its own decision on the frequency of update: it will continuously measure its own prediction errors and adjust its parameters. Hence, a really good adaptive system would adapt its own speed of adaptation!

Figure 5 illustrates the adaptation process. The *recent input* and *recent output* are taken from very recent history, and historic data is used to construct and train the prediction module. The adaptation module would take the recent input and output and, if necessary, adapt the parameters of the prediction module to decrease the prediction error — in other words, to adapt the prediction module to changes in the environment so it can make better predictions in the future. This was essential component for Adaptive Business Intelligence system for car distribution problem,

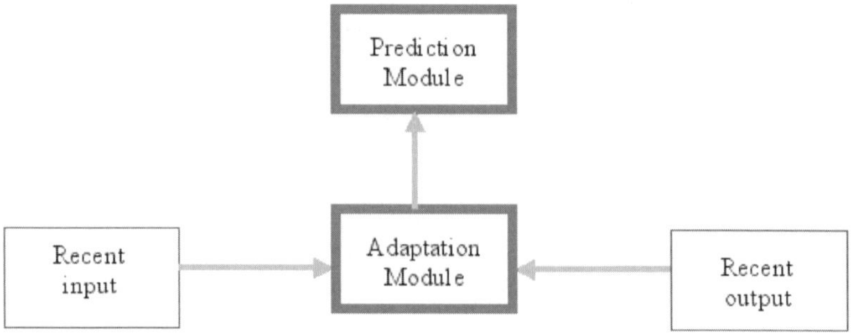

Fig. 5. Adaptation and prediction modules

as — on daily basis — new sales data were coming in, so it was possible to compare the predicted and real values and make appropriate adjustments.

Such a system was implemented for the car distribution problem presented in section 2. The system is in daily production use since 2002 and is running on Pentium 4 1GHz PC. The total running time for optimization is around 1.5 hours, but the system needs additional time to load all required input files (including the whole US inventory).

The system receives the inventory of the cars to be distributed from the Inventory Management System that the client maintains. The system automatically downloads from the ftp server and processes the input files early in the morning. The input files contain data about the cars that need to be distributed daily; the current, up-to-date inventory of the cars at the auctions and on the way to the auctions; and the information about the cars that have been sold (the latter is used to adapt the price prediction module).

The optimization starts automatically once the input files have been loaded and processed. By the start of the business day the optimization is completed and the proposed solution ready. However, the solution is implemented only after a remarketing officer checks the results and (possibly) makes some small adjustments. Less than 1% the cars from the recommended distribution are changed manually — these changes usually correspond to last-minute actions based on newest information (e.g., an approaching ice-storm). The final output files are sent to the system that manages the transportation of the cars to the auctions.

The implemented system helps the remarketing team with decisions on daily, near-optimal distribution of cars. As discussed earlier, the problem is extremely complex and the implemented software addresses the issues of transportation, volume sensitivity effect, price depreciation, recent istory, current inventory, risk factors, and dynamic market changes. The system recommends (on daily basis) the best distribution of cars and generates a multi-million dollar net lift per year. And parameter control methods, described in section 4 of this chapter, contributed to this lift in a significant way!

There were a few possibilities to validate this lift. For example, a daily sample of 4,000 can be divided into two sets of 2,000 cars each (with almost identical distribution of make/models in these two sets). One set is distributed by the "old" business model, whereas the other set by the system described above, and the results are compared later when all cars are sold. Alternatively, on selected days of the week (e.g., Mondays, Wednesdays, and Fridays) the old system is used, whereas on remaining days (e.g., Tuesdays and Thursdays) — a new system. Again, the results are compared later, when all cars are sold. Both methods were used to evaluate our new system.

7 Conclusions

Adaptive Business Intelligence systems address two fundamental questions that are of the utmost importance for people, businesses, and government agencies. These are:

- What will happen next?
- What is the best decision right now?

These two questions deal with daily lives of individuals and various organizations, and express the desire to make "good" decisions. Everyone would like to know what is likely to happen next (prediction) and everyone would like to make the best decision under uncertainty and risk (optimization). For example, when people drive to work in the morning, they try to predict the traffic pattern and select the best driving route. On a larger scale, banks may need to predict future currency exchange rates, and based upon these predictions, make decisions that will maximize profit.

Effective solutions for complex business problems require software that is capable of operating in time-changing environments, detecting current trends, and using optimisation to recommend a near-optimum decision. Further, self-learning modules continuously adapt to improve future predictions and recommendations. Through the concept of adaptability, a software system is able to learn and adapt by evaluating the actual outcome of each decision made. This gives enterprises the ability to monitor business trends, evolve and adapt quickly as situations change, and make intelligent decisions on uncertain and incomplete information [8]. All these activities require adaptation of many parameters in the prediction, optimization, and adaptation modules.

References

1. De Jong, K., "The Analysis of the Behavior of a Class of Genetic Adaptive Systems," PhD Dissertation, Department of Computer Science, University of Michigan, Ann Arbor, Michigan, 1975.
2. Goldberg, D.E., *Genetic Algorithms in Search, Optimization and Machine Learning*, Addison-Wesley, 1989.
3. Grefenstette, J.J., Optimization of Control Parameters for Genetic Algorithms, *IEEE Transactions on Systems, Man, and Cybernetics*, Vol.16, No.1, pp.122-128, 1986.
4. Jin, Y. and Branke, J,. Evolutionary Optimization in Uncertain Environments - A Survey, *IEEE Transactions on Evolutionary Computation*, Vol.9, No.3, pp.303-317, 2005.
5. Makridakis, S., Wheelwright, S.C., and Hyndman, R.J., *Forecasting. Methods and Applications*, Wiley, 1998.

6. Michalewicz, Z. and Fogel, D.B., *How to Solve It: Modern Heuristics*, 2nd edition, Springer, 2004.
7. Michalewicz, Z., Schmidt, M., Michalewicz, M., and Chiriac, C.,A Decision-Support System based on Computational Intelligence: A Case Study, *IEEE Intelligent Systems*, Vol.20, No.4, July-August 2005, pp.44-49.
8. Michalewicz, Z., Schmidt, M., Michalewicz, M., and Chiriac, C., *Adaptive Business Intelligence*, Springer, Berlin, 2006.
9. Rechenberg, I., *Evolutionsstrategie: Optimierung technischer Syseme nach Prinzipien der biologischen Evolution*, Frommann-Holzboog, Stuttgart, 1973.
10. Wagner, N., Michalewicz, Z., Khouja, M., and McGregor, R.R., Time Series Forecasting for Non-static Environments: the DyFor Genetic Program Model, to appear in *IEEE Transactions on Evolutionary Computation*, 2006.
11. Weiss, S.M. and Indurkhya, N., *Predictive Data Mining*, Morgan Kaufmann, San Francisco, 1998.
12. Wolpert, D.H. and Macready, W.G., No Free Luch Theorems for Optimization, *IEEE Transactions on Evolutionary Computation*, Vol.1, No.1, pp.67-82, 1997.

Parameter Adaptation for GP Forecasting Applications

Neal Wagner[1] and Zbigniew Michalewicz[2]

[1] Department of Mathematics and Computer Science, Augusta State University, Augusta, GA 30904, USA nwagner@aug.edu

[2] School of Computer Science, University of Adelaide, Adelaide, SA 5005, Australia, Institute of Computer Science, Polish Academy of Sciences, ul. Ordona 21, 01-237 Warsaw, Poland, and Polish-Japanese Institute of Information Technology, ul. Koszykowa 86, 02-008 Warsaw, Poland zbyszek@cs.adelaide.edu.au

Summary. Genetic Programming (GP) has been applied to time series forecasting often with favorable results. However, for forecasting tasks several open issues concerning parameter settings exist. Many real-world forecasting tasks are dynamic in nature and, thus, static parameter settings may lead to inferior performance. This paper presents the results of recent studies investigating non-static parameter settings that are controlled by feedback from the GP search process. Specifically, non-static settings for population size and training data size are explored.

Applications based on Evolutionary Algorithms (EA) rely on numerous parameters (e.g., population size, mutation and recombination rates, etc.) to guide and control algorithm execution. The values of these parameters have a significant effect on the performance of an EA. A common practice is to tune the parameters, that is to make several pre-experiment runs with various parameter settings in order to find good values. These values are then used for the actual experiment and typically remain fixed for the duration of the experiment. Manually tuning parameters is problematic because parameters are not necessarily independent of each other and, thus, trying all possible combinations requires an enormous amount of experiments.

For many applications, especially those used in dynamic environments, optimal parameter settings may vary over the course of a run. For such applications the use of static parameters can lead to inferior performance. *Adaptive* parameter control refers to the use of feedback from the evolutionary search process to adjust (adapt) parameter values during a run [34].

EA and other biologically inspired methods (e.g., Neural Networks) have often been applied to forecasting tasks [4, 7, 8, 11, 13, 17, 20, 27, 37, 38, 40]. Genetic Programming (GP) is a sub-category of EA that is also commonly used for forecasting [1, 5, 6, 14, 15, 16, 19, 21, 22, 23, 24, 25, 32, 35, 36, 41]. While GP forecasting applications often yield good results, a number of open issues concerning parameter settings

N. Wagner and Z. Michalewicz: *Parameter Adaptation for GP Forecasting Applications*, Studies in Computational Intelligence (SCI) **54**, 295–309 (2007)
www.springerlink.com

exist. Kaboudan [26] recognizes some of these issues including: 1) how to determine the training data size (i.e., the number of data to be utilized for training) and 2) how to determine the population size. A few GP forecasting studies have examined good (static) values for training data and population sizes empirically (for example, [10, 12]), but there is still no agreement in the literature on what may be optimal. This paper presents the results of two recent studies which investigate adaptive parameter control for training data size and population size, respectively. Additionally, we will discuss the another aspect of the GP search process for forecasting, that of finding an optimal evaluation (fitness) function.

The rest of this paper is organized as follows: section 1 gives a brief description of the application of GP to forecasting tasks, sections 2 and 3 detail algorithms for adaptively controlling the training data and population sizes, respectively, section 4 discusses the search for an optimal fitness function, and section 5 concludes.

1 The Use of GP for Forecasting

In order to have a clear picture of how GP is used for forecasting, it is necessary to understand the different kinds of forecasting problems that exist.In general, there are three types:

1. classification problems,
2. regression problems, and
3. time series problems.

In classification problems the goal is to predict the *class* or *category* of a newly occurring event by examing old events and their corresponding classes. For example, credit card companies need to classify newly occurring credit card transactions into one of two classes, fraudulent or non-fraudulent. For this type of problem, the issue of time is of secondary importance and either does not play a part in the classification process or is incorporated as one of many features of the process.

Regression problems try to learn about the relationship between several independent (predictor) variables and a dependent (criterion) variable. For regression, the desired output is a number. For example, business analysts may wish to predict a manager's salary given knowledge of his or her position, number of supervised employees, number of years at the position, education level, etc. As for classification problems, the issue of time is either non-existant or is included as a feature of the problem.

For time series problems the dominant feature is time and the data to be analyzed presents itself in the form of numerous observations/measurements given in sequential order of time. The goal of this type of problem is to predict a future (numeric) value of some variable by examing past values of that variable and/or past values of other related variables. For example, investment firms try to predict the future value of a particular stock based on past values of that stock and past values of other related variables such as trading volume, interest rates, etc.

GP has been used for all three of the above-listed forecasting problems, however we will focus on time series problems as GP applications for this type of problem are prevalent.

In GP solutions are represented as tree structures. Internal nodes of solution trees represent appropriate operators and leaf nodes represent input variables or constants. For time series forecasting applications, the operators are mathematical functions and the inputs are lagged time series values and/or explanatory variables. Each solution tree represents a candidate forecasting model. Figure 1 gives an example solution tree for time series forecasting. Variables x_{t1} and x_{t2} represent time series values one and two periods in the past, respectively. Crossover is performed by

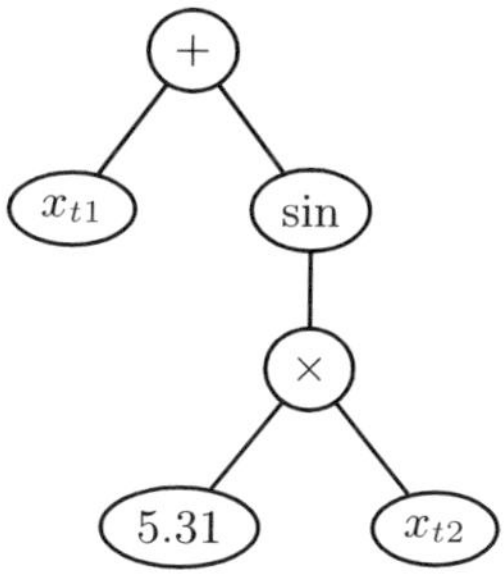

Fig. 1. GP representation of forecasting solution $x_{t1} + \sin(5.31x_{t2})$

exchanging subtrees from two parent solution trees. Mutation is performed by selecting a subtree of a single solution tree and replacing it with a randomly-constructed subtree.

GP starts by creating an initial population of candidate forecasting models. A candidate model is produced by (randomly) constructing a tree made up of operators and inputs. Each model is ranked based on its prediction error over a set of training data and new populations are generated by selecting fitter models and applying the crossover and mutation operations. New populations are created until the fittest model has a sufficiently small prediction error, repeated generations produce no reduction of error, or some limit for maximum generations has been reached. GP was developed by Koza [29] as a problem-solving tool with applications in many areas. He was the first to use GP to search for model specifications that can replicate patterns of observed time series.[3]

2 Adapting the Training Data Size

The primary objective of time series forecasting is to find a model that accurately represents the underlying data generating process and use this model to forecast the future. In order to accomplish this, GP requires a set of training data that sufficiently represents the data generating process in question. If the training data size is too small, forecasting models evolved by GP are unlikely to accurately represent the data generating process and, thus, will have poor forecasting performance. If the training data size is too large, GP models may suffer from "over-fitting" [9]. Over-fitting

[3] In [29] Koza refers to this as "symbolic regression."

refers to forecasting models that fit both the underlying data generating process and extent noise in the time series. The problem of setting the training data size is compounded when a time series is produced by a non-constant process (i.e., one that varies over time). In such a case different segments of a time series may have different underlying processes. If data from two different segments is specified for GP training, forecasting models evolved may be skewed. Consider the subset of time series data shown in figure 2. Suppose this represents recent historical data and has

$$\underbrace{22, 33, 30, 27, 24, 20, 21, 20, 20,}_{\text{segment1}} \underbrace{23, 26, 29, 30, 28, 29, 32, 30, 31}_{\text{segment2}} \mid \underbrace{\ldots}_{\text{future}}$$

Fig. 2. Time series containing segments with differing underlying processes.

been chosen as training data for GP. Suppose further that the subset consists of two segments each with a different underlying process. The second segment's underlying process represents the current process and is valid for forecasting future data. The first segment's process represents an older environment that no longer exists. Because both segments are analyzed, evolved forecasting models will be distorted.

A recent study by Wagner and Michalewicz [42] investigates a novel algorithm for adapting the training data size by using "windows" of training data that slide with time and expand or contract based on feedback from the GP search. In this GP forecasting application (called "Dynamic Forecasting GP" or "DyFor GP"), training starts at the beginning of the available historical time series data. Some initial windowsize (training data size) is set and several generations are run (in the manner described in the previous section) to evolve a population of solutions. Then the data window slides to include the next time series observation. Several generations are run with the new data window and then the window slides again. This process is repeated until all available data have been analyzed up to and including the most recent historical data. Figure 3 illustrates this process. In the figure, | marks the end of available historical data. The set of several generations run on a single data window is referred to as a "dynamic generation." Thus, a single run of DyFor GP includes several dynamic generations (one for each window slide) on several different consecutive data windows.

DyFor GP adjusts the data windowsize using feedback from the GP search process. This is accomplished in the following way.

1. Select two initial windowsizes, one of size n and one of size $n + i$ where n and i are positive integers.
2. Run a dynamic generation at the beginning of the historical data with windowsize n.
3. At the end of the dynamic generation, use the best evolved solution (forecasting model) to predict a number of future data and then measure the prediction's accuracy.
4. Run another dynamic generation also at the beginning of the historical data with windowsize $n + i$.
5. At the end of the dynamic generation, predict future data and measure the prediction's accuracy. Note which windowsize generated the better prediction.

$$\underbrace{22, 33, 30, 27, 24, 20, 21, 20, 20}_{\text{window1}}, 23, 26, 29, 30, 28, 29, 32, 30, 31 | \underbrace{\ldots}_{\text{future}}$$

$$22, \underbrace{33, 30, 27, 24, 20, 21, 20, 20, 23}_{\text{window2}}, 26, 29, 30, 28, 29, 32, 30, 31 | \underbrace{\ldots}_{\text{future}}$$

•
•
•

$$22, 33, 30, 27, 24, 20, 21, 20, 20, \underbrace{23, 26, 29, 30, 28, 29, 32, 30, 31}_{\text{windowi}} | \underbrace{\ldots}_{\text{future}}$$

Fig. 3. A sliding data window.

6. Select another two windowsizes based on which windowsize had better accuracy. For example if the smaller of the 2 windowsizes (size n) predicted more accurately, then choose 2 new windowsizes, one of size n and one of size $n - i$. If the larger of the 2 windowsizes (size $n + i$) predicted more accurately, then choose windowsizes $n + i$ and $n + 2i$.
7. Slide the data windows to include the next time series observation. Use the two selected windowsizes to run another two dynamic generations, predict future data, and measure their prediction accuracy.
8. Repeat the previous two steps until the the data windows reach the end of historical data.

Thus, predictive accuracy is used to determine the direction in which to adjust the windowsize toward the optimal. Successive window slides bring the windowsize closer and closer to this destination.

Consider the following example. Suppose the time series given in figure 2 is to be analyzed and forecasted. As depicted in the figure, this time series consists of two segments each with a different underlying data generating process. The second segment's underlying process represents the current environment and is valid for forecasting future data. The first segment's process represents an older environment that no longer exists. If there is no knowledge available concerning these segments, automatic techniques are required to discover the correct windowsize needed to forecast the current setting. DyFor GP starts by selecting two initial windowsizes, one larger than the other. Then, two separate dynamic generations are run at the beginning of the historical data, each with its own windowsize. After each dynamic generation, the best solution is used to predict some number of future data and the accuracy of this prediction is measured. Figure 4 illustrates these steps. In the figure **win1** and **win2** represent data windows of size 3 and 4, respectively, and **pred** represents the future data predicted.

The data predicted in these initial steps lies inside the first segment's process and, because the dynamic generation involving data window **win2** makes use of

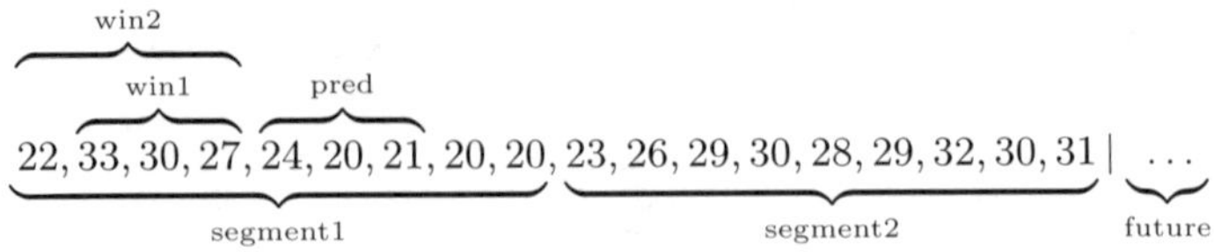

Fig. 4. Initial steps of window adaptation.

a greater number of appropriate data than that of **win1**, it is likely that **win2**'s prediction accuracy is better. If this is true, two new windowsizes for **win1** and **win2** are selected with sizes of 4 and 5, respectively. The data windows then slide to include the next time series value, two new dynamic generations are run, and the best solutions for each used to predict future data. Figure 5 depicts these steps. In the figure data windows **win1** and **win2** now include the next time series value, 24, and **pred** has shifted one value to the right.

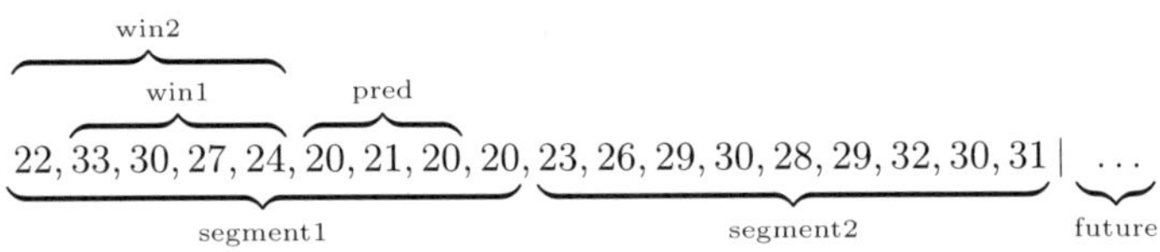

Fig. 5. Window adaptation after the first window slide. Note: **win1** and **win2** have size 4 and 5, respectively.

This process of selecting two new windowsizes, sliding the data windows, running two new dynamic generations, and predicting future data is repeated until the data windows reach the end of historical data. It may be noted that while the prediction data, **pred**, lies entirely inside the first segment, the data windows, **win1** and **win2**, expand to encompass a greater number of appropriate data. However, after several window slides, when the data windows span data from both the first and second segments, it is likely that the window adjustment reverses direction. Figures 6 and 7 show this phenomenon. In figure 6 **win1** and **win2** have sizes of 4 and 5, respectively.

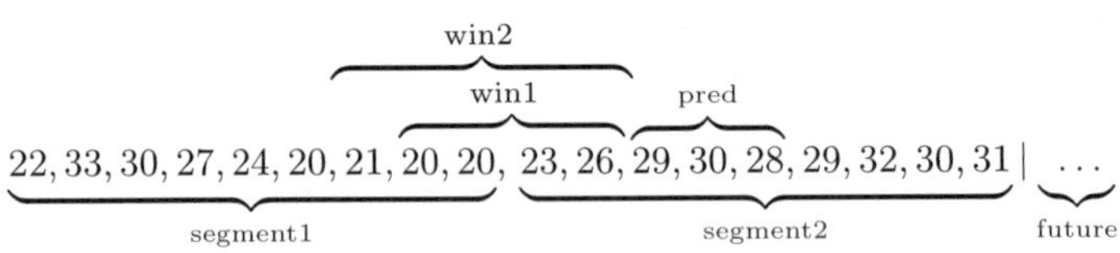

Fig. 6. Window adaptation when analysis spans both segments. Note: the smaller window, **win1**, is likely to have better prediction accuracy because it includes less erroneous data.

As the prediction data, **pred**, lies inside the second segment, it is likely that the dynamic generation involving data window **win1** has better prediction accuracy

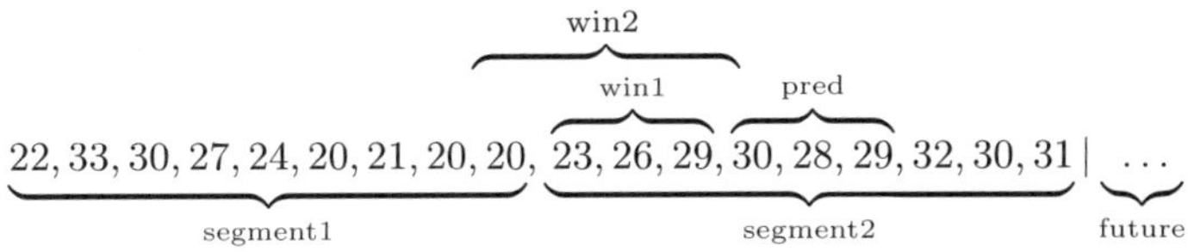

Fig. 7. Window adaptation when analysis spans both segments. Note: **win1** and **win2** have contracted to include less erroneous data.

than that involving **win2** because **win1** includes less erroneous data. If this is so, the two new windowsizes selected for **win1** and **win2** are sizes 3 and 4, respectively. Thus, as the data windows slide to incorporate the next time series value, they also contract to include a smaller number of spurious data. In figure 7 this contraction is shown.

After the data windows slide past the end of the first segment, they expand again to encompass a greater number of appropriate data. Figures 8 and 9 depict this expansion.

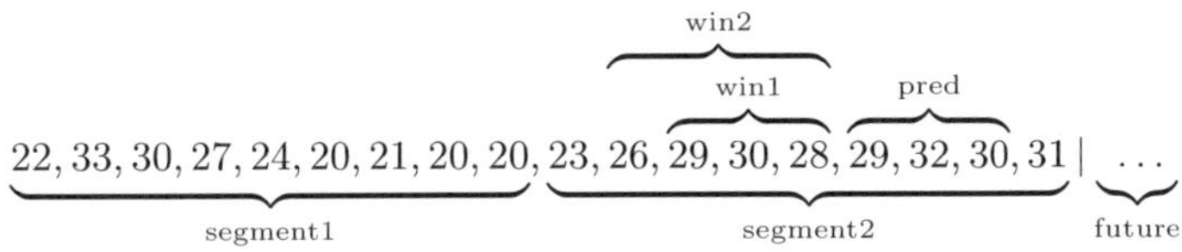

Fig. 8. Window adaptation when analysis lies entirely inside the second segment. Note: the larger data window, **win2**, is likely to have better prediction accuracy because it includes a greater number of appropriate data.

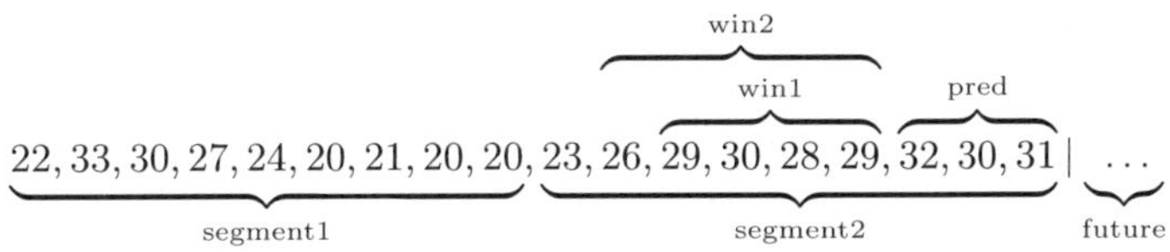

Fig. 9. Window adaptation when analysis lies entirely inside the second segment. Note: **win1** and **win2** have expanded to include a greater number of appropriate data.

As illustrated in the above example, DyFor GP uses predictive accuracy to adapt the size of its training data automatically towards the optimal. Automatic determination of an appropriate training window is beneficial for forecasting concerns in which the number of recent historical data whose underlying data generating process corresponds to the current environment is not known. Furthermore, this adaptation takes place *dynamically*. This means that as DyFor GP moves through (analyzes) historical time series data, the training window expands or contracts depending on the environment encountered. In the study conducted by Wagner and Michalewicz [42], DyFor GP with adaptive training data size is compared to conventional GP (with static training data size) for forecasting performance. Experimental results show that DyFor GP yields the more efficient forecasts. The following section describes how population size can be controlled for GP forecasting applications.

3 Adapting the Population Size

Bloat in GP is the tendency for solution trees to grow large as they approach the optimal [30, 31]. Solutions may become so large that they exhaust computer resources. Additionally, bloat hinders a GP model's adaptability as solutions become too specialized to adjust to changing conditions. Bloat is a problem for any GP model regardless of the application [3]. GP evolves a population of solutions for a given problem and, thus, must contain some method to control the number and size of solutions in the population. The standard method of GP population control is due to Koza [29] and uses a static population cardinality and a maximum tree depth for solutions.[4] However, this method does not protect a GP model from bloat. If numerous solutions in a population have full or nearly full trees of depth close to the maximum, available resources may be exhausted. Additionally, the artificial limit for tree depth prohibits the search process from exploring solutions of greater complexity, which, especially for many real-world problems, may be solutions of higher quality.

An alternative method for GP population control is presented Wagner and Michalewicz [41] that allows natural growth of complex solutions in a setting that more closely emulates the one found in nature. In nature the number of organisms in a population is not static. From generation to generation, the population cardinality changes depending on the quality and type of individual organisms present. The proposed natural non-static population control (NNPC) is based on a variable population cardinality with a limit on the total number of tree nodes present in a population and no limit for solution tree depth. This method addresses the following issues:

1. allowing natural growth of complex solutions of greater quality,
2. keeping resource consumption within some specified limit, and
3. allowing the population cardinality to vary naturally based on the make-up of individual solutions present.

By not limiting the tree depth of individual solutions, natural evolution of complex solutions is permitted. By restricting the total number of tree nodes in a population, available resources are conserved. Thus, for a GP model that employs NNPC, the number of solutions in a population grows or declines naturally as the individual solutions in the population vary. This method is described in more detail below.

NNPC works in the following way. Two node limits for a population are specified as parameters: the soft node limit and the hard node limit. The soft node limit is defined as the limit for adding new solutions to a population. This means that if

[4] Koza [29] used population sizes of 500, 1000, and 2000 and maximum solution tree depth of 17 in his early GP experiments.

adding a new solution to a population causes the total nodes present to exceed the soft node limit, then that solution is the last one added. The hard node limit is defined as the absolute limit for total nodes in a population. This means that if adding a new solution to a population causes the total nodes present to exceed the hard node limit, then that solution may be added only after it is repaired (the tree has been trimmed) so that the total nodes present no longer exceeds this limit. During the selection process, a count of the total nodes present in a population is maintained. Before adding a new solution to a population, a check is made to determine if adding the solution will increase the total nodes present beyond either of the specified limits.

Wagner and Michalewicz [41] compare a GP forecasting model with NNPC to one with the standard population control (SPC) method introduced by Koza [29]. Observed results indicate that the model with NNPC was significantly more efficient in its consumption of computer resources than the model with SPC while the quality of forecasts produced by both models remained equivalent. The following section discusses the search for an optimal fitness function for GP forecasting applications.

4 The Search for an Optimal Fitness Function

Evolution-based techniques such as GP require that some fitness function be used to measure the quality of candidate solutions. However, it may not be clear how to select such a measure for a particular problem. It may be that a single measure performs well under certain conditions but badly in others. Most GP forecasting applications use a mean squared error (MSE) fitness function for model evolution. To date, there have been no significant studies investigating alternative fitness functions for GP forecasting applications. One alternative measure that might be considered is the mean absolute deviation (MAD). Comparing the MSE and MAD measures, it can be seen that the error value of MSE grows quicker than that of MAD when outlier data are present. Thus, outliers tend to influence analyses based on MSE more than they do analyses based on MAD. An outlier datum can represent one of two possibilities: noise (which should be ignored or have reduced impact on model construction) or new information representing a shift in the underlying process. For series in which outlier data represent noise, MAD might be the more effective measure. For series in which outlier data represent a process shift, MSE might be preferable. The question of which measure to employ would depend upon the characteristics of the time series to be forecast.

A novel fitness function is presented by Wagner and Michalewicz [42] that incorporates aspects of both the MSE and MAD measures. The idea is to use MSE when data encountered is not considered an outlier and MAD when data encountered is considered an outlier. This new measure (called "CF" for "combined fitness") requires a user-specified parameter, Ω, to determine which data are outliers and which are not. Figure 10 gives a graphical depiction of the CF measure as a function of the relative error. From the figure, when the relative error is within the threshold given by Ω, CF measure values follow those of the squared error. However, when the relative error falls outside of the Ω threshold, CF measure values follow those of the absolute deviation.

A number of new GP forecasting experiments were conducted on real data to compare the MSE, MAD, and CF fitness functions. Here, we forecast the U.S. Gross

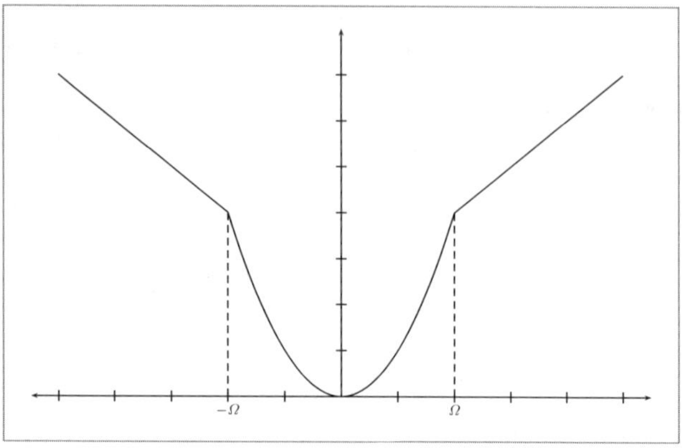

Fig. 10. The CF measure as a function of the relative error.

Domestic Product (GDP) using several relevant economic variables. The following three sections describe the GDP time series, experimental setup, and observed results, respectively.

4.1 Forecasting the U.S. Gross Domestic Product

According to the U.S. Department of Commerce [39], the GDP is defined as "the market value of goods and services produced by labor and property in the United States." The GDP is a metric frequently employed as a measure of the nation's economy. The GDP series was selected because it is a widely-studied, non-linear time series with a well-known set of explanatory variables. Figure 11 gives a graphical depiction of the quarterly GDP (growth) time series. In the figure real GDP growth is calculated as a quarter-over-quarter annualized percent change. A contemporary

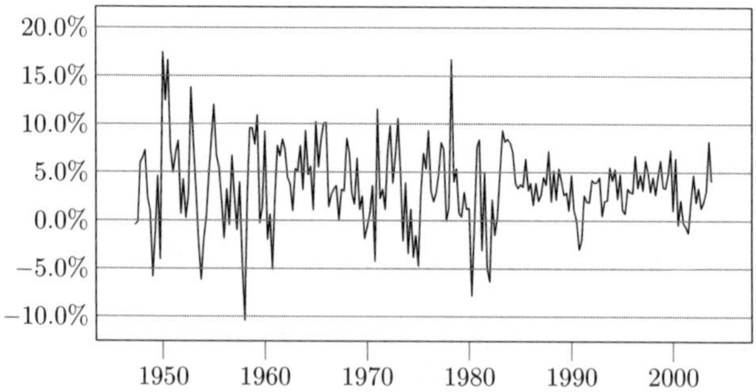

Fig. 11. Gross Domestic Product (growth): 1947-2003.

study conducted by Kitchen and Monaco [28] forecasts the GDP, a time series with quarterly frequency, using multiple economic indicators that are measured monthly. Thirty indicators are utilized in all and can be subdivided into the following categories: employment (6), financial (4), survey (6), production and sales (12), and other (2). The results of their study show that forecasting models constructed using these indicators provided efficient forecasting performance for the period of 1995Q1 through 2003Q1.

4.2 Experimental Setup

For this set of GP forecasting experiments, three fitness functions are compared: MSE, MAD, and the CF measure described in section 4. Recall that the CF measure requires that a parameter, Ω, be specified representing the threshold between outlier and non-outlier data. For these experiments three different settings of Ω are tested: one that is 5.0% of the median level of the time series, one that is 7.5% of the median level, and one that is 10.0%.

29 of the 30 economic indicators listed in the Kitchen and Monaco's GDP forecasting study [28] are utilized as inputs[5] by all competing models and the outputs are one-step-ahead, quarterly forecasts for the current quarter when only one month of historical data for that quarter is available. Historical GDP data dating back to 1965Q1 is used for training the GP and one-step-ahead forecasts for 1995Q1 through 2003Q1 are produced. Table 1 gives the GP parameter values used by all competing models.

Table 1. GP parameter settings.

Parameter	Value
crossover rate	0.9
reproduction rate	0.0
mutation rate	0.1
population size	1000
max. no. of generations	51
elitism used?	yes
fitness function	MSE/MAD/CF

All parameter values listed in table 1 were selected to match those used by Koza [29] with the following exceptions.

1. Elitism (reproduction of the best solution of the population) is used.
2. Parameter values for "reproduction rate" and "mutation rate" were exchanged. This was done for two reasons: 1) increasing the mutation rate allows for greater search-space exploration [33] and 2) decreasing the reproduction rate to zero was

[5] One of the indicators, "Business Week Production Index," could not be obtained at the time of the experiments.

not thought to harm the effectiveness of the evolutionary process since elitism is used.

Because the GP search process is a stochastic one, a set of GP runs is executed rather than just a single run. Setsize = 20 is used for all GP forecasting experiments executed here.

4.3 Results

As mentioned in the previous section, a set of runs is executed (setsize = 20) for each competing model. For a single run, forecasting performance is measured by calculating the MSE of all forecasts. For a set of runs, forecasting performance is measured by calculating the mean and standard deviation of MSE values over all (20) runs. Table 2 gives the observed results for the GDP experiments. In the table, the first column shows the fitness function used to drive the GP process. Note that three separate experiments employing the CF measure are shown, one for each of three Ω settings (5.0%, 7.5%, and 10.0%, respectively).

Table 2. GDP forecasting results.

GP fitness function	mean MSE	std. dev.
MSE	4.96	1.62
MAD	4.46	1.42
CF($\Omega = 5.0\%$)	3.99	1.19
CF($\Omega = 7.5\%$)	3.69	0.70
CF($\Omega = 10.0\%$)	3.96	1.19

Table 2 reveals some interesting results. All three CF fitness functions outperform both the MSE and MAD fitness functions with the one utilizing a 7.5% setting for Ω giving the best performance overall. Between the MSE and MAD fitness functions, MAD yields the better forecasts.

The MAD fitness function's better performance when compared to MSE may mean that for the GDP series outlier data more frequently represents noise rather than a shift in the underlying data generating process. The fact that the CF fitness function outperforms both MSE and MAD may mean that it is better able to distinguish noise from true process shifts and, thus, can lessen the effect of noisy data and still respond to changes in the process.

The following section draws conclusions and discusses potential areas for future work.

5 Conclusions and Future Work

In this paper, parameter adaptation for GP forecasting applications is discussed. Novel algorithms from recent studies are presented that seek to control two significant GP parameters, training data size and population size. Additionally, GP fitness functions MSE and MAD are detailed and compared. A new fitness function from a recent study that combines aspects of both the MSE and MAD measures is described and new experiments are conducted that compare the performance of this new measure, called the CF measure, with that of the MSE and MAD measures for a real time series of U.S. GDP data. The CF measure is shown to give better forecasting performance than MSE and MAD measures for the GDP series.

This CF measure contains a parameter, Ω, that represents a threshold between outlier and non-outlier data in a time series. The experiments presented in this study test several different settings for Ω. Further studies might examine the CF measure for other time series and, perhaps, develop some algorithm for adapting the Ω parameter toward its optimal setting.

Although the algorithm detailed here for adapting the training data size is only applicable to GP forecasting applications, the described algorithm for adapting the population size may be appropriate for other GP applications as well. Future investigations could compare this adaptive population size to the commonly-used static population size for several prevalent GP applications such as circuit design, database query optimization, etc.

References

1. Andrew M. and Prager R. 'Genetic programming for the acquistion of double auction market strategies.' *Advances in Genetic Programming*, vol. 1 (1994), pg. 355-368.
2. Angeline P. 'Genetic programming and emergent intelligence.' *Advances in Genetic Programming*, vol. 1 (1994), pg. 75-98.
3. Banzhaf W. and Langdon W. 'Some considerations on the reason for bloat.' *Genetic Programming and Evolvable Machines*, vol. 3 (2002), pg. 81-91.
4. Chambers L., editor. *Practical Handbook of Genetic Algorithms: Applications.* CRC Press, 1995.
5. Chen S. and Yeh C. 'Toward a computable approach to the efficient market hypothesis: an application of genetic programming.' *Journal of Economics Dynamics and Control*, vol. 21 (1996), pg. 1043-1063.
6. Chen S., Yeh C., and Lee W. 'Option pricing with genetic programming.' *Genetic Programming 1998: Proceedings of the Third Annual Conference*, vol. 1 (1998), pg. 32-37.
7. Chiraphadhanakul S., Dangprasert P., and Avatchanakorn V. 'Genetic algorithms in forecasting commercial banks deposit.' *Proceedings of the IEEE International Conference on Intelligent Processing Systems*, vol. 1 (1997), pg. 557-565.
8. Deboeck G., editor. *Trading on the Edge: Neural, Genetic, and Fuzzy Systems for Chaotic and Financial Markets.* John Wiley and Sons, Inc., 1994.
9. Diebold F. *Elements of Forecasting.* International Thomson Publishing, 1998.
10. Fernandez T. and Evett M. 'Training period size and evolved trading systems.' *Genetic Programming 1997: Proceedings of the Second Annual Conference*, vol. 1 (1997), pg. 95.
11. Gately E. *Neural Networks for Financial Forecasting.* John Wiley and Sons, Inc., 1996.

12. Gathercole C. and Ross P. 'Small populations over many generations cand beat large populations over few generations in genetic programming.' *Genetic Programming 1997: Proceedings of the Second Annual Conference*, vol. 1 (1997), pg. 111-118.
13. Goto Y., Yukita K., Mizuno K., and Ichiyanagi K. 'Daily peak load forecasting by structured representation on genetic algorithms for function fitting.' *Transactions of the Institute of Electrical Engineers of Japan*, vol. 119 (1999), pg. 735-736.
14. Hiden H., McKay B., Willis M., and Tham M. 'Non-linear partial least squares using genetic programming.' *Genetic Programming 1998: Proceedings of the Third Annual Conference*, vol. 1 (1998), pg. 128-133.
15. Iba H. and Sasaki T. 'Using genetic programming to predict financial data.' *Proceedings of the Congress of Evolutionary Computation*, vol. 1 (1999), pg. 244-251.
16. Iba H. and Nikolaev N. 'Genetic programming polynomial models of financial data series.' *Proceedings of the 2000 Congress of Evolutionary Computation*, vol. 1 (2000), pg. 1459-1466.
17. Jeong B., Jung H., Park N. 'A computerized causal forecasting system using genetic algorithms in supply chain management.' *The Journal of Systems and Software*, vol. 60 (2002), pg. 223-237.
18. de Jong E., Watson R., and Pollack J. 'Reducing bloat and promoting diversity using multi-objective methods.' *Proceedings of the Genetic and Evolutionary Computation Conference (GECCO 2001)*, vol. 1 (2001), pg. 11-18.
19. Jonsson P. and Barklund J. 'Characterizing signal behavior using genetic programming.' *Evolutionary Computing: Lecture Notes in Computer Science*, vol. 1143 (1996), pg. 62-72.
20. Ju Y., Kim C., and Shim J. 'Genetic based fuzzy models: interest rate forecasting problem.' *Computers and Industrial Engineering*, vol. 33 (1997), pg. 561-564.
21. Kaboudan M. 'Forecasting with computer-evolved model specifications: a genetic programming application.' *Computer and Operations Research*, vol. 30 (2003), pg. 1661-1681.
22. Kaboudan M. 'Genetically evolved models and normality of their residuals.' *Journal of Economics Dynamics and Control*, vol. 25 (2001), pg. 1719-1749.
23. Kaboudan M. 'Forecasting stock returns using genetic programming in C++.' *Proceedings of 11th Annual Florida Artificial Intelligence International Research Symposium*, vol. 1 (1998), pg. 502-511.
24. Kaboudan M. 'Genetic programming prediction of stock prices.' *Computational Economics*, vol. 6 (2000), pg. 207-236.
25. Kaboudan M. 'Genetic evolution of regression models for business and economic forecasting.' *Proceedings of the Congress of Evolutionary Computation*, vol. 2 (1999), pg. 1260-1268.
26. Kaboudan M. 'A measure of time series' predictability using genetic programming applied to stock returns.' *Journal of Forecasting*, vol. 18 (1999), pg. 345-357.
27. Kim D. and Kim C. 'Forecasting time series with genetic fuzzy predictor ensemble.' *IEEE Transactions on Fuzzy Systems*, vol. 5 (1997), pg. 523-535.
28. Kitchen J. and Monaco R. 'Real-time forecasting in practice.' *Business Economics: the Journal of the National Association of Business Economists*, vol. 38 (2003), pg. 10-19.

29. Koza J. *Genetic Programming: On the Programming of Computers by Means of Natural Selection.* MIT Press, 1992.
30. Langdon W. 'The evolution of size in variable length representations.' *1998 IEEE International Conference of Evolutionary Computation*, vol. 1 (1998), pg. 633-638.
31. Langdon W. and Poli R. 'Fitness causes bloat.' *Soft Computing in Engineering Design and Manufacturing*, vol. 1 (1997), pg. 13-22.
32. Lee D., Lee B., and Chang S. 'Genetic programming model for long-term forecasting of electric power demand.' *Electric Power Systems Research*, vol. 40 (1997), pg. 17-22.
33. Michalewicz Z. *Genetic Algorithms + Data Structures = Evolution Programs.* Springer-Verlag, 1992.
34. Eiben A., Hinterding R., and Michalewicz Z. 'Parameter control in evolutionary algorithms.' *IEEE Transactions on Evolutionary Computation*, vol. 3 (1999), pg. 124-141.
35. Mulloy B., Riolo R., and Savit R. 'Dynamics of genetic programming and chaotic time series prediction.' *Genetic Programming 1996: Proceedings of the First Annual Conference*, vol. 1 (1996), pg. 166-174.
36. Neely C. and Weller P. 'Predicting exchange rate volatility: genetic programming versus GARCH and RiskMetrics^{TM}.' *The Federal Reserve Bank of St. Louis*, (2002).
37. Smith K., Gupta J. *Neural Networks in Business: Techniques and Applications.* Idea Group Pub., 2002.
38. Trippi R., Turban E., editors. *Neural Networks in Finance and Investing: Using Artificial Intelligence to Improve Real-World Performance.* Irwin Professional Pub., 1996.
39. U.S. Department of Commerce Bureau of Economic Analysis. http://www.bea.doc.gov/bea/glossary/glossary_g.htm, 2004.
40. Venkatesan R. and Kumar V. 'A genetic algorithms approach to growth phase forecasting of wireless subscribers.' *International Journal of Forecasting*, vol. 18 (2002), pg. 625-646.
41. Wagner N. and Michalewicz Z. 'Genetic programming with efficient population control for financial times series prediction.' *2001 Genetic and Evolutionary Computation Conference Late Breaking Papers*, vol. 1 (2001), pg. 458-462.
42. Wagner N., Michalewicz Z., Khouja M., and McGregor R. 'Time series forecasting for dynamic environments: the DyFor genetic program model.' To appear in *IEEE Transactions on Evolutionary Computation*, 2006.

Index

Printing: Krips bv, Meppel
Binding: Stürtz, Würzburg